Chicago Lincoln Statue, Lincoln Square

WESTINGHOUSE ELECTRIC RAILWAY TRANSPORTATION

WESTINGHOUSE ELECTRIC RAILWAY TRANSPORTATION

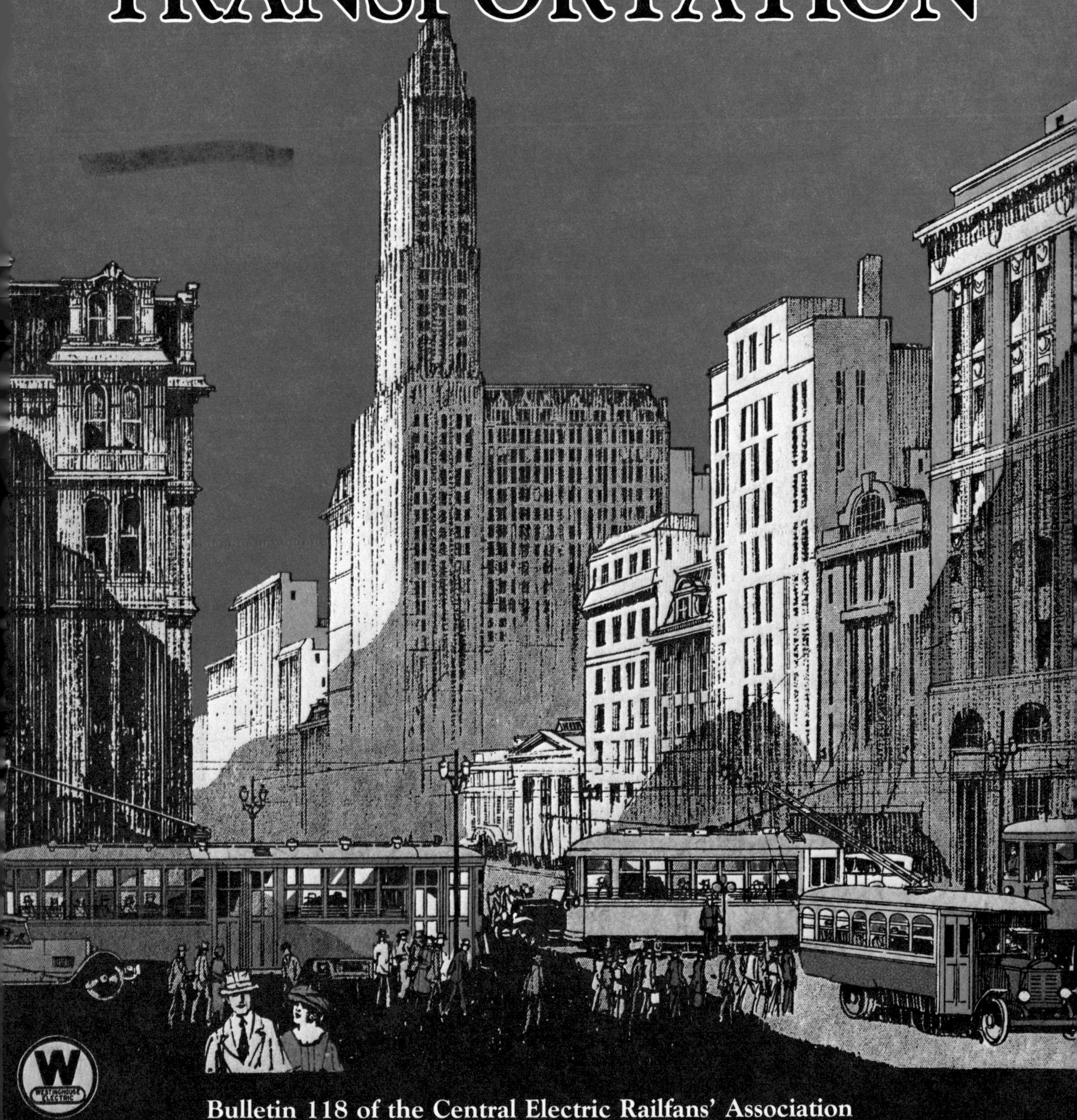

Bulletin 118 of the Central Electric Railfans' Association

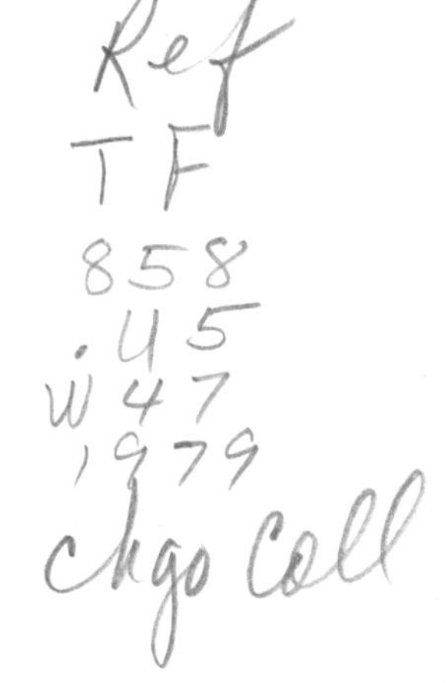

WESTINGHOUSE ELECTRIC RAILWAY TRANSPORTATION

BULLETIN 118 OF CENTRAL ELECTRIC RAILFANS' ASSOCIATION

Published 1915, 1917, 1922, 1924, 1929 and 1936
by Westinghouse Electric & Manufacturing Co.
Published 1915 by Electric Railway Journal

Reprinted 1979 with the Permission of
Westinghouse Electric Corporation and
McGraw-Hill Publications Company by
Central Electric Railfans' Association
An Illinois Not-For-Profit Corporation
P.O. Box 503, Chicago, Illinois 60690

Library of Congress Catalog Card No. 78-74493
International Standard Book Number 0-915348-18-7

CERA Bulletins are technical, educational references prepared as historical projects by members of Central Electric Railfans' Association, working without salary in the interest of the subject as a hobby. This Bulletin is consistent with the stated purpose of the corporation: To foster the study of the history, equipment and operation of electric railways. If you the reader can provide any unknown information or are of the opinion that certain information is incorrect please send your information, documented by source material where possible, to: Curator of Corrections, Central Electric Railfans' Association, P.O. Box 503, Chicago, Illinois 60690, U.S.A.

HD

CERA wishes to express its appreciation to Westinghouse Electric Corporation and McGraw-Hill Publications Company for granting us permission to reprint this material on street railway, mainline and interurban electric railway transportation. Many of these articles had a very limited printing and have been out-of-print for many years. We wish to thank Westinghouse and McGraw-Hill for granting us the opportunity to give wider circulation to these most interesting technical and historic publications.

CONTENTS

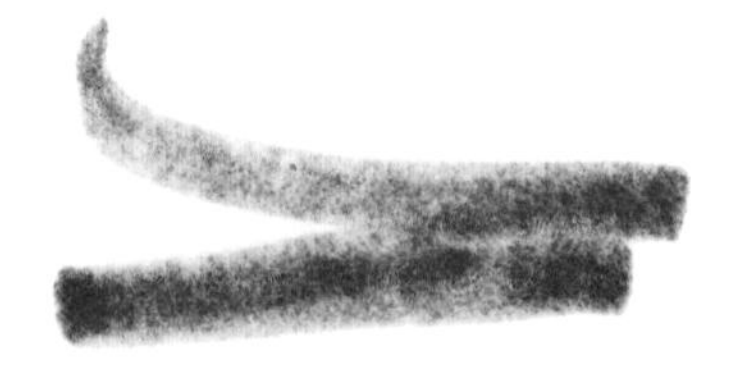

RESULTS
of
ELECTRIFICATION

Heavy Traction Department

Westinghouse Electric & Manufacturing Company

East Pittsburgh, Pa.

Circular No. 1505 April, 1915

Overhead Construction—Six-Track Trunk-Line Electrification

Results of Electrification

THE Electric Locomotive is no longer an experiment in its application to steam railroad operating conditions.

Its first introduction into the highly developed sphere of steam railroading was somewhat compulsory, with the object of overcoming the smoke nuisance of the steam locomotive. Its application has, however, proven conclusively that for reliability of operation, economy in maintenance, flexibility of control and availability for service, it is superior to the steam locomotive. In development it has kept pace with the recent rapid advance of the steam locomotive, until to-day, single Electric Locomotives are being designed with axle loads of 65,000 pounds and continuous capacity ratings of 4,000 horsepower.

The name Westinghouse has always been synonymous with railroading. The close affiliation of the Westinghouse Electric & Manufacturing Company with the Westinghouse Air Brake Company and the Baldwin Locomotive Works has been a large factor in the success achieved in Electric Locomotive operation.

The Electric Locomotive has now "won its spurs" in nearly every class of steam railroad operation, including tunnel, main line, grade, suburban and branch line application.

The railroad world now has an opportunity to judge the results obtained by electrification from an economic, as well as from a service standpoint in the Paoli suburban service of the Pennsylvania Railroad at Philadelphia, the New York, Westchester & Boston Railroad and the Elkhorn Grade Division of the Norfolk & Western Railroad and many others, all of which are installations of Westinghouse Railway Equipment.

Hoosac Tunnel Electrification, Boston & Maine Railroad—11,000-Volt Line Construction With Double Insulation

Tunnel Work

TUNNELS usually form the throat of railroad systems. The delays incurred while waiting for the smoke to clear, and the inability to see signals form a serious obstacle to the free operation of freight trains with steam engines in long tunnels.

These difficulties, coupled with the disagreeable features as regards passenger traffic have long been recognized by steam railroad officials, and the advantages of Electric Operation in service of this class have been realized in several notable instances.

Typical Passenger Train in Electric Operation at Hoosac Tunnel. The Steam Locomotives are Hauled Through with Banked Fires

Hoosac Tunnel Electrification

Boston & Maine Railroad

THE famous Hoosac Tunnel of the Boston & Maine Railroad, built in 1875, was always an obstructing feature to the traffic of this railroad until its electrification by the Westinghouse Single-Phase System in 1911. The delays caused by steam locomotive smoke and gases made this tunnel the limiting section of the division. Passenger trains now go through the tunnel with windows open in the summer time, refreshing the patrons with the cool breezes.

Under Steam Operation the entire tunnel was a block five miles long, as no other train was permitted to enter while a passenger train was passing through, now several at one time pass through.

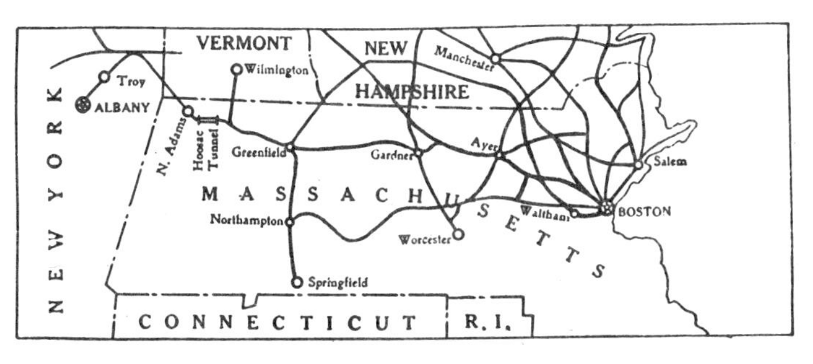

Map—Boston & Maine Railroad, Showing Relative Location of Hoosac Tunnel

A 130-Ton Baldwin-Westinghouse Electric Locomotive Hauling Freight Trains in Hoosac Tunnel Service

THE freight trains are no longer packed three and four deep at each portal waiting for a chance to get through. Three trains are allowed on each track at one time, the steam locomotive being hauled through with banked fires. These trains, averaging 1600 tons trailing load, including steam engine, make the trip in about eighteen minutes and consequently the capacity of the tunnel is now largely determined by the size of the power house, the number of electric locomotives and the number of blocks, all of which can be increased as traffic demands.

The maintenance cost of the roadbed, tunnel arch, and cars was greatly reduced in this tunnel by the change to electric locomotives. The old troubles of the rock and brick work being shattered by the engine exhaust, the telephone and signal lines being corroded by acid gases, and bad rails caused by condensing steam have all disappeared.

The track gangs can also work with safety and dispatch in making repairs, where formerly only two hours' effective work in a day shift was possible.

Simple Inspection Shed for Electric Locomotives in Hoosac Tunnel Service

Substation for Operation of Tunnel Service

IN electrifying this system the crossing of the 600-volt lines of the city railway, by the 11,000-volt single catenary construction of the Boston & Maine Railroad was accomplished.

This was one of the many interesting engineering features encountered in the installation of the overhead construction, all of which was supplied by the Westinghouse Company.

The entire traffic through this tunnel, reaching 77 trains per day, is handled by a total of six Baldwin-Westinghouse single-phase locomotives having a total weight of 130 tons each, with 102 tons on the drivers.

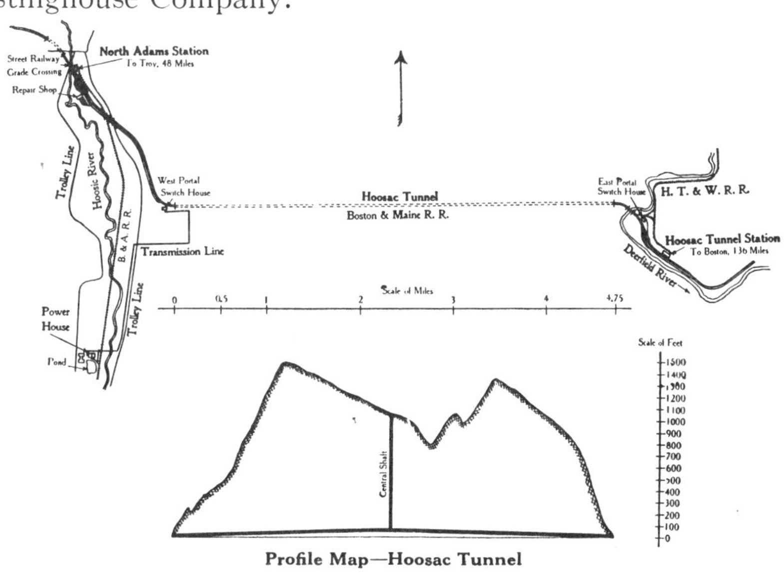

Profile Map—Hoosac Tunnel

The St. Clair Tunnel is a Cast Iron Tube 40 Feet Below St. Clair River and is Whitewashed and Electrically Illuminated

St. Clair Tunnel Electrification

Grand Trunk Railway

THE officials of the Grand Trunk Railway System were among the first to realize the advantages of electric locomotives for tunnel operation. In 1908 they inaugurated electric service in the St. Clair Tunnel which connects the town of Port Huron, Mich., with Sarnia, Ont., by passing under the St. Clair River. The operation of this section has been so successful with the Westinghouse Single-Phase System and locomotives that the yearly reports are now summed up by stating "Everything is going about the same old way. No troubles and no delays."

The grade to the tunnel approaches and the incline portions of the tunnel are all two per cent, necessitating high tractive forces to operate the freight trains on these approach grades. This required, prior to electrification, special anthracite coal burning locomotives, in order to reduce the smoke to a minimum.

A 162-Ton Baldwin-Westinghouse Locomotive and 1000-Ton Train. No Pusher Service is Used

THE maximum tractive effort of these steam locomotives, however, limited the weight of trains to about 760 tons, and the speed up the 2 per cent grades was very slow. Being a single tube tunnel operated on the Staff System of train dispatching, the increasing traffic prevented efficient operation.

Electrification not only eliminated the smoke and bad gases, but increased the capacity of the tunnel about 100 per cent. The double unit 132-ton Baldwin-Westinghouse electric locomotives have developed a dynamometer draw-bar pull of 90,000 pounds, on reasonably dry sanded rails. The normal draw-bar pull of these locomotives is, however, 50,000 pounds at 10 miles per hour, and the maximum speed is 35 miles per hour. A remarkable decrease in the number of train breakages followed the inauguration of electric service. This was effected by the gradual increase in draw-bar pull possible during acceleration.

The power house which was installed by the Westinghouse Company is located over the tunnel, and the power lines run direct to it. It supplies power for pumping the drainage system, and also for the machine shops and roundhouses, in addition to operating the electric locomotives.

Results of Electrification

Double Track Approaches to St. Clair Tunnel, Showing Transmission Line and 3300-Volt Single-Phase Trolley Overhead Construction

Electric Locomotives Emerging From Tunnel

Passenger Train Entering Pennsylvania Terminal Station, New York City

Trunk-Line Terminal Work

Electrification of the New York Extension

Pennsylvania Railroad

THE electric locomotive was an important factor in the successful solution of New York Terminal problem of the Pennsylvania Railroad. It rendered possible the use of tube tunnels under the North River for the purpose of reaching the heart of New York City, and increased the attractiveness of the magnificent Pennsylvania Station by eliminating the noise, smoke and fire danger of the steam locomotive.

There is no steam locomotive operation in the electrified zone, the entire service of 500 trains daily being handled with Westinghouse railway equipment.

Changing Locomotives—Penna. R. R., Manhattan Transfer

AT Manhattan Transfer, where the change is made from steam to electric locomotives on trains to and from Pennsylvania Station, four minutes is allowed.

Changing Locomotives—Penna. R. R., Manhattan Transfer

THE entire operation including the necessary testing of air brakes has been performed in two minutes.

Running Gear, Showing Two 2000-H.P. Westinghouse Motors—Pennsylvania Railroad Locomotive

THESE double articulated locomotives have a maximum rating of 4000 horsepower each, being equipped with two Westinghouse 600-volt, direct-current, field-control motors. They not only handle the heavy through trains on the two per cent grades in the tubes, but operate at passenger speeds of 60 miles per hour, and also work on switching service, rendering valuable assistance in increasing the yard capacity over that which would be possible with steam locomotives.

In regular service these locomotives start 850-ton trains on the two per cent grade and have a recorded draw-bar pull of 79,000 pounds. While designed to start and accelerate a 550-ton train in addition to the locomotive, trains of 14 all-steel cars, weighing over 1000 tons, have frequently been handled by one locomotive. Such diversity of service would be impossible with steam locomotives.

The electric zone extends from Harrison, N. J., through the Pennsylvania Station, New York City, to Sunnyside Yards where trains are made up. The route is 14.9 miles with 105 miles of single track electrified with third-rail.

The service rendered by these electric locomotives shows a remarkable efficiency. The monthly mileage is 4500 per locomotive, while they only require a four-hour overhauling period every 2500 miles, aside from the daily road inspection customary in all railroad apparatus before going into service.

Western Portal to Tunnel Under the Hudson River—Pennsylvania Railroad

Operating Record of Pennsylvania Locomotives During 1912

Number of locomotives	33
Total miles	994,592
Average miles per locomotive	30,139
Total train-minute delay, due to all causes	66
Total miles per train-minute delay due to all causes	15,070
Total train-minute delay due to electrical causes	43
Total miles per train-minute delay due to electrical causes	23,100
Total train-minute delay due to control	10
Total miles per train-minute delay due to control	99,459

THIS constant availability for service forms one of the chief advantages of the electric over the steam locomotive. It means reduction in round house, coaling, watering and hostlering expense. The general maintenance of the Pennsylvania electric locomotives is about one-half of the steam or 7.2 cents against 14.72 cents.

The results obtained by the Pennsylvania Railroad in the electrification of the New York terminal have led this company to contract with the Westinghouse Electric Company for the installation of electric service in their Philadelphia terminal.

By using multiple-unit electric trains for suburban traffic, the congestion now existing in Broad Street Station will be greatly relieved, due to the adaptability of multiple-unit cars which require no turning or shifting.

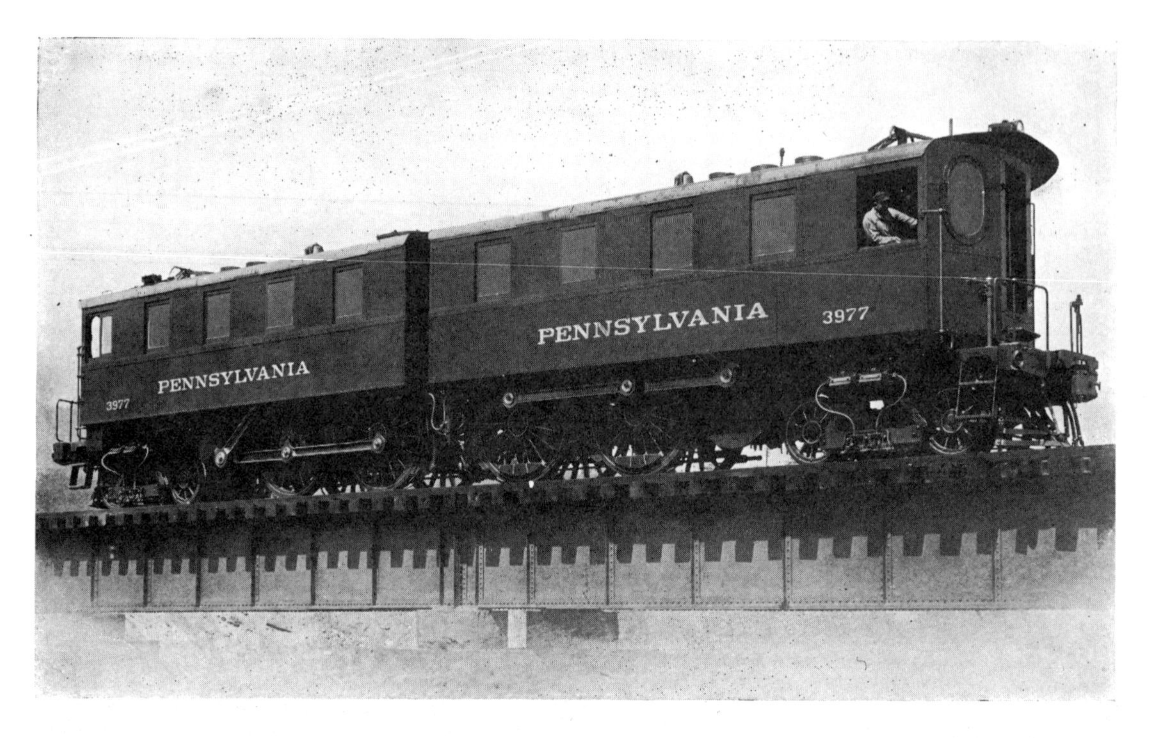

One of Thirty-Three 156-Ton 4000-H.P. Electric Locomotive Used in the Electrification of the New York Extension—Pennsylvania Railroad

ON November 28th, 1914, these locomotives completed four years of service with the following phenomenal record:

Locomotive-miles	3,974,746
Total engine failures	45
Total minutes detention to trains	271
Locomotive-miles per detention	88,328
Locomotive-miles per minute detention	14,667

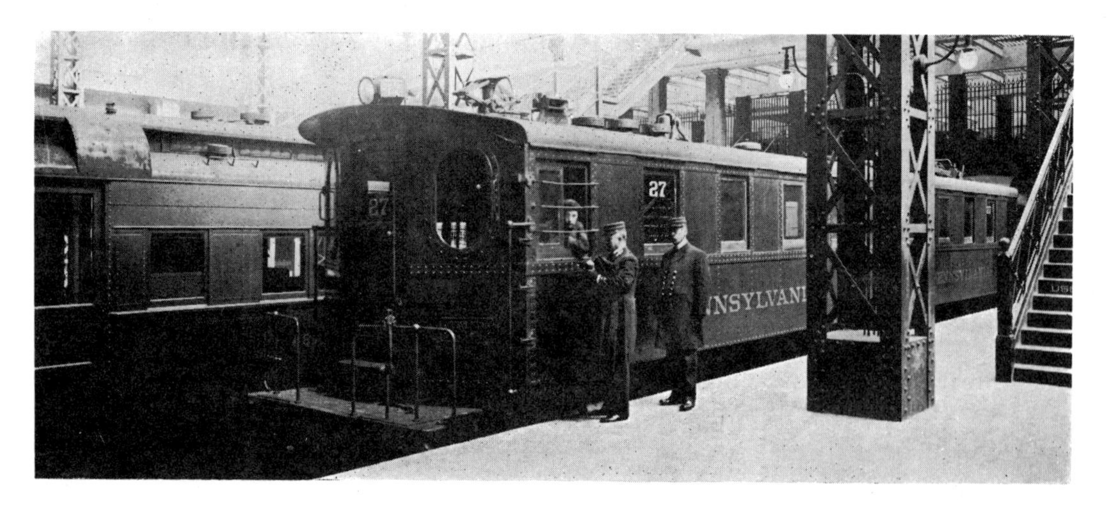

Electric Locomotive About to Leave With Train

A 110-Ton Baldwin-Westinghouse Locomotive and Train in Trunk-Line Freight Service

Trunk-Line Work

New York, New Haven & Hartford Railroad

THE New York, New Haven & Hartford Railroad electrification is the largest and most important trunk-line electrification in the world. The initial work which extended from Woodlawn to Stamford, Conn., a route distance of 21.4 miles, has now been continued to New Haven, a route of 73 miles from New York City. In addition the Harlem River Division, and the Oakpoint and Westchester freight yards are operating with electric locomotives, making a total trackage in the electric zone of approximately 550 miles.

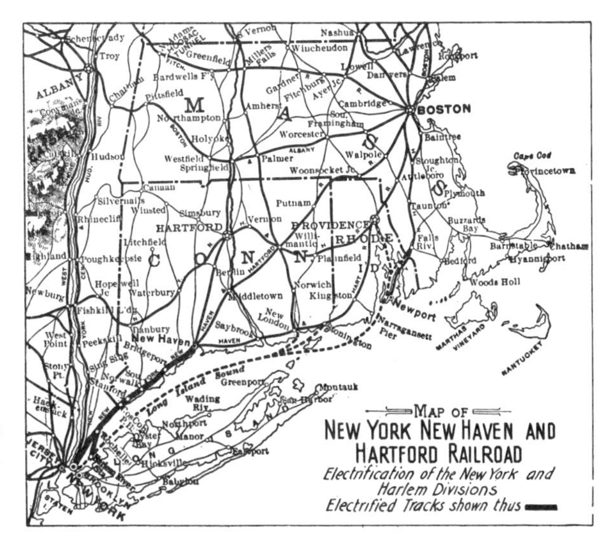

Double Heading Electric Locomotives in Passenger Service—New York, New Haven & Hartford Railroad

THE success of the initial installation on this railroad system is emphasized by the extensions made. With a daily traffic of approximately 240 passenger trains in addition to the heavy freight haulage, the amount of coal burned for operating trains has been reduced one-half.

Over 100 Baldwin-Westinghouse locomotives are now in use on this system.

Cos Cob Power House—N. Y., N. H. & H. R. R. 36000 Kw. in Westinghouse Turbo-Generators Installed

A Typical Way-Station, Pelham Park—New York, New Haven & Hartford Railroad Electrification

11,000-Volt Overhead Line Construction, Oakpoint Classification Yards

A Familiar View Found at All Steam Railroad Terminals Having Suburban Service

Suburban Terminal Service

THE stand-by losses of steam locomotives in large terminals form a considerable factor in the operating cost of this service. To this should be added the time and extra track capacity necessary for shifting, turning and trips to the round house for coal and water.

Stand-By Losses

Overhead Construction—Main-Line Electrification at St. Davids—Paoli Division, Pennsylvania Railroad

Philadelphia Electrification Paoli Division

11,000-Volt Single-Phase Alternating Current

THE limit of the physical capacity of Broad Street Station of the Pennsylvania Railroad, at Philadelphia, is fast being reached. This condition is being relieved by electrifying the suburban service, which will eliminate a large number of empty train and locomotive movements. The principal point of congestion, is at the "throat" of the yard where six suburban routes meet.

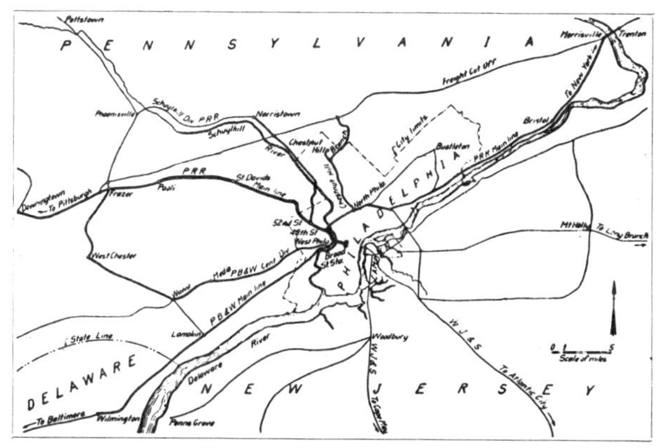

Map—Paoli Electrification

Under steam operation it was necessary to make a large number of movements through this "throat" after trains had been unloaded. By electric operation when trains run into the station loaded, all unnecessary shifting and turning of locomotives will be eliminated.

One of 93 Multiple-Unit Cars Equipped With Westinghouse Motors and Unit Switch Control

THUS, it is figured that electrification will save from 16 to 20 per cent of the total "throat" movements, which will result in a corresponding relief at the rush hour periods, and in addition there will be a gain in station track capacity equivalent to two station tracks.

This electrification will thus afford the same relief from congestion as considerable station enlargement, which could not be accomplished by any other means than a large investment for real estate.

The initial electrification consists of the Broad Street-Paoli service only, which includes 20 miles of four-track route.

This service will require 93 motor cars operating in multiple-unit trains from the single-phase overhead trolley system with 11,000 volts on the trolley wire.

Each car is equipped with two 225-horsepower Westinghouse single-phase motors and electro-pneumatic control, geared for a balancing speed of 60 miles per hour.

As soon as the operation of the Paoli branch is electrified, work will be started electrifying the Chestnut Hill branch with the same system.

An Eleven Car Multiple-Unit Motor Car Train. Each Car is Equipped With Two Westinghouse Motors and Unit Switch Control

The Long Island Railroad Electrification

THE Long Island Railroad was one of the first steam railroads to operate extensively by electricity.

It uses the Pennsylvania Terminal Station in New York, operates 200 trains per day and on its 200 miles of electrified track handles an annual service of 600,000,000 passenger-miles.

The results of this electrification are best shown by the following extracts from a paper read by Mr. J. A. McCrea before the New York Railroad Club.

"To anyone connected with the operating department of the Long Island Railroad, the use of anything else but electricity as motive power in connection with the western terminals could hardly be considered. This is entirely due to the rapid changes in the conditions under which American travel is handled, that have taken place in the last five or six years, traffic having developed along lines made possible by multiple-unit service.

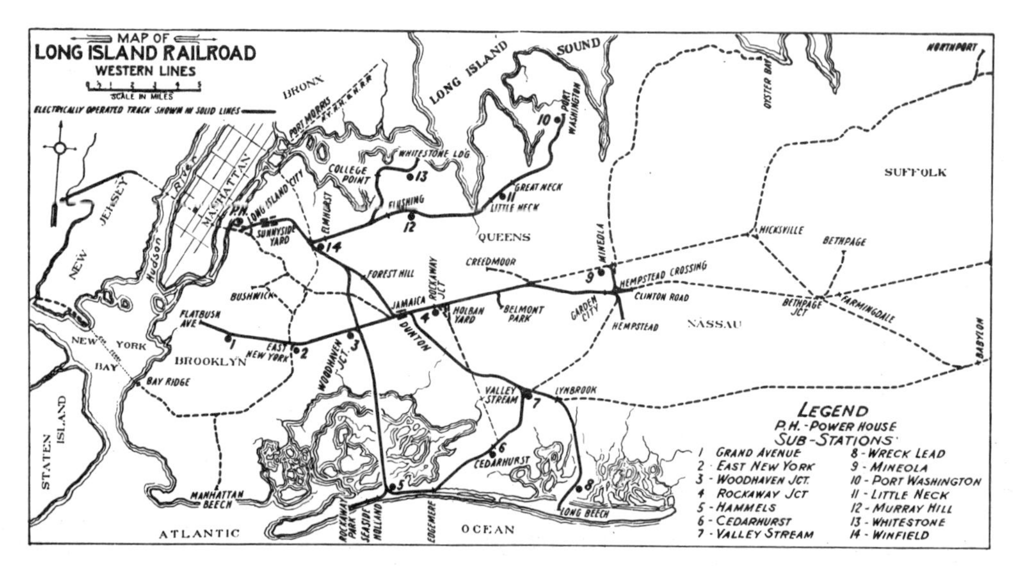

Map—Long Island Railroad, Electrified Lines

"WERE we today forced to abandon electricity as a motive power it would mean a complete revolution in our train service and either the enlarging of our Flatbush, Brooklyn, Station to two or three times its present size, at an almost prohibitory expense, or what is more probable it would be necessary to abandon much of the service not possible under the restrictions of steam operation.

"It is in the operation of a terminal that the ideas of a man accustomed to a steam situation must change radically. In the average terminal throughout the country operated by steam, it is not possible to place, load, and dispatch more than five or six trains per hour from any one track. In fact, this would be a high average, unless the location of the coach yard was particularly advantageous and special attention was given to the operation; but with multiple-unit equipment, even under adverse circumstances, it is possible to obtain the number of movements out of one track up to eight or ten trains per hour, the equipment of some four or five of which being that of trains that have come in and unloaded their passengers on that track. This performance is further emphasized by the fact that the time of the day the terminal is taxed to its capacity is usually during the rush hours late in the afternoon, and it is necessary to fill out by adding cars to all trains that come into the terminal, which, in the case of multiple-unit operation, must be done on the station track and not in the coach yard as with steam."

Dutton Inspection Shed for Multiple-Unit Cars

Main Concourse and Stairways to Platforms—Pennsylvania Station, New York City, Showing Arrangement of Electrified Tracks—Western Terminal for Long Island Railroad Suburban Trains

"THE points in multiple-unit electric operation which have impressed those who have been identified with steam operation are:

First—Regularity and reliability of service.

Second—The possibilities of running with close headway and the great increase of speed in local service.

Third—The very marked reduction in amount of switching and lay-over time at a terminal.

"Relative to regularity and reliability; in five years there has been but one serious delay in electric operation on the Long Island which was due entirely to something over which we had no control, and the same sort of an accident might have crippled the steam service.

"A multiple-unit shifting crew makes but half the number of movements as compared with steam service. A crew consists of usually a motorman and conductor; sometimes a helper or yard brakeman, or car inspector is added. This crew is very flexible, being able to move with great rapidity from one track to another and easily accomplishing the work of two yard engines. It is very important to avoid as far as possible, special make-ups on trains; i. e., the location of the smoking cars and baggage cars should be optional as to the head end or rear end of a train, as regular assigned make-up of a train causes delay. The equipment should be pooled, for the same reason."

High-Speed Multiple-Unit Car Operating During Non-Rush Hour Period

New York, Westchester & Boston Railway

THE New York, Westchester & Boston Railway is a striking example of suburban service by electric train under steam operating conditions. The main line is a four-track system with 6.8 route miles, extending to Columbus Avenue, where it divides into two double-track branches, running to New Rochelle, two miles beyond and to White Plains, 9.4 miles from Mt. Vernon. The main line connects with the Harlem River Division of the New Haven Railroad, tracks of which are used for three miles to the Harlem River Station.

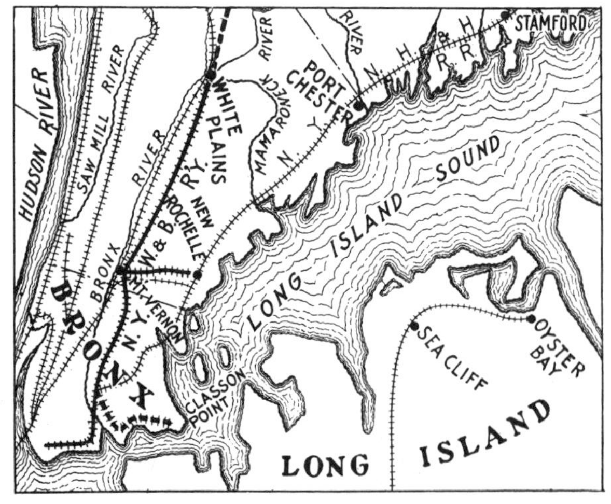

Map—New York, Westchester & Boston Railway

11,000-Volt Catenary Line Construction on Harlem River Branch of the New Haven System

THESE trains consist normally of single cars for local service which operate on a schedule of 22 miles per hour, making a 15-second stop every mile. The express trains which alternate with the locals in this service maintain a schedule of 37 miles per hour, including a 25-second stop every 2½ miles.

A Ten-Car Multiple-Unit Train

High-Voltage Single-Phase Trolley Construction Under Road Crossing Bridges

THE equipment consists of 30 steel cars weighing 60 tons each and seating 78 people. Each car is equipped with two Westinghouse single-phase motors of 170 h.p. capacity each and the electro-pneumatic control.

The best demonstration of the high grade service which can be rendered by electric equipment in congested railroad suburban traffic is shown in the following tabulation published by the N. Y. W. & B. Ry. Co.:

DELAYS FOR 1913

Month 1913	Brakes No.	Brakes Min.	Electrical Equipment No.	Electrical Equipment Min.	Pantographs No.	Pantographs Min.	Miscellaneous No.	Miscellaneous Min.	Total Delays No.	Total Delays Min.	Miles Per Delay	Miles Per Min. Delay	Cars Out of Service	Total Miles
Jan.	2	8	4	34	2	32	2	10	11	84	10,129	1,326	8	111,421
Feb.	6	27	0	0	3	20	1	12	10	59	10,010	1,686	17	100,182
Mar.	3	10	3	23	5	72	0	0	10	105	11,865	1,127	8	118,608
April	3	19	2	4	1	15	0	0	6	38	19,527	3,083	5	117,166
May	1	6	3	16	0	0	1	5	5	27	25,889	4,779	2	129,449
June	2	7	2	7	3	20	1	1	8	35	16,174	3,690	7	129,184
July	1	12	1	9	1	22	0	0	3	43	43,258	3,018	7	129,776
Aug.	0	0	0	0	0	0	0	0	0	0	126,198	126,198	7	126,198
Sept.	0	0	1	8	1	8	0	0	2	16	60,956	7,619	5	121,913
Oct.	2	10	1	5	0	0	1	23	4	38	30,678	3,229	9	125,682
Nov.	6	49	1	11	0	0	0	0	7	60	16,737	1,952	10	119,835
Dec.	0	0	1	20	0	0	0	0	1	20	128,239	6,412	11	128,239
Totals	26	148	19	137	16	189	6	51	67	525	......	.....	96	1,457,653
Av. per month	2.1	12.3	1.6	11.4	1.3	15.7	0.5	4.2	5.6	43.7	41,638	13,343	8	121,471

Sectionalizing Bridge at Junction of New Rochelle and White Plains Branches

Shop Repairs Cost in Dollars

1913	Complete Electrical Equipment	Car Body	Trucks Wheels Axles	Gears and Pinions	Pantos	Air Brakes	Brake Shoes	Miscellaneous and Supervision	Totals	Cents Cost Per Car Mile
January	2074.27	116.05	87.87	40.94	122.31	168.71	123.82	659.37	3393.34	3.04
February	698.54	162.66	117.87	85.82	110.05	111.53	46.50	675.74	2008.71	2.00
March	822.52	418.54	103.81	23.81	216.02	80.18	170.75	1312.16	3147.79	2.61
April	949.51	208.94	74.32	181.83	119.48	75.45	62.91	699.54	2366.98	2.02
May	196.12	57.70	75.03	6.00	159.03	101.36	221.51	835.65	1652.40	1.27
June	293.19	107.90	180.65	39.07	219.00	97.47	140.25	605.64	1683.17	1.30
July	360.12	115.39	277.91	5.00	157.39	66.31	228.12	421.54	1632.78	1.25
August	321.29	161.59	282.26	29.05	160.83	74.44	162.41	436.03	1627.90	1.28
September	303.15	276.48	193.29	......	119.43	108.35	164.81	439.37	1604.88	1.31
October	332.44	139.08	165.57	13.25	95.37	82.33	152.14	366.72	1346.90	1.07
November	245.27	234.42	117.55	7.25	68.69	77.75	137.72	465.18	1353.73	1.12
December	446.47	213.55	112.27	28.99	169.65	60.91	187.78	300.00	1509.62	1.17
Average	586.31	183.95	149.04	38.50	143.10	92.05	149.06	601.49	1944.01	1.62

Running Inspection Cost in Dollars

1913	Complete Electrical Equipment	Car Body	Trucks Wheels Axles	Gears and Pinions	Pantos	Air Brakes	Brake Shoes	Miscellaneous and Supervision	Totals	Cents Cost Per Car Mile
January	186.69	40.50	37.30	16.15	31.15	29.75	15.25	150.00	506.79	.450
February	171.50	35.55	37.50	15.00	30.50	32.60	22.00	135.50	480.15	.480
March	186.25	38.90	40.25	12.00	32.50	34.75	24.23	125.00	493.88	.416
April	184.67	33.62	73.12	11.75	55.23	37.35	60.52	138.99	595.25	.508
May	192.87	50.00	76.41	44.42	75.00	41.41	28.86	156.10	665.07	.514
June	98.16	16.83	58.90	17.58	26.53	26.17	26.85	77.96	358.98	.288
July	134.07	20.00	73.93	2.85	50.59	58.96	38.60	103.68	482.74	.372
August	137.14	23.05	83.94	29.05	39.00	64.13	32.56	120.70	528.57	.418
September	81.13	45.01	44.01	3.00	34.00	75.13	32.11	110.90	425.29	.350
October	133.33	42.01	86.12	12.28	29.00	30.37	25.11	113:68	471.90	.376
November	129.49	30.04	63.32	7.23	36.74	53.00	26.00	111.40	457.27	.381
December	189.60	40.53	37.29	16.08	31.00	29.98	23.92	151.75	520.15	.500
Average	152.07	34.67	59.34	15.62	40.10	42.80	29.67	124.65	498.92	.410

Multiple-Unit Cars

A careful record of the power consumption of all cars is kept. Considerable interest in reducing this item to a minimum is created among the motormen by monthly prizes for the best records obtained.

Power Used and Road Inspection Cost

Month	Kilowatt-Hours Per Car Mile	Watthours Per Ton Mile	Lubrication Cents Per Car Mile	Oilers Cents Per Car Mile	Terminal Inspection Cents Per Car Mile	Total Cents Per Car Mile
January	5.60	93.3	.021	.0526	.269	.3426
February	6.00	100.0	.002	.0493	.240	.2913
March	5.42	88.3	.026	.038	.284	.348
April	4.76	79.3	.025	.055	.320	.400
May	4.55	75.9	.021	.375	.287	.3455
June	4.61	76.8	.026	.057	.277	.360
July	4.65	77.5	.018	.070	.280	.368
August	4.58	76.3	.016	.0487	.298	.3627
September	4.48	74.5	.016	.0613	.300	.3773
October	4.54	75.6	.022	.0772	.297	.3962
November	4.91	81.6	.017	.0738	.300	.3908
December	5.34	89.0	.023	.0447	.285	.3527
Average	4.94	82.3	.021	.0554	.286	.3626

Repair Shop and Inspection Shed

Heavy Mountain Grade Operation With Steam Locomotives—Showing the Possibilities of Electrification

Mountain Grade Work

THE constant increase of traffic on many steam railroads, necessitating longer and heavier trains, is making economical operation on grade divisions a serious problem with the steam locomotive.

The electric locomotive can be operated in multiple-units with one crew. It can exert the maximum tractive effort of the Mallet locomotive, at considerably higher speeds.

Its service capacity is increased by cold weather, when more power is needed.

Two notable examples of grade electrification are here given:

Giovi Division—
Italian State Railways

Elkhorn Grade—
Norfolk & Western Railway

Freight Train on 3½ Per Cent Grade With Three-Phase Electric Locomotives Using Regenerative Braking

Giovi Grade Division Electrification

Italian State Railways

THE application by the Westinghouse Company of the electric locomotive to the Giovi Grade Division of the Italian State Railways has resulted in a complete reorganization of this system. Forming the connection between the harbor of Genoa and Milan, the principal city of Northern Italy, this road crosses the Apennine Mountains on grades reaching 3½ per cent, and uses six tunnels, the longest of which is 2.02 miles. Increased traffic in 1889 necessitated the construction of a secondary line over the 14.4-mile mountain section with a 1.16 per cent grade.

Freight Yard, Italian State Railways, Showing Overhead Construction

A special type of steam locomotive for this service was later installed. These improvements were, however, not able to take care of the increasing traffic. The accumulated tonnage on the docks at Genoa would reach as high as 575,000 tons at times, causing large losses due to the rapid deterioration of perishable freight.

Realizing that the possibility of the steam locomotive had been exhausted, the Italian Government contracted for Westinghouse three-phase alternating-current type locomotives. These locomotives were designed with the maximum power allowed by the type of rolling stock used, and to operate at a constant speed of 28 miles per hour. The electric service was started in 1910, and the results have justified all the claims for it.

The operating speed of trains was increased 88 per cent on the up-grades, and 75.5 per cent on the down-grades.

2000 H.P. at 28 M.P.H.
Heavy Freight Locomotive

2600 H.P. at 62 M.P.H.
High Speed Passenger Locomotive

Three-Phase Locomotives—Italian State Railways

THE number of trains was increased from 28 per day with steam to 38 with electric locomotives. The number of cars delivered over the up-grade section in the last year of steam operation was 109,963. In 1910-11, the first year of electrification, the number reached 172,063, while the average weight of cars was also increased. Considering the up-grade section only, which is the limiting factor of the division, the comparative capacity was found by test to be as follows:

Maximum Steam	1912-13 Electric	Maximum Electric
2,574,800 Tons	3,825,000 Tons	5,000,000 Tons

The characteristics of the above steam locomotive with tender, and the electric locomotive show why this increased capacity is possible.

	Steam Locomotive	Electric Locomotive
Number of Axles	5	5
Weight of Locomotive	75 tons	60 tons
Weight of Tender	28 tons	....
Diameter of Drivers	53 inches	42.51 inches
Capacity on 3½ per cent grades	170 tons	190 tons
Speed on 3½ per cent grade	15.5 M.P.H.	28 M.P.H.
Horsepower Rating	900 H.P.	2000 H.P.

Three-Phase Line Construction, Using Two Overhead Conductors

ANOTHER item of interest in connection with this operation is the fact that a detailed record of the running expenses shows a saving by electrification of 22½ per cent, including interest and depreciation on the investment.

The Italian State Railways already have 110 Westinghouse locomotives of the Giovi type in service and is considering the electrification of other lines in preference to double tracking sections of same. This study has been greatly advanced by the success of the Giovi work, which has likely eliminated the necessity of building a new direct line involving a tunnel 11.8 miles long, for which $30,000,000 was voted at the time electrification was started.

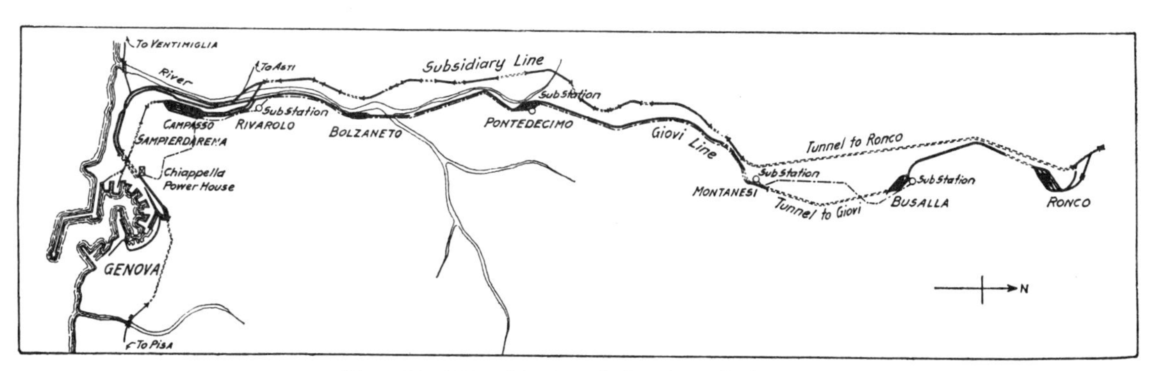

Map—Giovi Electrification—Italian State Railways

Results of Electrification

The Elkhorn Grade

Steam Operation—*Required three Mallet locomot 3250-ton train at 7 miles per hc*

Electric Operation—*Two Baldwin-Westinghouse Ele previous speed; namely, 14 miles*

A Two-Engine Train on Elkhorn Grade Consisting of Two 270-Ton Baldwin-Westinghouse Locomotives and 25 Gigantic 130-Ton Six-Axle Steel Coal Cars

Norfolk and Western Railway

quipped with superheaters and mechanical stokers to handle a this two per cent grade.

ocomotives are now handling the same tonnage trains at double the our.

In Descending the Elkhorn Grade the Front Locomotive Automatically Holds the Entire Train by Means of Electric Regeneration Without Use of the Air Brakes

**One of Twelve 270-Ton Baldwin-Westinghouse Induction-Motor Type Electric Locomotives—
Norfolk & Western Railway**

Elkhorn Grade Electrification

Norfolk & Western Railway

THE Norfolk & Western Railway Company is the American pioneer in substituting the electric locomotive for the Mallet steam locomotive in long mountain grade work. Realizing the great advantage of the three-phase locomotive for this work as demonstrated in Italy, they contracted with the Westinghouse Company for twelve locomotives with these characteristics.

The system of application, however, has been so improved that these locomotives will require only one contact wire, using the simple single-phase system of distribution instead of the double contact wires which the European three-phase distribution demands.

One Type of Steam Locomotive Being Replaced by Electric Locomotives

Two 270-Ton Baldwin-Westinghouse Induction-Motor Type Locomotives and Train

THE object of this electrification is to obtain an increase in the capacity of the existing tracks, by increased speed, by hauling heavier trains, and by eliminating congestion caused by delays for coal and water. Another object is the reduction of the over-all operating expenses, in maintenance, crews, and fuel.

Thirty miles of main line involving 85 miles of single track are being electrified.

Coal trains weighing 3250 tons make up the normal trains, the heaviest traffic being up grades of a maximum of 2 per cent.

Test trains of 3250 tons have been hauled up this two per cent grade with two electric locomotives. These same trains have been held by the automatic regeneration of the front locomotive with less than full-load current. The air brakes were not used, but maintained except as an emergency feature for additional safety.

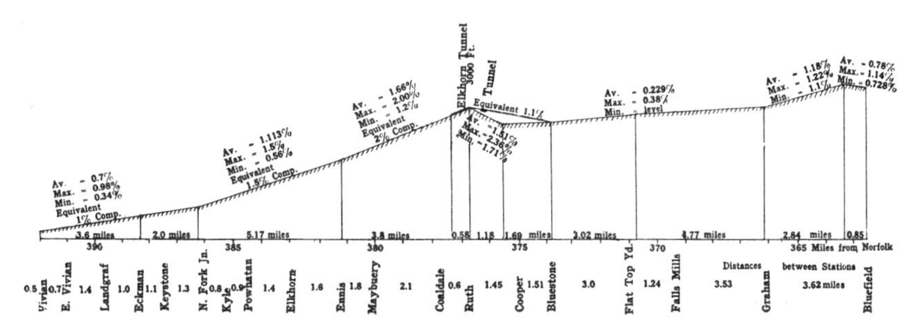

Profile of Line—Vivian to Bluefield

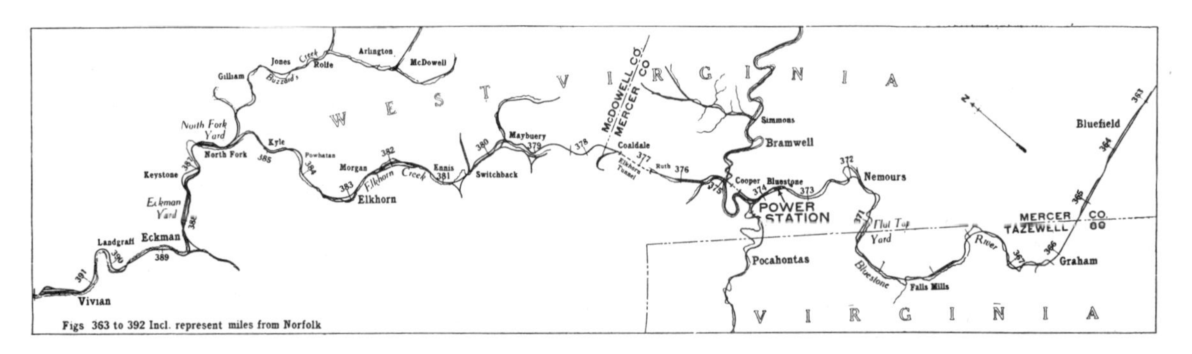

Outline Map Showing Location of Electrified Section of Norfolk & Western Railway

THE daily tonnage over this division reaches at times 50,000 tons, and with Mallet steam locomotives operating at 7 miles per hour on the grades, the road was badly handicapped by congested traffic. The improvements in service which will be effected are shown in the following table:

TABLE I—NORFOLK & WESTERN RAILWAY

Information on Freight Movement

	Vivian to Elkhorn Tunnel		Elkhorn Tunnel to Graham		Graham to Bluefield	
	East Bound	West Bound	East Bound	West Bound	East Bound	West Bound
Trailing load (tons)	3250	2000	3500	2000	3250	2000
Locomotives per train...	2	1	1	1	1	1
Speed on grades........	14	14	28	14	14	14
Maximum grade, per cent	+2	—2	+0.4	+1.10	+1.22	+1.14
H. P. on maximum grade	6350	0	3450	2560	3980	2640

TABLE II—TRACTIVE EFFORTS FOR DIFFERENT GRADES AND LOADS

	Train on 1.5 and 2% grades	Train on 1% grades	Train on 0.4% grades
Weight of train, tons................	3250	3250	3250
Locomotives per train................	2	1	1
Approx. speed, miles per hour........	14	14	28
Draw-bar pull per locomotive pounds:			
Uniform acceleration..............	91800	114000	79400
On 2 per cent grade................	75400		
On 1 per cent grade................		85800	
On 1. 22 per cent grade...........			
On 0.4 per cent grade.............			4600
Maximum guaranteed accelerating tractive effort per locomotive, pounds..	133000	133000	90000

The advantages of this type of locomotive for grade service lie in its constant speed characteristics for all loads and smoothness of acceleration, the latter is a safety factor for draw heads.

Regenerating power which means a saving in power cost, and a greater saving in wear and tear on brakes and running gear in descending grades.

Simple and sturdy design of mechanical construction.

Westchester Yard, Showing Overhead Construction for 11,000-Volt Trolley—New Haven System

Steam Railroad Switching Service

GREATER operating economies have been secured by the use of electric switcher locomotives than in any other class of service. To date not a single feature has developed in which the electric engine is not superior to the steam locomotive in this service.

One of Sixteen 80-Ton Baldwin-Westinghouse Switcher Locomotives in Classification Yard and General Switching Service—New York, New Haven & Hartford Railroad Electrification

Oakpoint Classification Yard, New York, New Haven & Hartford Railroad, Showing Baldwin-Westinghouse Switcher Locomotives at Work

Trunk-Line Classification Yard and General Switching

New York, New Haven & Hartford Railroad

THE sixteen Westinghouse switcher locomotives furnished the New York, New Haven & Hartford Railroad are doing the same work which formerly required double the number of steam locomotives.

The saving in coal as determined by actual test runs was demonstrated to be 66 per cent of the amount used by steam locomotives.

These single-phase locomotives weigh 80 tons, have a maximum three-minute tractive effort of 36,000 pounds up to six miles per hour and a continuous tractive effort of 15,000 pounds at 11.2 miles per hour.

Six of these locomotives do all the switching work between Stamford and Harlem River Station. They are kept in service 24 hours a day, each making an average of 140 miles daily.

The ease with which these Baldwin-Westinghouse locomotives are controlled, the elimination of stand-by losses, and those necessary where coal and water are used; the full capacity power in freezing weather, and availability for service at all times, have proven a source of satisfaction to the operators who formerly had to accomplish this work with steam power.

A Baldwin-Westinghouse Switcher Locomotive Unloading Car Floats—Oakpoint Yard, Harlem River Division

IN handling the work at the Oakpoint Yards on the Harlem River, practically all of the cars are transferred from floats, which work requires careful handling. The smoothness of the unit switch control on these electric switchers is a strong feature with the engineers.

Nine locomotives have made a monthly record of 38,000 locomotive-miles, operating between the Westchester Yards and the Harlem River docks, and handling 65,000 cars with a total weight of approximately one million tons.

As all this freight movement is within the corporate limits of New York City, the elimination of smoke has greatly benefited this locality.

The overhead construction used in these yards is composed entirely of steel cable, this being possible without excessive losses due to the 11,000-volt trolley system used. The construction, while exceedingly strong, is also inexpensive, and leaves no dangerous obstructions in the way of the yard crews.

An item of interest in connection with this property is that a large interconnecting railroad bridge is nearing completion at this point, which will make physical connection with the Pennsylvania and Long Island System, making an all-rail route between New England and the South and West.

A 60-Ton Baldwin-Westinghouse Locomotive and Train Ascending a 1½ Per Cent Grade—Niagara Junction Railway

Steam Railroad Freight Interchange Service

Car Spotting

Niagara Junction Railway

THE Niagara Junction Railway has completed the most extensive yard electrification for industrial switching work in America.

They interchange from 1800 to 2000 standard freight cars monthly between the industrial plants located on their 4-mile route and the steam railroads. In addition considerable work in local movements between plants is performed. The entire traffic is handled by two 60-ton Baldwin-Westinghouse locomotives operating on the standard 600-volt trolley system.

Two 60-Ton Baldwin-Westinghouse Locomotives Used in Steam Railroad Freight Interchange Service. Each Locomotive is Equipped With Four Westinghouse Commutating-Pole Field Control Motors and Unit Switch Control

THESE locomotives have been found to be well adapted to the service of spotting cars, where movements within six inches are easily made.

Each of these locomotives handles about 30,000 ton-miles of freight per month working on a 10 hour per day basis, with Sundays excepted. They are proving their superiority over the steam engines replaced in many ways. The availability of the market, permitting the purchase of cheap power affords an opportunity for fuel economy, which coupled with the greatly decreased maintenance cost of these locomotives, and crew wages has caused a larger reduction in operating expense.

The repair expense saved amounted to $1,931.59 on two locomotives.

The electric locomotives, in addition, eliminated the water expense and created a reduction in crew and labor wages of approximately $2,160.00 per year.

These locomotives are given a running inspection every other day, which takes one hour, while they are given a general inspection the first Sunday in each month.

A Three-Car Multiple-Unit Passenger Train—Spokane & Inland Empire Railroad

Single Track Systems

THE introduction of the high-voltage system which permit. operation over extended distances with only a small loss in power, has resulted in many new railroad promotions using the electricity as a motive power, which would formerly have been installed using the steam locomotive. The low cost of water power, the multiple-unit operation of passenger cars, and the electric locomotive for freight work result in reduced operating costs over steam operation.

Two 52-Ton Baldwin-Westinghouse Locomotives and Freight Train—Spokane & Inland Empire Railroad

A Two-Car Multiple-Unit Train Arriving Only 20 Minutes Late After Bucking Snow Drifts. Steam Roads in This Vicinity Suspended Operations on Account of the Storm

Spokane & Inland Empire Railroad

THE Spokane & Inland Empire Railroad is an instance of the contemporary development of the railroad along with the territory served.

This great system of the Northwest was built considerably in advance of the country through which it operates, and is consequently responsible to a great extent for the rapid development of this vast domain.

Starting in 1906, this road has constantly expanded, opening new country until it now covers 186 miles of route.

Standard three-car passenger trains operate over this entire system using the single-phase apparatus with 6600-volt trolley, with a total daily service of 22 trains.

The schedule speed of these trains is 30 miles per hour, making schedule stops every 4 miles and subject to an equal number of flag stops.

Cuts Cleared Out by Electric Locomotive Bucking Snow—Spokane & Inland Empire Railroad

THE rapid expansion and growth of this road was made possible without great additional expense by the initial installation of the high voltage trolley requiring only simple transformer substations.

They make a regular 180-mile freight run with all the local freight work incidental to it in 13 hours. The ton-mileage of these locomotives closely approximates 3,000,000 per month.

The electric locomotive is especially well adapted for cold weather and blizzards. It does not have the drawback of the steam engine due to low steam in cold weather, but has an overload capacity at all times which is of great advantage in bucking snow.

Two of the Westinghouse single-phase 52-ton locomotives behind an ordinary snow plow cleared 15 miles of track in February, 1913, without injury to the apparatus.

The snow-drifts in cuts reached a depth of 15 feet, while reports showed a fall of 54 inches over this entire section of country.

The power house equipped by the Westinghouse Company is built in the bed of the river as part of the dam, and is so located that the former bed of the river forms a tail race. In addition to operating the railroad, it supplies two 1500-kw., motor-generators and a 100-kw. industrial load.

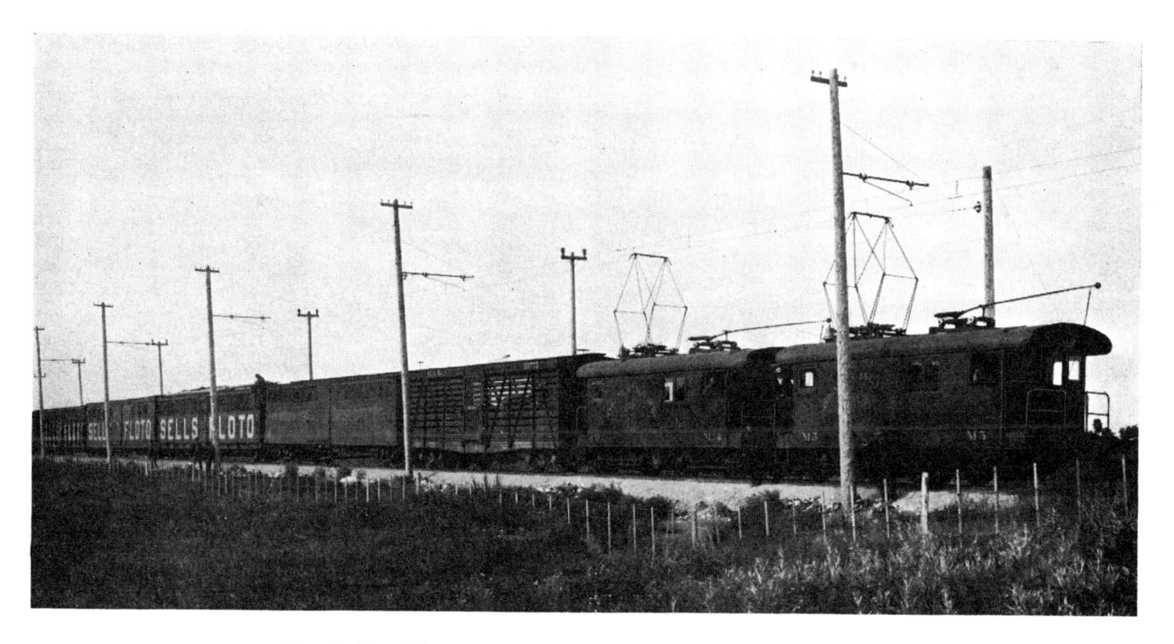

Two 52-Ton Baldwin-Westinghouse Locomotives and Circus Train

THE freight service is handled by double-headed 52-ton locomotives using one crew, which can handle a 450-ton train at approximately 17 miles per hour on a 2 per cent grade, while their maximum speed is 30 m. p. h. Also 72-ton locomotives are used for freight service on this system.

Loading Wheat at Palouse, Wash.

A Four-Car Multiple Unit Train—Spokane & Inland Empire Railway

Nine Mile Power Station

Located on the Spokane River nine miles from Spokane. 125 miles of the Inland Division are operated from this plant, as well as two 1,500 kw. railway motor-generator sets and approximately 1,000 k.v.a. industrial load.

Car Emerging From Redwood Peak Tunnel

Oakland, Antioch & Eastern Railway

THIS new development operates with Westinghouse Electric equipment on the same basis as modern steam railroad practice.

It forms a direct line between Oakland on San Francisco Bay to Sacramento, 84 miles distant, through a beautiful and fertile country which has not heretofore had adequate transportation facilities.

This road crosses the Coast Range mountains through Shepards Canon on a 3 per cent grade, piercing Vontra Cocto hills into the Redwood Canon by a 3500-foot tunnel of solid rock.

It affords a splendid example of the application of Westinghouse equipment for high-voltage direct-current operation. Passenger service is handled by multiple-unit and locomotive hauled trains.

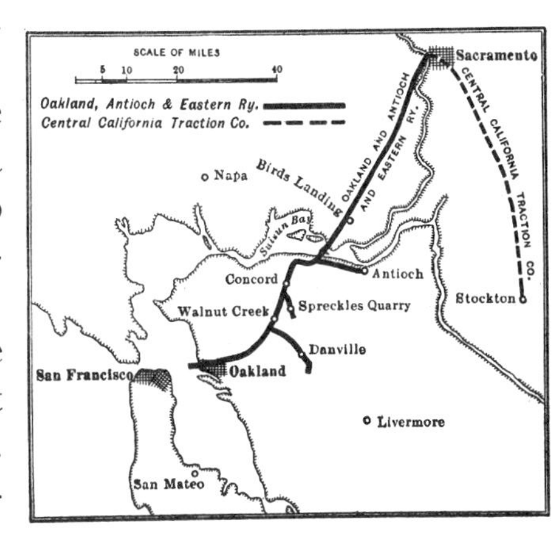

Map—Oakland, Antioch & Eastern Railway

View Showing 1200-Volt Direct-Current Line Construction

THE passenger cars weigh 43 tons and maintain a schedule speed of 30 miles per hour, reaching a free running speed of 50 miles per hour.

General View, Showing Right-of-Way

One of Two 47-Ton Baldwin-Westinghouse Freight Locomotives

THE Baldwin-Westinghouse passenger locomotives weighing 62 tons are provided with pony trucks and designed with all the fundamental principles which steam practice has proven essential for high speed work.

They haul trains of five coaches at a speed of 45 miles per hour and are designed for a maximum speed of 70 miles per hour.

The freight service is handled by 47-ton Baldwin-Westinghouse locomotives which operate at 27.5 miles per hour maximum speed and with a maximum tractive effort of 24,500 pounds.

The economies effected in the operation of this road together with its flexibility for ready expansion which so soon became necessary were both important factors of high speed and frequent electric service.

One of Two 62-Ton Baldwin-Westinghouse High-Speed Passenger Locomotives

A 55-Ton Baldwin-Westinghouse Locomotive and Freight Train

The Piedmont & Northern Lines

1500-Volts Direct Current

THE Piedmont & Northern Lines comprises a system of approximately 125 miles of electric railways, having at the present time its northern terminal at Charlotte, N. C., and its southernmost terminal at Greenwood, S. C. This system lies principally east of and parallel to the Piedmont Ranges of the Appalachian Mountain System in the states of North and South Carolina.

The Piedmont & Northern Lines is one of the representative railway systems of the South, forming a net work of lines which handle heavy steam railroad and electric railroad traffic. These lines constitute a railroad system with its private right-of-way, low grades, long radius curves, steel bridges, brick and concrete bridges, high-powered locomotives for freight service, high-speed, large-capacity motor cars for passenger and express service; in fact, combining the light frequent service of the interurban with the great capacity of the steam railroad; a complete railroad system using electricity as its motive power.

Electric Locomotive and Train of Standard Steam Railroad Rolling Stock

FOUR distinct types of train service are maintained on the Piedmont lines, as follows: (1) limited passenger; (2) local passenger; (3) light freight and express; (4) heavy freight. Limited and local passenger service is handled by high-speed interurban cars. Light freight service is handled with express car locomotives. Heavy freight is transported in standard freight cars hauled by Baldwin-Westinghouse electric locomotives.

The 55-ton Baldwin-Westinghouse locomotives used in the freight service regularly haul 800 tons trailing load, and occasionally pull as much as 1000 tons. These locomotives often handle solid through trains carrying all steel, electrically-lighted drawing room, sleeping cars, which make connections with Atlanta, and all points South and West.

A 55-Ton Locomotive and Coal Train

A Holiday Excursion Enroute—Four-Car Multiple-Unit Train

SUNDAY and holiday service is particularly heavy to recreation parks along the Piedmont & Northern Lines. The multiple-unit passenger train shown herewith is a train on its way to Recreation Park near Mt. Holly, N. C., on the Catawba River.

Special excursion trains are chartered by societies and fraternal organizations. The government of South Carolina has also used this road for the transportation of state troops to the annual encampment.

The package service in addition to the Southern Express Company's business often requires multiple-unit operation with express and passenger coaches on the run between Charlotte and Gastonia. A single express car schedule is also maintained. Heavy freight service is increasing on all sections, particularly over the 94-mile run from Spartanburg to Greenwood. The former place is the transfer point for both the Southern Railway and the Carolina, Clinchfield & Ohio Railway, and the latter with the Seaboard Air Line.

A typical combination passenger and freight station used in certain cities like Spartanburg, S. C., is shown herewith. These depots are combined with large cotton warehouses, cotton and coal forming a large percentage of the freight movements in this part of the country.

A Four-Car Multiple-Unit Train—Package Freight and Passenger Service

In addition to the regular local trains, parlor observation limited cars run between various points, and by the Seaboard Air Line connection at Greenwood, S. C., with the Spartanburg-Atlanta Limited.

Express-car locomotives are used for hauling light freight and express. In addition to carrying a load of express, or freight, these cars are capable of hauling a trailing load of 150 tons. A vestibule and motorman's operating equipment is provided at each end of the car for operation in either direction. The electrical equipment is similar to that used on the passenger cars. These express-car locomotives are 50 feet long, 9 feet 2 inches wide, and 13 feet 6 inches from rail to top of trolley board.

A Multiple-Unit Train—Combination Passenger and Freight Service

A 60-Ton Baldwin-Westinghouse Locomotive and Train—Southern Pacific Company, San Francisco

Southern Pacific Company

Pacific Coast Properties

600-1200-1500-Volts Direct Current

THE Southern Pacific Company in the electrification of certain sections of their system are using 15 Baldwin-Westinghouse locomotives of the above type. These locomotives operate in general freight service, hauling anything from a gravel train to a fruit train, operating on an average of 14 hours per day and making from 1700 to 2400 miles per month. The trains hauled on level track, sometimes weigh from 975 to 1300 tons, being made up of from 15 to 20 loaded cars. On grades a smaller number of cars are hauled, depending on the length and slope of grades to be encountered. During the fruit season, these locomotives prove an important factor in moving the fruit to market.

One of Ten 60-Ton Baldwin-Westinghouse Locomotives and Train—Pacific Electric Company, Los Angeles

HEAVY holiday service is sometimes taken care of by using the electric locomotives to haul passenger cars. On one occasion several steel motor cars, which were received too late to have the electrical equipment installed before a holiday, for which they were to be used, were placed in trains of three to seven cars each and hauled by electric locomotives.

The Pacific Electric Railway is one of the best examples of railroad electrification. It covers approximately 1000 miles of track. The average locomotive failure is one per 25,000 miles.

The electric locomotives are kept in service every day and are making a record for maintenance and reliability. These locomotives are always ready for service, and about the only time any of them are in the barn is when a run is finished.

The smooth operating control, ease of inspection, reliability and the small number of repairs required have made these locomotives a source of great satisfaction to the train crews and operating men.

Passenger service is handled by single car and multiple-unit train operation.

Steam vs. Electricity

Branch-Line and Feeder Systems

THE steam train carries a crew of five persons and costs sixty cents or more per mile to operate, giving infrequent service.

The electric car is handled by two men, and the cost of running rarely exceeds twenty cents per mile. It will stop every mile if necessary for passengers, and carry them quickly and comfortably into the city.

Frequent and superior service where such lines parallel usually result in the electric road getting the major portion of the business.

The steam railroads are now realizing these conditions and are investigating the electrification of their own branch lines.

Limited Train—Lackawanna & Wyoming Valley R. R.

11,000-Volt Catenary Line Construction on a Branch Line

Single-Phase Catenary Line Construction

Line Construction for Branch Line and Feeder Systems

1500-Volt Direct-Current Line Construction

Four-Car Multiple-Unit Train in Branch Line or Feeder Service

Erie Railroad—Rochester Division

Branch Line Electrification

ONE of the first steam railroad branch line electrifications in the United States, operating since 1907, is that of the Erie R. R., between Rochester and Mt. Morris.

Electric operation resulted in an increase of 50 per cent in the passenger travel based on local ticket sales.

This 34-mile section is maintained by three car inspectors, one lineman and a general supervisor, the right-of-way being handled by the steam division.

The usual service is handled by trains of one motor-car and one trailer. However, a multiple-unit train operates on special days.

The single-phase system using an 11,000-volt trolley permits heavy traffic without expensive feeder system.

It is a noteworthy fact that the entire steam history of this division fails to show a single perfect month.

The total operating cost of the electrified division is 15.6 cents per car-mile and 21.8 cents per motor-car-mile. This covers maintenance of overhead line, the rolling stock, the substation and power.

Three-Car Multiple-Unit Train

Lackawanna & Wyoming Valley Railroad

THE Lackawanna & Wyoming Valley Railroad is an example of competition resulting from an electric line paralleling the existing steam road. This road is a double track system extending from Scranton to Wilkesbarre, Pa., a distance of 19 miles. Although in direct competition with steam trains, the high class and frequent service furnished, results in a very heavy traffic. The flexibility of the electric system is shown by the fact that trains are operated, using from one to four cars as the traffic demands, the limited trains on hourly headway making the 19-mile run in thirty minutes, while local trains operating on 20-minute headway, make the run in forty-one minutes. There are 260 trains handled at each terminal daily. In addition to the passenger service, a large freight business is conducted, consisting of six freight trains each way daily, which service requires only two locomotives. These locomotives are equipped with four Westinghouse direct-current, 600-volt motors of 150 horsepower capacity each, and have a tractive effort of 23,500 pounds.

Two-Car Multiple-Unit Train

Rock Island & Southern Railway

THE Rock Island Southern is a practical example of high voltage railroading. It is a real railroad, and its economic location is good, especially for heavy freight traffic.

Here is a 50-mile route with the power plant at the middle of the line and a single trolley wire between terminals, without a feeder and without a substation. With 11,000 volts on the trolley a 500-ton train can be handled at the end of the line without a material drop in pressure at the train.

The transportation facilities which this new route furnish are unlocking the resources of this territory. Stockyards are going in, grain elevators are being built, towns are growing, and diversified industries are springing up. The reliable service furnished by the electric motors is the main factor in the development. The steam locomotive with its infrequent three-car local passenger trains and its slow moving local freight trains could not accomplish such results as the Rock Island Southern is obtaining.

Results of Electrification

THE simplicity of the equipment is emphasized by the fact that no corps of electrical experts is required to keep the "motors rolling." The general superintendent, previous to his coming to this property, had no experience with electric motive power. The master mechanic, the dispatchers, likewise the roadmaster are all steam experienced men. Necessarily, some of the men at the car barn are electrical men, but the bulk of the work at the barn is mechanical. Keeping the motors and control in commission is by no means the most expensive part of the barn work.

Steam road men watching the performance of the rolling stock on the Rock Island Southern cannot escape being impressed with the simple solution which high voltage trolley and low voltage motors and control give for steam-road branch-line work. These men see that this motive power is easy to apply and easy to operate; that it is simple and reliable; and that it can handle a heavy freight train as easily as it handles a single passenger train. Assuming that power can be bought at the trolley, as it actually is in the case of the Rock Island Southern, they see that the investment practically narrows down to the trolley, the motors and a modest addition to the shop and stores. This brings the investment down to a point where steam-road managements can begin to take interest in the subject of branch-line electrifications. Then, with examples of satisfactory performance, such as the Rock Island Southern electrical equipment affords them, the steam men can give branch-line electrification serious attention.

Motor Car for Branch Line or Feeder Service, Equipped With Westinghouse Motors and Unit Switch Control—Rock Island Southern Railway

Three-Car Multiple-Unit Train Equipped for 1200-Volt Direct-Current Operation

The Maryland Electric Railways

THIS road is considered as one of the oldest in the South, and among the first to appreciate the value of electric service, which was inaugurated in May, 1908. Prior to electrification, the steam service consisted of three trains per day each way; but now the attractive service of a train every half hour is offered. Operation is carried on by running a local train on the hour, and a limited on the half hour.

The Maryland Electric Railways, known as the Annapolis Short Line extends from Baltimore terminal to Annapolis, Md., terminal, a distance of 25.3 miles.

The twelve-hundred-volt direct-current system is used for supplying power on this road, and the rolling stock is all equipped with Westinghouse railway equipment.

Part of the freight, the greater portion of which is moved in car-load lots, is now handled by electric locomotive at a decided reduction in cost.

Electric service brought with it a consistent increase in travel and a handsome increase in population along the right of way. The water fronts became more attractive for summer outings and year-round homes; and apparently the two cities have been brought closer together.

A 47-Ton Baldwin-Westinghouse Locomotive and Train

The Visalia Electric Railroad

ELECTRIC operation commenced in March, 1908, on this road, which is located in Tulare County, California, in the heart of a rich citrus fruit district. This road extends from Visalia to Exeter, a distance of 10.1 miles, and then to Red Bank 18.8 miles farther, and in all there are approximately 40 miles of route including all branches.

Power is purchased from the Mount Whitney Power and Electric Company. This feeds at 35,000 volts to the three transformer substations, and these feed the trolley at 3300 volts.

The heavy freight traffic on this road takes place during the months of October, November, December and the first part of January. This time of year is the season for packing and shipping the citrus crop.

Multiple-Unit Car

DURING these months in the season of 1913-14 the one locomotive handled over 1200 cars of oranges and lemons. Throughout the fruit packing season positive and unfailing service is demanded from this locomotive; any failure which would cause it to miss a single scheduled trip would entail the loss of probably the entire shipment of fruit which was then prepared for movement. The growth of the freight business handled over this system has been very rapid and last season's tonnage was between three and four times the tonnage handled five years ago.

Overhead Tangent Construction

The present passenger schedule of this road provides for a total of two round trips between Visalia and Lemon Cove and seven round trips between Visalia and Woodlake, two of these latter trains running on to Red Banks. It is possible with the equipments in service to reduce considerably the scheduled running time, but since the schedule is largely determined by connections with steam trains, there is no necessity for doing so. The average daily car mileage made by the passenger equipment is 545.

Figures compiled from the operating records for the last quarter of 1913 show that the average maintenance cost of equipment was as follows:

Cost of electrical and mechanical repairs and renewal of parts of passenger and combination cars, including motor and control equipment, air-brake equipment, trucks and bodies, 1.839 cents per car-mile.

Cost of electrical and mechanical repairs and renewal of parts of locomotive, including motor and control equipment, air-brake equipment, cab and trucks, 1.134 cents per locomotive-mile.

Total cost per car-mile of both passenger and locomotive equipment, including the cost of inspection, oiling, car cleaning, etc., 2.487 cents per car-mile. Average power consumption at the switchboard is 77 watt-hours per ton-mile for all service.

Overhead Yard Construction

Mechanical Parts of Ten Electric Locomotives Shipped 350 Miles in One Train From the Baldwin Locomotive Works, Philadelphia, to the Westinghouse Electric & Mfg. Co., East Pittsburgh, Pa., for Installation of Electrical Equipment

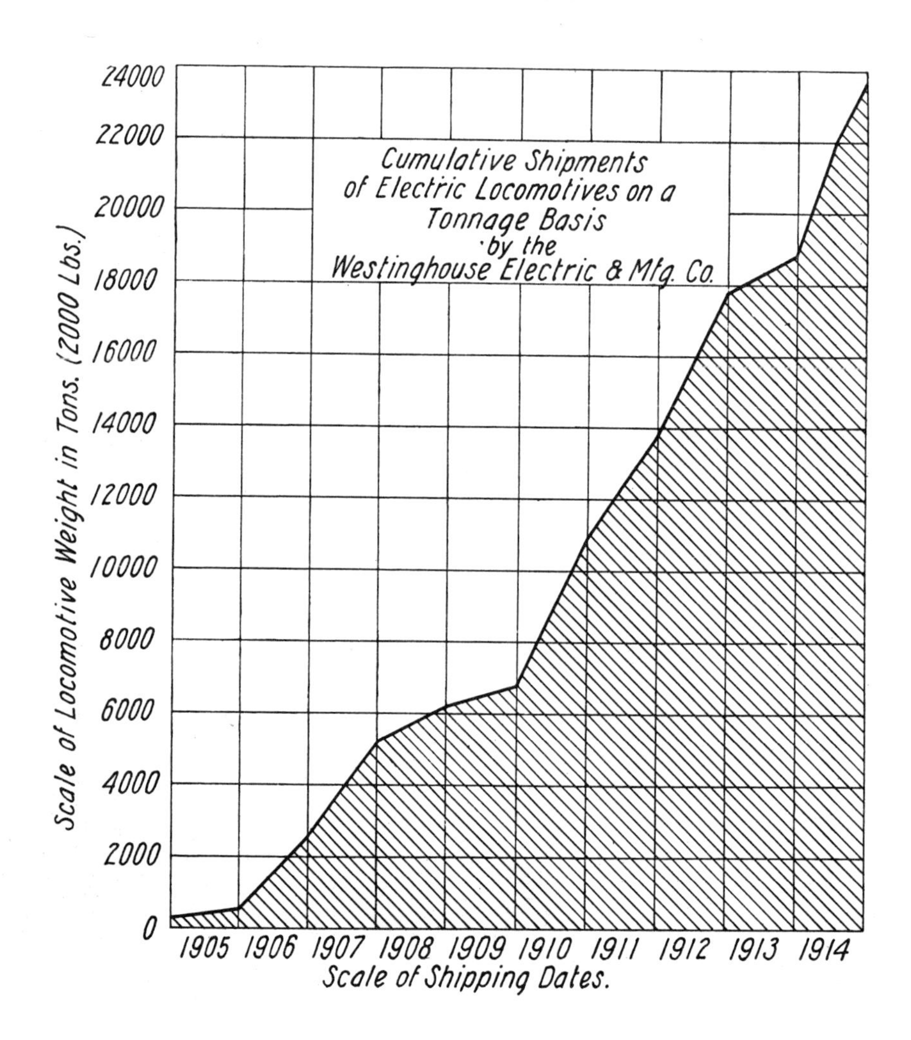

Electrified Steam Railroads in the United States of America

No.	Road	Date	Miles of Single Track	Miles of Route	Locomotives	Cars
1	Baltimore & Ohio R. R.	1895	7½	3	6	..
2	Northern Pacific R. R.	1903	32	20	...	29
3	*N. Y. N. H. & H. R. R.	1905	550	88	100	70
4	*Long Island R. R.	1905	250	100	...	393
5	New York Central R. R.	1906	235	53	61	200
6	W. Jersey & Seashore R. R.	1906	150	60	...	68
7	*Erie Railroad	1907	40	34	...	8
8	Oneida Railroad	1907	118	44	...	23
9	*Annapolis Short Line	1908	30	25	...	12
10	*Grand Trunk Railway	1908	12	3½	6	..
11	*Visalia Electric Railroad	1908	45	31	1	8
12	Great Northern R. R.	1909	6	6	4	..
13	*New York Terminal—Pennsylvania R. R.	1910	132	20	33	8
14	*Rock Island Southern Rwy.	1910	78	50	2	6
15	Michigan Central R. R.	1910	19½	4	10	..
16	*Wash. & Old Dominion R. R.	1911	56	52	...	10
17	*Boston & Maine R. R.	1911	22	8	6	..
18	*Southern Pacific R. R.	1911	48	25	15	45
	Southern Pacific R. R.	1911	48	25	...	65
19	Butte, Anaconda & Pacific R. R.	1913	90	30	21	..
20	*Norfolk & Western Ry.	1914	85	30	12	..
21	*Philadelphia Terminal—Pennsylvania R. R.	1914	90	20	...	93
22	*Denver Interurban R. R.	1908	57	44	...	8
23	*Jamestown, Westfield & Northwestern R. R.	1913	42	32	1	10
24	*Niagara Junction Railway	1913	12	3	2	..
	Total		2255	810.5	280	1056
	***Westinghouse Electric & Mfg. Co.**		**1549**	**565.5**	**178**	**671**

THE above list includes all of the steam railroad track which has been electrified in the United States. The roads marked with an asterisk designate the portion of this work which has been done with Westinghouse apparatus.

In addition to this list, there have been a large number of new roads inaugurated which were installed with an initial equipment for electric operation. These roads, however, operate on a steam railroad basis and would have used the steam equipment had it not been for the great advance in electric operation, as proven whenever installed. These roads are such as the Spokane & Inland System comprising 160 miles of track with 28 motor cars and 11 electric locomotives; the Piedmont Traction Company with 280 miles of track, and the New York, Westchester & Boston Railroad; also, many others operated with Westinghouse equipment.

Westinghouse
Electric Railway Equipment
for
Speedy and Comfortable Transportation Service

SPECIAL PUBLICATION, 1863
OCTOBER, 1929

Westinghouse Electric & Manufacturing Company
East Pittsburgh, Penna.

FOREWORD

SINCE the beginning of street railway operation, hamlets have grown to cities and cities to metropolitan centers of production, wealth and culture; expanding under the beneficial influence of low cost transportation.

In spite of the constantly increasing number of automobiles, the electric railway industry continues to furnish the citizens of our centers of population with safe, reliable, frequent and inexpensive service.

Today, traffic congestion in our arterial streets grows steadily worse and the remedy seems to be restricted parking, signal control and more and more speed for all classes of vehicles. In fact, this is the age of speed; records in all modes of transportation are being broken almost daily.

Included in this situation is the street car which must be equipped to hold its place in the never ending race of getting countless passengers somewhere.

With the modern equipped street car, higher accelerating and running rates of speed can be accomplished with passenger comfort comparable with that of the latest model automobile. Actual operation of these cars results in increased patronage for the railway industry.

Westinghouse railway engineers, long experienced in transportation equipment matters, have cooperated with operating engineers and developed the motive power apparatus described in this book. This apparatus will enable railways to furnish higher speed and more comfortable service to an ever increasing number of passengers.

In addition to the description of modern street car equipment, data are included on the latest practices and apparatus for train operation, rapid transit, higher speed interurban service, trolley and gas electric busses, and electric locomotives for freight haulage.

Cleveland Public Square During the Rush Hour at Five O'Clock in the Afternoon

A Modern Car for Rapid City Service in Brooklyn

THE MODERN STREET CAR

THE modern street car has become the butterfly of urban transportation. Red or yellow exteriors have changed to color combinations which excite the envy of the modern roadster. Hard, slat seats have given place to thick-cushioned ones with leather covering, individual chairs with form fitting backs or others equally remote from the type all too common in the past.

Realizing that pleasing color schemes and comfortable seats, important as they are, are not sufficient to place the street car on the plane its fundamental characteristics warrant, the Westinghouse Electric and Manufacturing Company has made certain notable developments which improve the performance of the car. These developments are:

Type VA (Variable Automatic) Control for hand or foot operation.

High-Speed Motors and W-N Double Reduction Drive.

High-Speed Motors for Timken and other Worm Drives.

High-Speed Standard Single Reduction Motors for 50 and 100-hp. ratings.

The use of high motor speeds permits maximum capacity for minimum weight. Therefore, for any fixed car weight, greater motor capacity can be provided. This greater motor capacity may be used to increase rates of acceleration, rates of braking (if dynamic braking is employed), free running speed, or to afford a combination of the three. The additional power provided results in higher schedule speeds. The new drives have further advantages of less unsprung weight, less noise, the elimination of axle bearings, the lubrication of gears with clean oil, and, in the case of W-N drive, small diameter wheels and low floors.

Non-automatic control has been favored in the past for the varied conditions of traffic and grades met on city streets. The earlier forms of automatic control functioned on a constant current basis. When the accelerating current was set for the maximum permissible rate on the level with a lightly loaded car, the resulting rates with fully loaded car and on grades were entirely too low. To overcome this difficulty it was necessary to provide means of advancing the control progression, that is, the control was made non-automatic. Even with carefully de-

signed resistors, non-automatic control does not afford the desired degree of smoothness with extremely high rates of acceleration.

Type VA control overcomes the objections to both non-automatic control and the older forms of automatic control. A number of accelerating points are provided. Various rates of acceleration can thus be employed and all rates are under the control of the current relay and are, therefore, automatic. This insures smoothness of acceleration over a wide range

Curve A 3.0 Miles per hour per second rate of accelerating and braking.
Curve B 1.5 Miles per hour per second rate of accelerating and braking.

Chart I (Curves A and B)

of operating conditions. The exact values of the rate provided depends on the particular operating conditions to be met. If the rates provided on the level

vary from 1.25 to 6 miles per hour per second the rates available on a five per cent grade are .25 to 5 miles per hour per second.

Chart I (Curves A and B) illustrates what can be accomplished under various frequencies of stops by increasing average accelerating and braking rates from 1.5 miles per hour per second to 3.0 miles per hour per second, all other conditions remaining the same.

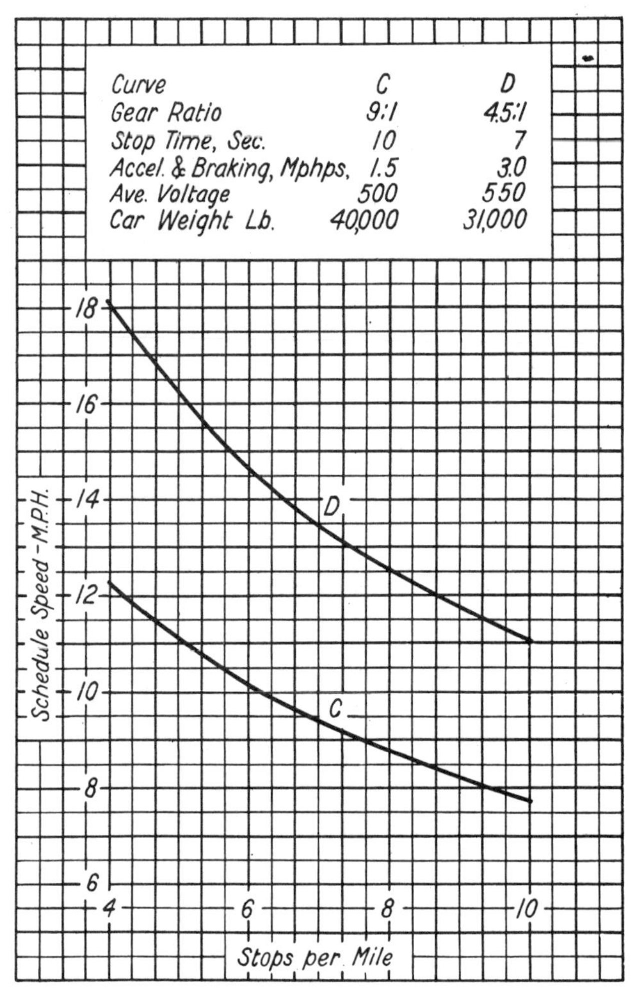

Chart II (Curves C and D)

If we assume that Curve C, on Chart II, shows the schedule speeds possible with a typical street car of old design weighing 40,000 pounds, the performance shown on Curve D may be obtained with a 31,000-pound modern car by raising the average line voltage, gearing for higher free running speed, increasing accelerating and braking rates, and reducing the duration of stop.

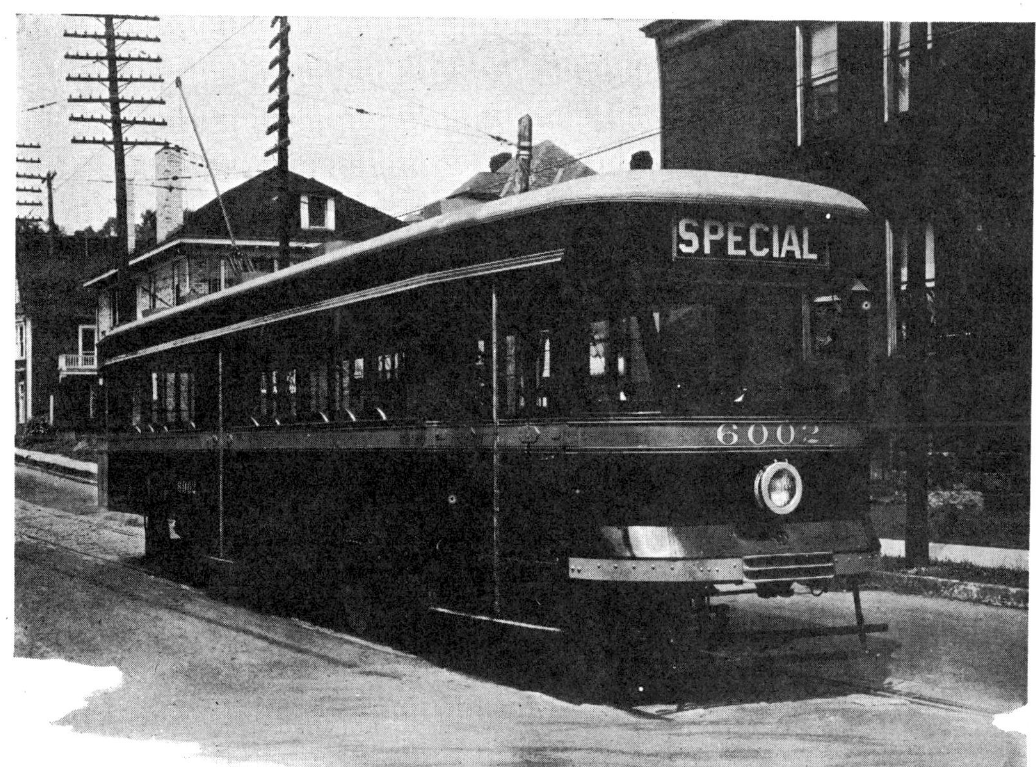

The New Pittsburgh Railways Light-Weight City Car with Foot-Operated VA Control and Dynamic Braking

MODERN CITY STREET CAR CONTROL

TRAFFIC congestion, due to the constantly increasing number of automobiles that use our streets, is a factor of great concern to those responsible for the operation of adequate railway service in any city. Municipal authorities in meeting the situation have placed most arteries used by street cars under "Stop" and "Go" signal control and have encouraged higher operating speeds for all vehicles. This has resulted in the realization by railway managers and equipment engineers that it is necessary for street cars to be able to travel at higher speeds when breaks or openings in traffic permit.

In frequent stop service, cars must make higher rates of acceleration and deceleration as well as increased free running speeds. This must be accomplished with comfort and safety for the passenger. In order to assure smooth yet higher rates of acceleration under all conditions of operation, equipment engineers recommend automatic control for city street cars with provision for variation in the rates of change of speed. Westinghouse engineers have developed the type VA (Variable Automatic) control to fulfill the requirements of increased comfort and speed in city street car operation under all conditions of service.

A typical bill of material for Westinghouse type VA control for four 50-hp. motors arranged for cabinet installation is given on the following page. This control may also be furnished with main switching parts designed for installation in boxes under the car floor. Foot operation is specified but hand operation is optional. Electro-pneumatic unit switches are also specified. Magnetic contactors can be furnished upon specification.

Operating Features of VA Control

Variable tractive effort is essential in order that the operator may be able to accelerate slowly in very

Bill of Material for Foot-Operated, Electro-Pneumatic Control, Variable Automatic Acceleration.

No Req.	Apparatus	Weight Lb.
1	US-24 Trolley Complete	115
1	Type 220-A-3 Fuse Box	11
1	Type 806-H Line Switch	80
1	Type 371-M-19 Overload Relay	14
8	Type UP-21 Electro-pneumatic Contactors	136
1	Type XD-524 Reverser	40
1	Type TK-752 Main and Motor Cutout Switch	8
1	Set of Type M Resistors	70
1	Type XM-14 Foot Controller	30
1	Type XS-116 Sequence Switch	90
1	Type 379-M Limit Relay	6
1	Set Pneumatic Details	35
1	Set Insulating Details	10
1	Set Main and Control Cable	120
1	Set Knuckle Joint Connectors	8
1	Type TC-2 Control Switch	3
	Total Weight	781

dense traffic and rapidly when the street or corner is clear. He also must be able to sustain the rate of acceleration when the load or grade increases, particularly at rush hours. Similar provision also must be made for most efficient operation over different rail conditions which may vary widely during the same day over different parts of the route.

Variable tractive effort is obtained by having a series of positions on the master controller which correspond to a predetermined accelerating rate. A current limit relay is used to govern the rate of acceleration. This relay has a shunt coil on it which is used in connection with the master controller. The position of the master controller determines the amount of resistance in series with the coil. Changing the current flow through the shunt coil by manipulation of the master controller, determines the setting of the relay. That is to say, the first accelerating notch on the master controller may give an accelerating rate up to full parallel of 1 mphphs. The second notch may give an accelerating rate up to full parallel of 1½ mphps. The sixth notch may give an accelerating rate up to full parallel of 3½ mphps. The rates of acceleration for each notch on the controller may be adjusted to suit the schedule requirements.

Motormen soon become skilled in operating the master controller to secure the car speed required to meet various conditions of traffic.

Operators Station of a Car Equipped with VA Control, Showing the Foot Pedals which Actuate the Master Controls for Both Acceleration and Braking

Quick response is necessary that the car respond immediately for any change in the position of the controller lever. With an initial movement of the controller to the "on" position power is applied without time lag to the motors. All automatic controlling devices instantly go to the "off" position when the controller is moved to that position. If, when coasting at 12 or 15 miles per hour, the controller is moved to a running position, the motors are thrown on the line without loss of time.

Quick response is obtained by using a fast-acting sequence switch properly controlled from the master controller. The sequence switch carries the interlock fingers which control the accelerating switches. It is driven by a balanced pressure air engine which has two opposing pistons. The cylinders are filled with air and the engine is moved by releasing air from either cylinder, depending upon the direction to be used. Due to the small torque required to rotate the drum which handles control circuits only, the operation is positive with no tendency to overrun.

Smooth and rapid acceleration are required to make increased schedule speeds and at the same time not to disturb the free movement of the passengers.

The smoothness of acceleration depends on three things; the type of control, whether automatic or non-automatic; the number of notches; and the effectiveness of each notch. The most important of these three factors is the application of automatic notching as it produces even steps. Smooth acceleration enables the use of high rates of acceleration without annoying the passengers. The limit of accelerating rates under this condition seems to be the point where the rates begin to be so high as to stop the passengers from moving in the car after boarding or before alighting. If the passengers do not adjust themselves promptly in the car, there will be delays in loading and unloading which cancel any saving in time gained by high acceleration rates.

Smooth acceleration is accomplished principally by the application of automatic notching. Each notch is taken at the proper time interval. The second factor which effects smoothness is the correct number of equally effective notches. The number of notches can be too few or on the other hand, they can be too many. A large number of notches ineffectively applied will not produce any better notching than a fewer number of notches effectively applied and tends to increase time of operation as it complicates the equipment.

Effective notching provides such a number of control notches as may be effective in obtaining smooth acceleration. This makes for simplicity of the control which is desirable from the standpoint of installation, operation, and maintenance. With Westinghouse VA control, when the controller is thrown on, the notching of the car is started immediately so that smooth acceleration is obtained with 11 notches. The resistance is designed so that each notch is an effective one and, therefore, the car is smoothly accelerated without loss of time.

Holding is desirable to enable the operator to stop the progression of the control at any point and hold this speed.

All automatic equipment provides a holding position at the master controller, the purpose of which is to hold up the progression of the control at any point. This enables the car to follow up in slow traffic. For ease of manipulation, the "holding" begins immediately on a slight back-up movement of the controller handle rather than the return of the handle to the specific point on the controller before the definite hold is effective.

This is accomplished by having the holding point on the master controller de-energize the "off" magnet on the sequence switch. This causes a sequence switch to stop at any point to which it has advanced. However, the "holding" begins immediately on a slight back-up movement of the controller handle.

Adaptability of method of operation to either foot or hand control is desirable to meet local requirements of service. One man operation is simplified by foot operation with certain types of cars. Either foot or hand controller produces the same car performance as related to the electrical equipment.

Safety devices of some form are desirable for one man operation and with foot control are arranged so that the operator does not need to have both feet on the operating pedals at the same time. This enables the operator to remove one foot from the pedals at a time for rest and comfort.

Simplicity and reliability are essential features for high grade schedule performance as regards continuity of service. This is assured in VA control by the use of standard unit parts, the satisfactory operation of which is proven by many years of service throughout the electric railway industry.

Emergency operation is provided so that two motors can be cut out with a switch located inside the car and near the operator. Automatic operation will notch up for 2 motors the same as for 4 motors. This is accomplished by properly locating the limit relay in the main circuits at the cut out switch. This type of control is adapted very well for the purpose of pushing or pulling a loaded crippled car off the road. In fact, it probably has a slight advantage over hand control in this respect. This is due to the use of variable tractive efforts, with the notching automatic and uniform. The standard equipment for type VA control is a hand operated reverser arranged with an emergency dynamic brake position. It requires only one action by the operator to secure braking through the motors, which braking is independent of the line voltage and the direction of the car.

Performance

The experimental car No. 6002 of the Pittsburgh Railways Company, as shown in the illustration, is equipped with Westinghouse type VA control and four type 1426-BT, 50-hp. motors. Due to the

Type VA Control-Panel for Cabinet Mounting

extensive use of aluminum alloy in the body, the complete weight is 27,000 lb. ready for service. The car is operated from a foot controller and the magnetic contactors are installed in a cabinet above the car floor.

The car has a free running speed of 45 mph. on level tangent track and 28 mph. on a 5.9 per cent grade with straight track. Remarkably smooth acceleration is experienced by passengers at the 5 mphps. maximum rate attained by this equipment and lower rates are obtained with equal comfort. Average smooth acceleration rates of 3 mphps. may be accomplished with ease in regular city service.

This car is also equipped with automatic dynamic braking operated by the foot controller. Smooth decelerations are made at rates up to 5 mphps. The dynamic braking operation includes the variable feature for use with different conditions of adhesion. Dynamic braking is considered to be in an experimental stage of development and is now applied on a limited number of trial street cars.

Standard HL Control

Many city street cars are operating in a very satisfactory manner with standard Westinghouse electro-pneumatic unit switch control. Specifications for this hand control apparatus will be furnished upon request. HL control is used extensively in Cleveland, Los Angeles, and San Francisco in congested city service.

Drum Control

The hand operated drum series parallel controller of the K type continues to be applied on many properties. The table below gives principal characteristics of this line of controllers.

Characteristics of Drum Controllers

Controller	No. of Motors	Max. Allowable Capacity of Each Motor		Number of Points		Wt.-Lb.
		Hr. Hp.	Cont. Amp.	Series	Parallel	
K-35-KK	4	65	60	5	3	225
K-63-BR	2	40	39	4	3	135
K-64-A	4	110	105	6	4	450
K-68-A	2	70	66	4	3	233
K-75-A	2	75	75	5	3	143
K-75-A	4	50	50	5	3	148

Westinghouse Line Switches

Line switches are recommended with all K controller installations on account of superior performance of the following primary functions:

To take a large percentage of the arcing from the controller fingers during normal operation.

To operate as a circuit-breaker, thus giving protection to the motors and control apparatus.

To function as an automatic disconnect for the motors and the control apparatus, when the controller is in the "off" position.

To remove all power current interruption from the car platform. This eliminates the danger of controller explosions and violent opening of the circuit-breaker, which have always been a source of fright and considerable danger to passengers.

Two different types of line switches are available, one magnetically operated, the other pneumatically operated. The table gives principal operating characteristics:

Line Switch Rating Data

Type	Max. Hp.	Net Wt.-Lb.	Remarks
UM-2-I	260	95	Magnetic switch with overload relay
UM-2-H	260	100	UM-2-I with pneumatic trip
806-J-4	260	110	Pneumatic switch with overload relay
806-J-7	260	115	806-J-4 with pneumatic trip

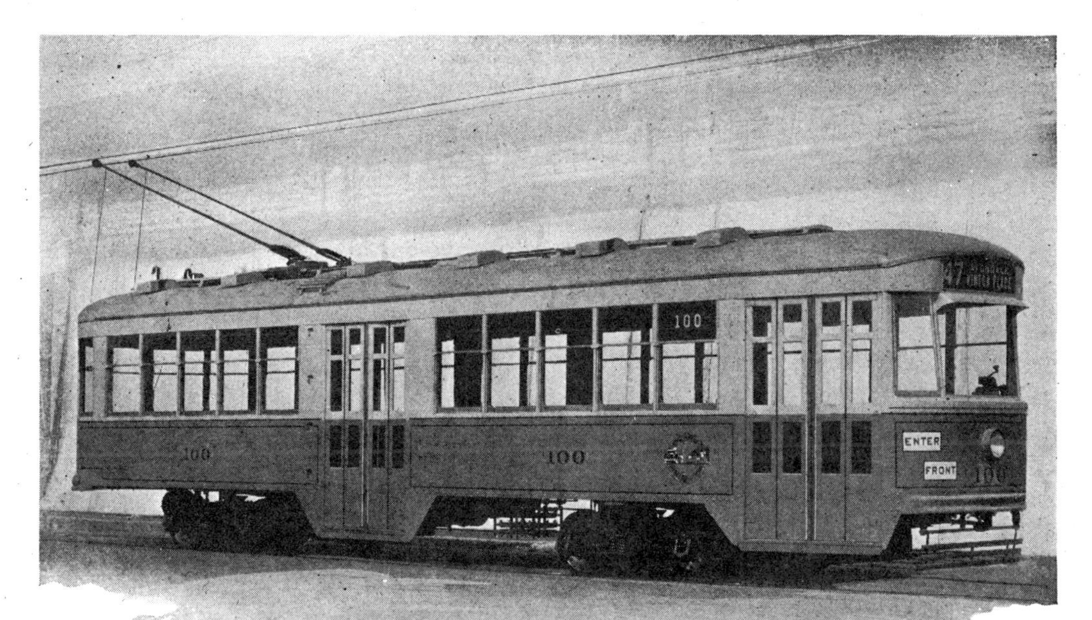

A Light-Weight, Low-Floor City Car of the Cincinnati Street Railway Company

STANDARD STREET RAILWAY MOTORS

IN parallel with the great improvements in street car body design in recent years, remarkable advances have been made by Westinghouse engineers in the design and application of standard single reduction motors as effecting passenger comfort and higher schedule speeds.

This has been evidenced in all of the following ways:

The use of greater rated horsepower of motors per ton of car weight either through reducing the total car weight for a given equipment or by the application of larger rated motors than formerly to a car of a stated weight and capacity.

This has been evidenced by the decided tendency of operating engineers to purchase quadruple thirty-five hosrepower motors instead of quadruple twenty-five horsepower motors for a given weight of street car. Furthermore, decided attempts have been made to reduce car body and truck weights in spite of the increasing tendency to luxurious appointments. Recent purchases specify quadruple fifty horsepower motors as essential for operation in two large American cities.

Two-motor equipments are no longer applied as new equipment for operation in congested city street car service but are superseded by four-motor equipments which permit higher speed and smoother, quieter, and higher rates of acceleration.

The development of street railway motors for all commercial ratings up to 100 horsepower for application with 28-inch or less diameter wheels, resulting in stream line body design, lower floor height, fewer steps, easier boarding and alighting, reduced stop time, and higher schedule speeds.

The design and perfection of motors up to 100 horsepower rating capable of withstanding higher average armature speeds resulting in weight reduction for the rating and faster transportation service.

Improvements in design details as to lubrication, ventilation, armature construction, ease of inspection, etc., which permit higher speed car operation with increased reliability of performance and reasonable maintenance cost. Some of the features developed by Westinghouse engineers are:

(a) Oil sealed armature and axle bearings, result-

ing in better lubrication and longer periods between oiling inspections.

(b) Dual ventilation, resulting in longer brush and commutator life.

(c) Forged steel fans, which are immune to the effects of crystallation, almost perfectly balanced and which do not break in service.

(d) Chrome-nickel steel alloy shafts, which withstand successfully the higher stresses.

(e) All armatures are dynamically balanced and winding insulation is tested with high frequency energy before assembly into the motors.

(f) Non-resonant gearing, which reduces operating noises due to motor equipment.

(g) Underside brushholder inspection is optional on 35 and 50-hp. sizes of motors.

(h) Through bolts on axle caps, which make high standard maintenance easy and inexpensive.

Dual Ventilation

The Westinghouse Electric and Manufacturing Company from the beginning of electric railway operation down to the present time, has been actively identified with the design, manufacture, application, and maintenance of motors for this important industry. Westinghouse engineers have been leaders in the inauguration of many new and desirable improvements which have contributed greatly to the progress of the art.

Among these improvements may be mentioned, in particular, the first commercially successful single reduction type motor known as the Westinghouse No. 3 and the first low-floor car motor, Westinghouse

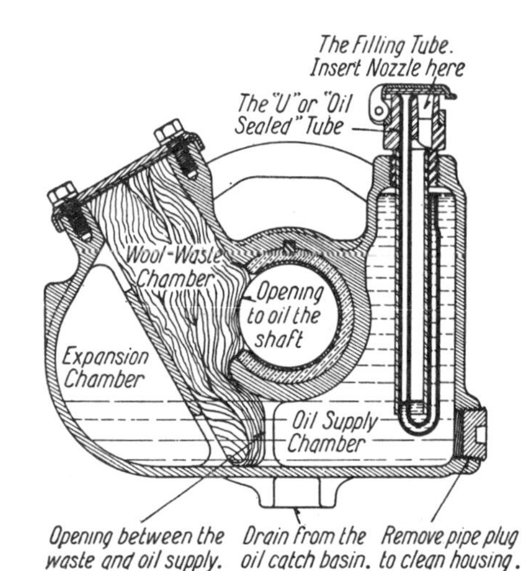

The Oil Sealed Housing has Tripled the Life of Bearings

No. 328, which revolutionized street car design. Westinghouse also produced the first light-weight car motor, No. 505, for application to the "Birney" street car which also led to advanced types of rolling stock of both single and double trucks.

Over 100,000 Westinghouse motors are in transportation service almost everywhere that street cars are in operation.

The Westinghouse Electric and Manufacturing Company has a complete standard line of box frame, self ventilated, commutating pole railway motors for operation on 600-volt direct-current power. A table of standard ratings in general use with speeds, weights, etc., is given below for application on standard gauge trucks.

Single Reduction Street Railway Motors

Type	Hr. Hp.	Rating Rpm.	Wheel Dia. In.	Gear Ratio	Std. Axle Dia. In.	Complete Wt.-Lb.	No. in Service
508-E	25	1285	26	13:74	4	1035	5300
508-C	25	1285	30	14:96	4½	1100	166
510-E	35	1085	26	13:69	4	1475	7846
510-C-2	35	1085	30	13:89	5	1590	251
514-A	40	760	24	15:58	4½	1650	5538
514-C	40	760	33	15:76	5	1770	1697
516-A	50	900	26	14:53	5	2185	252
532-A	50	670	26	15:[illegible]	5	2250	4447
532-B	50	670	33	15:69	5	2325	1697
535-A	60	795	26	15:57	5	2400	321
535-B	60	795	33	15:69	5½	2475	48
306-CV-4	65	695	33	15:69	5	2700	5983
539-A	100	1105	28	15:57	5½	2665	40
539-B	100	1105	33	15:69	5½	2730	2
548-C-8	100	805	33	16:73	6	3175	376

Non-Resonant Helical Gear

Detailed specifications, dimensions, and application data may be obtained from any Westinghouse office. Westinghouse engineers are available to make studies and recommendations as to the proper size motors to apply to any transportation service.

Construction Details

Westinghouse motors differ slightly in design features among the various sizes and types. There are over 8000 type 510—35-hp. motors in street car service throughout the world. Features of this motor are:

Frame
- Cast steel box type.
- Nose suspension.

Axle Bearings
- Bronze.
- Pinion end and commutator end interchangeable.
- Standard 4-in. diameter by 7 in. long.
- Through bolts hold axle caps.

Brushholders
- Holders interchangeable.
- Porcelain sealed on tube.
- Adjustable steel spring.
- Flat contact tips.
- Heavy flexible shunt.
- Brushes 5.8 in. by $1\frac{1}{4}$ in.

Housings
- Secured in frame by driving fit.
- Two jack holes for removing.

Lubrication
- Oil and waste method of oil seal type.
- Oil box cover bolted.

Ventilation
- Dual ventilation.
- Parallel type ventilation.
- Single piece forged steel fan.
- Air intake cast in frame.

Field Coils
- Wound with rectangular ribbon.
- Impregnated and baked
- Held by stiff flat springs.
- Flexible cable leads.

Armature Bearings
- Bronze with babbitt lining.
- Pinion end $2\frac{3}{4}$-in. diameter by $6\frac{1}{2}$ in. long.
- Commutator end $2\frac{3}{8}$-in. diameter by $4\frac{1}{2}$ in. long.

Armature
- Wound with round wire.
- No short cross-overs.
- No sharp bends to commutator.
- U-shaped insulation at ends of slots.
- Complete armature dipped and baked.
- Open rear end armature winding.

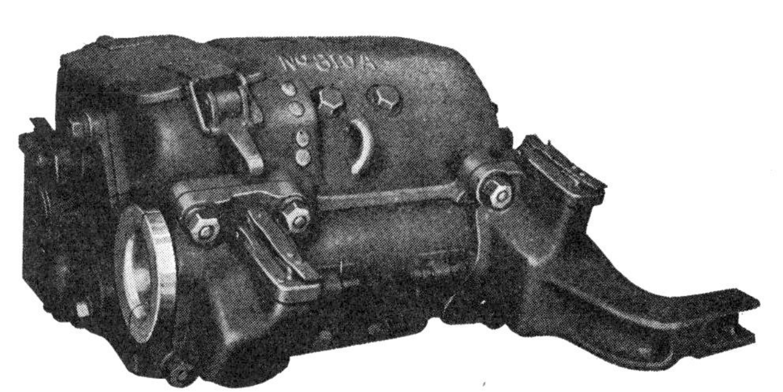

Eight Thousand Type 510 Motors Power More Than Two Thousand Modern Cars

Commutator
- V-ring held by ring nut.
- Mica undercut $\frac{3}{64}$-in.
- Moulded mica V-rings.

Gearing
- Helical $7\frac{1}{2}$-deg. angle.
- Grade BP-forged steel.
- Short addendum gear teeth.
- Long addendum pinion teeth.
- 4-D. P. 4-in. face.
- Non-resonant gearing optional.

Gear Case
- Two-point suspension.
- Drawn steel.

A Monongahela West Penn Light-Weight City and Interurban Car Equipped with W-N Drive

HIGH-SPEED MOTORS AND W-N DRIVE

TO the electric railway industry movement for more comfortable and speedier street car service, the Westinghouse Electric and Manufacturing Company contributes the high-speed motor and W-N drive.

This improved equipment for street car propulsion consists of four powerful high-speed spring borne, low-voltage motors each flexibly coupled to a self contained unit of double reduction gears through which the car axles are driven.

High-speed motors and W-N drive are designed for use on light-weight, low-floor city cars in quadruple equipments where fast schedule speeds and quiet, smooth operation are highly important requisites of service.

Unprecedented improvements in car construction and operation have resulted from this new development. The most important improvements are:

Low car floors are permitted by use of 22-inch and 24-inch wheels resulting in efficient stream line design of body, elimination of ramps and low steps, which makes boarding and alighting easy for passangers and expedites schedules.

Substantial weight reduction results in greater ratio of horsepower per ton of car weight which increases the acceleration rates and free running speeds enabling the car to keep ahead of other traffic.

Rates of from 3 mphps. to 5 mphps. are easily obtainable without discomfort to passengers.

Motor is spring supported, reducing force of hammer blows at rail joints. About one-half of the weight of the gear unit is also spring supported. This car truck with motors and gear unit has minimum unsprung weight and the car rides smoother than old designs.

Extreme quiet operation.

Maintenance of equipment is reduced due to the elimination of oil and waste lubrication, axle bearings, longer gear life, less unsprung weight, and better clearance under motor.

For the same schedule, energy consumption should be less approximately in proportion to the weight reduction.

Improved motor operation of 300-volt motors due to better inherent ability of this design to commutate heavy surges of current and withstand flashing at high peripheral speeds. Low-voltage results in better commutation and less flashing at critical speeds.

The 300-volt motor armature has less tendency to have broken leads due to the conductors being twice the size of normal 600-volt leads and on

W-N Drive Unit

account of the sturdy armature construction.

Dynamic braking may be applied where desirable with 300-volt motors, two in series, with the same connections as for a two-motor equipment. This works out very successfully from a control standpoint. Dynamic braking is still in the stage of development and has not been generally applied by the railway industry.

Westinghouse high-speed motors and W-N drive may be applied to the trucks of any street railway car manufacturing company. The low-wheel truck design results in some additional weight reduction.

The type of control used may be Westinghouse type VA, HL, or K with the line switch as may be desirable to meet local requirements.

The Monongahela West Penn Public Service Company has had in revenue service at Parkersburg, West Virginia, during the past 12 months, 5 city street cars and 5 suburban cars each equipped with quadruple type 1425-A, 35-hp. motors and double-end K-75 control. The city cars have averaged 40,000 car miles and the suburban cars 60,000 car miles of service each. These cars are demonstrating all of the forementioned desirable features and continue to furnish the very highest form of service.

Comparative Data between Standard Single Reduction and Double Reduction Motors.

Type	1425-A W-N	510-E	1426 W-N	516-D
Hour Hp.	35	35	50	50
Hour Rpm.	2100	1085	1700	900
Motor Wt. Bare	570	1285	850	1910
Coupling and Gear Unit	470	....	610	....
Gearing and A-C	....	190	....	275
Total Weight	1040	1475	1460	2185
Relative Per Cent	100	142	100	150
Wt. of 4 Motors	4160	5900	5840	8740
*Wt. Saving	1740	....	2900	....
Wheel Diameter—Inches	22	26	24	26
Clearance—Inches	4½	3½	4½	3½
Relative Per Cent	100	78	100	71
Safe Max. Rpm.	5000	2400	4000	2400
Max. Reduction	10:1	5.3:1	9.1:1	3.78:1
Min. Reduction	3.33:1	3.15:1	3.83:1	2.25:1

*Further weight reduction possible due to smaller diameter wheels.

Car and Performance Data—Monongahela West Penn Public Service Company.

Class of Car	Suburban	City
Seating Capacity	48	48
Weight Total	34,000	32,000
Car Body	16,200	14,200
Trucks	11,000	11,000
Motor and Control	5,300	5,300
Airbrake Equipment	1,500	1,500
Type of Motor	4–1425	4–1425
Total Hp.	140	140
Gear Ratio	W-N 5.44:1	W-N 8.45:1
Control Type	D.E. K-75	D. E. K-75
Wheel Dia.	22 in.	22 in.
Car Body Mfgr.	Kuhlman	Kuhlman
Truck Mfgr.	Brill	Brill
Air Brake Mfgr.	Westinghouse	Westinghouse
Schedule Speed	15.98	8.54
Stops per Mile	2.47	6.3
Duration of Stop	20.1 sec.	11.4 sec.
Average Voltage	525	525

Dimensions:

Length Over Bumpers	45 ft., 3 in.
Truck Center Distance	23 ft., 4 in.
Truck Wheel Base	4 ft., 10 in.
Rail to Step	11 5/16 in.
Step to Platform	9 in.
Wheel Diameter	22 in.

Type 1425 High-Speed Motor

The Allegheny Valley Street Railway Company operating a suburban service between Pittsburgh, Penna., and New Kensington, Penna., has recently ordered 12 new cars each weighing 30,000 lb. and equipped with 4—1425 motors, W-N Drive, and K-75 control.

Westinghouse engineers are prepared to make studies and recommendations regarding the application of high-speed motors and W-N drive to any electric railway car.

The Original Worm Drive City Car—Springfield (Mass.) Street Railway

HIGH-SPEED MOTORS FOR WORM DRIVE

IN the quest for methods of improving the mechanism and operation of the electric street car, equipment engineers have carefully studied the design practices successful in the automotive industry. As a result the transportation utility industry has a new tool in the quiet running high-speed street car equipped with amply powered motors and worm drive trucks.

The propulsion apparatus of these modern cars consists of four high-speed, spring-borne, low-voltage motors each connected through two universal couplings, and a shaft, to the self-contained worm gear reduction units with wheels comprising each car axle drive. Westinghouse high-speed motors are designed particularly for this application and are in successful operation on many worm drive cars. Dimensions and characteristics permit the application of the motors to any worm drive truck.

The type of control may be Westinghouse VA, HL, or K with a line switch as required to meet operating conditions.

Some of the desirable features of worm drive motor equipment are as follows:

Lessens noise.

Reduces weight due to high-speed motors.

Possibilities of energy saving are in proportion to weight reduction for the same schedule. There is practically no increase in energy consumption for higher schedule speeds due to weight reduction.

Higher schedule speeds are possible due to greater horsepower per ton of car weight.

Electrical equipment maintenance may be reduced on account of elimination of oil and waste lubrication, axle bearings, greater clearance under motor, and motors being spring borne.

Improved motor operation due to 300-volt motors having inherently better commutating characteristics at high critical speeds.

Dynamic braking, although still in the experimental stage of development, may be applied where desirable with 300-volt motors.

The pioneer street car built in the United States with worm drive was constructed in 1927 for the Springfield, Massachusetts, Street Railway Company This car is shown in the illustration as equipped with Timken drive and powered with four vehicle type, high-speed Westinghouse motors and type HB control. It has a body constructed of aluminum alloy and weighs complete for service 25,300 lb. It seats 45 passengers and has operated approximately 40,000 car miles of revenue service during the past two years.

The second worm drive car was built in the same year for the Chicago and Joliet Electric Railway Company and was also equipped with Timken drive and Westinghouse quadruple motors and control. This car was the first modern street car using dynamic braking with resistors arranged to utilize the accelerating and decelerating currents for car heating. This was also an aluminum car with a seating capacity of 50 and weighed 23,722 lb. ready for service.

Timken Worm Drive Truck and Westinghouse Type 1426-BT Motors

Data on Motors Applied with Worm Drive

Type	1425-B	1426-B	1426-BT
1 Hr. Hp.	35	50	50
1 Hr. Rpm.	2100	1700	1700
Motor Wt.	675	855	945
Wheel Dia.	26	26	26
Clearance Under Motor	6½	5	5
Safe Maximum Rpm.	5000	4000	4000
Max. Reduction	10:1	10:1	10:1
Min. Reduction	3:1	3:1	3:1
Shaft Extension for Brake Disk	No	No	Yes

Type 1426-B High-Speed Motor

The Pittsburgh Railways Company is equipping seven city street cars with Westinghouse type VA foot control, quadruple type 1426-BT, 50-hp., motors and Timken drive.

One line will be completely operated with these cars, affording an opportunity of demonstrating the actual possibilities of improving city transportation service with this type of apparatus.

Practically all recent worm drive applications have included Westinghouse quadruple type 1426-BT, 50-hp., motors to insure increased schedule speeds and improved accelerating rates.

A Two-Car Multiple-Unit Train of the Pittsburgh Railways Company

MULTIPLE-UNIT OPERATION

DUE to constantly increasing traffic congestion, many large city electric railway systems in recent years have made studies and plans for re-routing cars on certain lines through the downtown districts in order to permit operation with greater facility and less delay.

Higher accelerating and braking rates, improved voltage, loading platforms, and traffic control have all contributed to improving the service but on account of limited street space and the increasing number of automobiles, it has become extremely difficult in many instances to operate additional street cars through the congested area.

Multiple-unit operation, already standard practice in rapid transit and interurban train movement, has been successfully applied to heavy street car service in many cities. Multiple-unit street cars are the most flexible and efficient means of providing transportation service in city streets under various traffic conditions. Operated singly this type of car can provide frequent service on the lighter lines during the entire day and on the heavier lines during the non-rush hours. At periods of heavy travel, the cars can be operated in trains of two or three cars. Remote control is being increasingly applied for single car operation to secure improved performance so that it should not be objectionable for train operation.

Motor car and trailer two-car trains are used on many properties but the following summary of the advantages incident to the operation of multiple-unit cars will be of interest:

Good riding qualities on all cars of a train, thus insuring a better distribution of load.

Elimination of wheel slippage, due to application of motors on all axles. This gives accelerations which are smooth and rapid.

Maximum flexibility and individual operation of any car are obtained by having motors on all cars. There are no cars which cannot be moved under their own power.

Satisfactory operation in both directions is practicable on lines which do not have loops.

Train speed and accelerating characteristics are the same as for single cars.

Multiple-Unit Street Car Operation Data

Property	No. Cars Per Train	Seating Capacity Per Train	Wheel Dia. In.	Total Wt.—Lb.	No. Motors	Total Hp.	Control	Balancing Mph.	Schedule Mph.	Stops Per Mile
Pittsburgh Railways Co.	2	100	26	76,000	8–514	320	HL	32.0	9.40	9.60
Boston Elevated Railway Co.	3	154	26	133,200	12–514	480	AL	38.5	10.00	7.50
Pacific Electric Railway Co.	2	110	26	111,600	8–532	400	HL	39.0	12.23	5.26
Cleveland Interurban Ry.	4	184	26	162,000	12–340	600	HL	42.5	15.00	4.00

Small wheel cars with low-level floors can always be used with the size of motor that is capable of handling one car.

Cars each have the correct motor capacity and are not wasteful of power when operated singly The investment for sufficient cars to perform identical service is very little more than that required for motor cars and trailers.

Operating savings are realized in power and labor, which pays an attractive return on any slight increase in investment.

Multiple-unit operation results in the greatest movement of passengers with the most efficient use of street space.

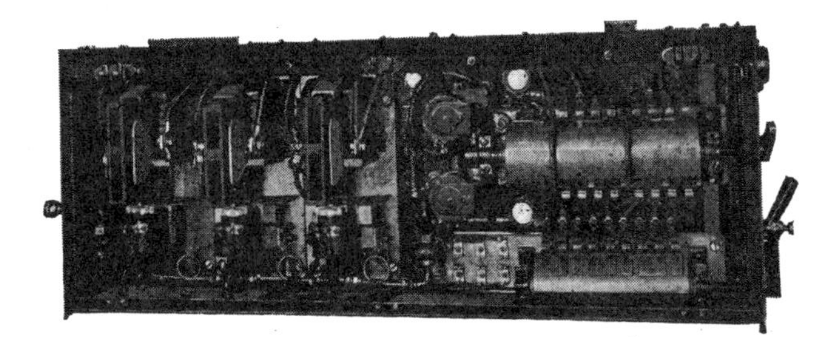

Switch boxes for Type 806 Control

Multiple-Unit Operation on the Pacific Electric Railway

Westinghouse electro-pneumatic HL unit switch control with multiple-unit features for train operation and standard 600-volt commutating-pole ventilated, series motors have established a record in all sections of this country for dependable efficient operation.

Type 516-D Motor—Two Hundred and Forty in Service in Chicago

Westinghouse engineers are prepared to make studies and recommendations regarding the desirability of operating this type of equipment on any property.

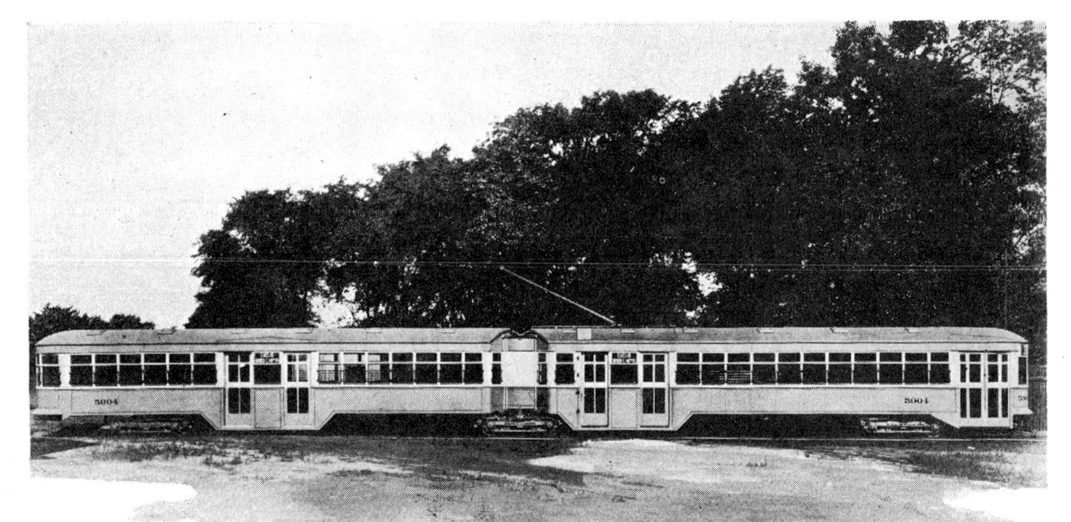

An Articulated Car of the Cleveland Railway Company in City Operation

ARTICULATED CARS

IN many cities electric railway travel on the main surface artery lines is sufficiently heavy to warrant two-car train service all day. In some cases the flexibility of multiple-unit car operation is not particularly desirable in that single cars are not required for any part of the schedule.

Motor cars and trailer units are used in some cities for this class of service but the most efficient type of rolling stock is the permanently connected two-car articulated unit with three trucks and four or six motors, depending upon local schedule requirements.

Very decided reductions are made in the weight of the unit and the arrangement of the body is greatly improved. The performance of the articulated unit in service is also better for passenger comfort and schedule making ability.

Experience has shown that articulated cars have a high degree of riding comfort and that there is freedom from the nosing effect common to the operation of motor car and trailer or multiple-unit cars. Passengers have a tendency to avoid riding in trailers due to the better riding qualities of motor cars. The articulated car overcomes this difficulty and insures an even distribution of load. It is the smoothest riding vehicle which can be obtained and, in this respect, is superior even to a multiple-unit train. Another important factor in connection with load distribution is the passage between the bodies of an articulated unit.

Type 306 Motor

In some cities consideration has been given to the use of a de luxe service for the convenience of long haul passengers. It is true that the bulk of city travel consists of short haul riders but a substantial percentage occupy cars for periods ranging from twenty-five to fifty minutes. There is a field among these long haul passengers for special service which will attract patrons that have been lost to the automobile.

The articulated car is a very satisfactory means of providing de luxe service since such units will meet the requirements of both the short haul and long haul passengers. It is the most economical operating unit that can be obtained for the purpose, from the standpoint of the cost of platform labor, maintenance, and power.

Articulated Car Unit Operation Data

Property	Car Bodies	Trucks	Seating Capacity	Wheel Dia. In.	Total Wt.—Lb	No. Motors	Total Hp.	Con-trol	Bal. Mph.	Sched. Mph.	Stops per Mile	Service
United Railways & Elec. Co. of Baltimore	2	3	87	33	66,070	4–306	260	HL	36	10.1	8.0	City
Cleveland Railway Co.	2	3	104	26	77,000	6–340	300	HL	35	10.2	5.4	City
Boston Elevated Ry. Co.*	2	4	88	26	88,800	8–514	320	MU	35	14.0	2.7	Subway
Washington, Baltimore & Annapolis Elec. R. R. Co	2	3	112	36	114,000	4–333	500	HL	53	28.2	0.11 2.53	Interurban
Brooklyn-Manhattan Transit Corp.	3	4	160	34	207, 000	4–584	840	ABF	45	25.0	0.73	Subway

*Permanently Coupled

The articulated car is not new or untried. Such units are operating very successfully in a number of cities and others have tried them for various kinds of service. The most important applications on city surface lines are in Cleveland, Baltimore, and Milwaukee.

An Articulated Car of the Washington, Baltimore & Annapolis Electric Railroad Company

Extensive application of articulated three-car units has been made recently in rapid transit subway service by the Brooklyn-Manhattan Transit Corp. where relatively long trains are required, even during the non-rush hours. The principal advantage of their operation is a material reduction in track maintenance on account of the elimination of practically all "nosing" effect of the cars. This results in much better riding qualties. Permanently coupled cars are used also in subway service, but the advantages are not as great with respect to track maintenance and riding qualities as in the case of articulated cars.

The Washington, Baltimore & Annapolis R. R. operate successfully high-speed two-car articulated units in inter-city service between Washington and Baltimore.

Westinghouse electro-pneumatic HL unit switch control and four to six standard 600-volt motors are performing satisfactory service on articulated car units in all of these applications.

A summary of the advantages of articulated cars, which have been demonstrated in actual service, is as follows:

Improved riding qualities, due to a negligible amount of side motion.

Passage between car bodies of a unit, thus insuring a distributed load.

Availability of a section that can be equipped for de luxe service to accommodate long haul passengers.

Reduced track maintenance incident to the elimination of practically all nosing.

Less train friction, resulting in quieter operation and a saving in energy consumption.

Improved schedule performance.

A Multiple-Unit Train of the Brooklyn-Manhattan Transit Corp.

RAPID TRANSIT

THE wonderful growth of New York, Chicago, Philadelphia, and Boston has demonstrated that electric rapid transit lines are essential to the expansion and development of metropolitan centers. Subway, elevated, and rapid transit systems are successful in fulfilling the exacting requirement of transporting large masses of people from one section of a city to another. This is accomplished safely at a high rate of speed for an extremely low rate of fare. The number of stops per mile is limited but due to the number of trains operating, the waiting time between trains is very short. Due to the frequency of service any slight delay caused by equipment failures, may tie up the entire system. Electrical equipment as now available functions in this exacting duty to furnish extremely reliable transportation service.

The City of Philadelphia recently inaugurated subway service under Broad Street. This rapid transit

Rapid Transit Car and Schedule Data

Property	Motor Car Wt.—Lb.	Trail Car Wt.—Lb.	Max. Ratio Motor and Trailer Car	Control	Motors Per Motor Car	Total Hp. Per Car	Hp. Per Ton
Philadelphia Rapid Transit Co.	110,000	......	1 M	ABF	2-581	420	7.64
Boston Elevated Railway Co.	85,900	......	1 M	ABF	2-577	400	9.32
Interborough Rapid Transit Co.	83,200	55,000	7 M3T	ABF	2-577	400	7.50
Interborough Rapid Transit Co.	83,200	55,000	7 M3T	ABF	2-577	400	7.50
Brooklyn-Manhattan Transit Corp.*	207,200	......	1 M	ABF	4-584	840	8.12
Philadelphia Rapid Transit Co.	91,200	......	1 M	ABF	2-333	250	5.48
Boston Elevated Railway Co.	67,850	......	1 M	ALFM	2-301	350	10.30
Interborough Rapid Transit Co.	73,780	......	1 M	ABF	2-333	250	6.97
Brooklyn-Manhattan Transit Corp.	71,500	......	1 M	ABFD	2-300	440	12.30
Chicago Rapid Transit Co.	76,900	64,000	3 M2T	ABF	2-567	340	15.80
Chicago Rapid Transit Co.	76,900	64,000	3 M2T	ABF	2-567	340	5.80

*Triplex Cars—3 Car Bodies on 4 Trucks.

service is operated with one hundred and fifty, 110,000-lb. cars, each powered with two Westinghouse type 581-A-1, 210-hp. motors, and double-end electro-pneumatic HL unit switch control.

The Westinghouse Electric and Manufacturing Company has been a pioneer in the development of electric motive power apparatus for this application and today occupies a dominant position in the number of installations operating in rapid transit service in America and abroad.

Motors

Motors for rapid transit service require the highest form of engineering and manufacturing ability. These machines combine high electrical efficiency and great mechanical strength and operate under severe limitations as to space and weight. Special care is taken to ventilate the motor windings and at the same time to exclude brake shoe dust which is prevalent in subways. Westinghouse motors are built to develop the high rates of acceleration and the running speeds demanded in this service, and at the same time to operate reliably with low maintenance.

Control

Westinghouse electro-pneumatic unit switch control is now the standard of the world for heavy-duty multiple-unit electric train operation. Automatic acceleration provides for a rapid predetermined rate of acceleration, independent of the operator, resulting in positive, safe, and smooth performance of the train in attaining full speed. Absolute control as to direction of operation of all cars in trains, shutting off of power from trains, and protection of equipment against overload or electrical short circuit is provided. In case of incapacity of the operator, the deadman's handle operates to stop the train at once. It is the usual practice to interlock all doors with the control so that the train cannot be started until all the doors are closed; nor can the doors be opened until the train is stopped.

Among the specially desirable features of Westinghouse electro-pneumatic control for subway and elevated trains are:

Independent low-voltage source of power (storage batteries) which enables the motorman to manipulate the control apparatus isolated from the main 600-volt power.

Reliable and adequate power (compressed air) to close the main current carrying switches that control the application of power to the motors, with powerful springs to insure the opening of these switches when the power is cut off.

Individual and separated main unit switches which are manipulated directly and automatically by the motorman, insuring smooth, automatic acceleration of trains and dependable release of power.

Every unit switch is provided with individual, powerful, magnetic blowout, with capacity capable of disrupting extremely high current arcs, thus providing additional security against sustained short circuits.

Electro-pneumatic control makes it possible to coordinate door and signal control to greater advantage and simplicity for operation by motorman and train crews.

The electro-pneumatic switch for car or train use affords the simplest and most reliable form of "circuit-breaker" or safety valve known to the art.

The master controller construction is very simple and extremely accessible.

Rapid Transit Car and Schedule Data

Seating Capacity	Service	Mph. Schedule	Av. Run Miles	Accel. Rate	Brkg. Rate	Mph. Bal. Speed	Division
50	Subway Local	15.5	0.540	1.8	2.00	47	Broad Street
72	Subway Local	21.4	0.750	1.7	2.00	48	Cambridge
48	Subway Local	14.8	0.340	1.5	2.00	44	Broadway Seventh
48	Subway Express	23.6	1.200	1.5	2.00	44	Broadway Seventh
158	Subway Local	15.0	0.459	1.7	2.00	44	Brighton
51	Elevated Local	16.9	0.450	1.3	2.00	42	Frankfort
48	Elevated Local	16.5	0.500	1.5	1.50	46	Atlantic Ave.
48	Elevated Local	13.7	0.300	1.5	1.75	41	
53	Elevated Local	15.8	0.330	1.5	1.75	47	
52	Elevated Local	15.6	0.320	1.3	1.75	44	Forest Pk. Loop Local
52	Elevated Express	20.8	0.830	1.3	1.75	44	Wilmette Loop Express

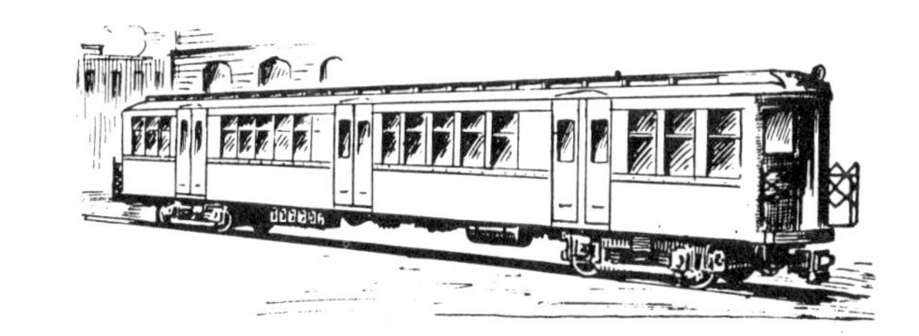

A Four-Wheel A.C.F. —Westinghouse Gas Electric Bus of the City of Detroit, Dept. of Street Railways

GAS ELECTRIC BUSSES

PERHAPS no utility vehicle since the beginning of history has undergone such rapid changes in design within a very few years as has the passenger bus. This remarkable development has been due in large measure to the desire of the automotive industry, with almost unlimited engineering and manufacturing facilities, to build "better busses".

In the transportation industry, the chief competitor of either busses or electric street cars is the privately owned automobile. Therefore, in order to attract and retain regular patrons, the public service vehicle must make available to the fare-paying passenger more than the popular sensation of riding in an automobile. The beautiful, yet serviceable chassis and body must not only be amply powered but also safely controlled by a polite and courteous operator. It must have quiet and smooth, yet rapid acceleration, relatively high free running speed, and a rapid but gradual deceleration. Quietness of operation and easy riding qualities are very important. All of these desirable characteristics are possible to a high degree with the application of electric control and transmission.

The advantages of gas electric equipment for busses and coaches are:

Simplifies the duties of the operator in starting and manipulating the vehicle, due to the elimination of the clutch and gear transmission.

Accomplishes rapid yet smooth acceleration, resulting in greater passenger comfort.

Increases schedule speed.

Results in quieter operation.

Less vibration and shock transmitted throughout the body, due to elimination of gear shifting.

Provides maximum starting effort with minimum abuse of the engine.

Minimizes engine racing and stalling.

Reduces possibility of accidents through simpler control and reduction of operator fatigue.

Affords additional safety by electric braking.

Provides a simple means of transmitting power from a single prime mover to two or more driving axles if desired.

Ease of operation affords greater opportunity for courtesy to passengers.

Operation of gas electric bus easier for new employee to master.

Fewer engine revolutions.

Electrical Equipment

The application of gas electric control and transmission involves the substitution of a generator and one or two motors and control in place of the clutch,

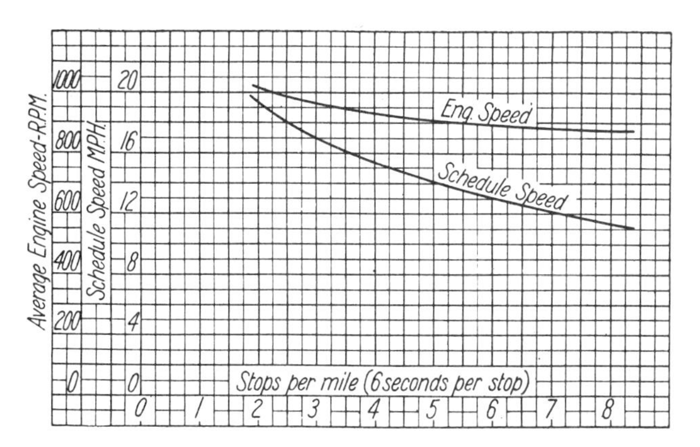

The Schedule Speed Performance of a Six-Wheel Bus with Full Load Determined from Actual Tests. The Corresponding Low Average Engine Speeds are Also Indicated

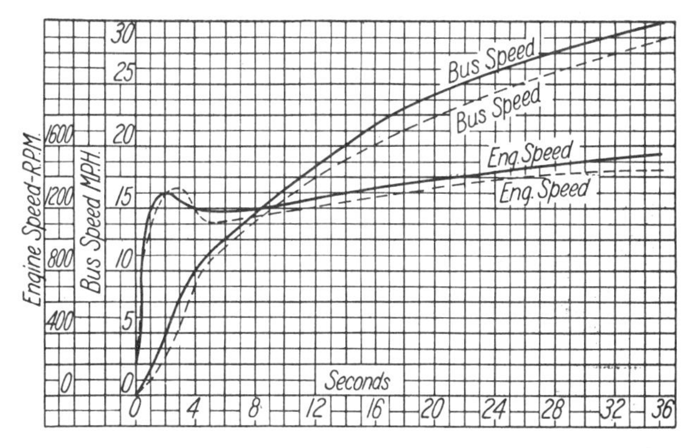

Acceleration Performance of a Six-Wheel Bus with Light and Full Load, Showing Corresponding Engine Speed Control

gear shift mechanism, and transmission commonly applied with the gasoline engine on the bus.

The Westinghouse Electric and Manufacturing Company is prepared to furnish specifications and prices for gas electric bus control and transmission equipment to be applied to standard 4-wheel, 6-wheel, and 8-wheel chassis for single-deck or double-deck bodies. Westinghouse engineers will cooperate with any bus manufacturing company in the application of the equipment to the chassis.

Gas electric drive is recommended for application to all heavy, frequent-stop, large city bus service as affording the maximum of speed, safety, and comfort, to passengers and operators.

The performance of a gas electric bus is illustrated in the accompanying curves as obtained by test. The acceleration is smooth and at a rapid rate desirable for duty in frequent-stop heavy congested service. The engine operates at minimum speed for the

Type 189-F Generator

service. The free running speeds are sufficiently high to result in high schedule speeds and the entire operation cycle is accomplished without interruption in application of power to the driving wheels. The result is extreme ease of operation and a maximum of comfort for the passenger with minimum stresses on the mechanism.

Generator

The generator used is of the shunt type designed for coupling to the variable speed gas engine. The characteristics are such as to deliver the horsepower

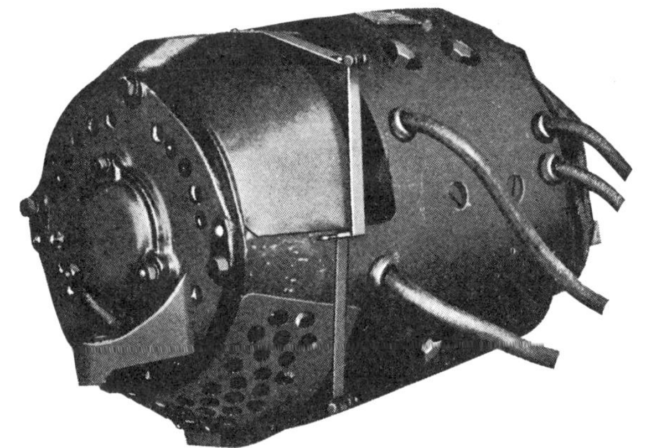

Type V-95-C Motor

of the gas engine to the motors in the form of electrical power in the most efficient manner.

The shunt winding gives a drooping characteristic which is necessary for proper engine speed control. To assure a quick building up of the voltage and high initial acceleration of the bus, a teaser winding is supplied which is excited from the regular starting battery and controlled through the accelerator switch and battery field relay.

Motors

The motors are of the series type, which from long experience, have been found to be especially suited for transportation vehicles. The characteristics of

Data on Typical Gas Electric Busses in Operation

Type	Exit	Seating Capacity	Approx. Wt.-Lb.	Approx. Loaded Wt.-Lb.	Engine Gross Hp.	Rpm.	Generator Type	Motor Type
4-Wheel	Center	34	17,000	21,700	122	1600	180F	1-V97
4-Wheel	Front	30	16,600	20,800	96	1850	188	2-V92-C
4-Wheel	Rear	40	19,000	24,600	96	1850	188C	2-V94-C
6-Wheel	Center	37	20,000	25,000	122	1600	189F	2-V95-F
6-Wheel	Center	37	18,500	23,000	114	1850	177F	2-V92-F
8-Wheel	Center	44	22,000	28,000	114	1850	178	2-V67

the motors are such that they are able to deliver high torques with low losses. As the power available is limited, efficient operation of the motors and the entire electrical apparatus is necessary.

Control

Two types of control may be applied to gas electric busses; drum or contactor.

The drum type, hand operated, makes all connections in the main controller for forward or reverse directions, and braking. It is designed to be conveniently located beside the driver's seat.

The contactor type of control employs a small direction or selector switch and three magnetic contactors. The selector switch is so constructed that the contactors are open in shifting from one position to another, thus eleminating arcing from the selector switch.

Bill of Material for Electric Control and Transmission for a Typical Modern Bus

No.	Item	Approx. Wt.—Lb.
1	Generator	985
2	Motors	900
1	Controller	75
1	Foot Accelerator Switch	5
1	Foot Operated Field Resistor Switch	5
1	Field Resistor Panel	10
1	Braking Resistor	20
1	Battery Field Relay	20
1	Motor Cutout Switch	20
1	Engine Test Terminal	10
	Total Weight	2050

Operation

The operation of a gas electric bus, equipped with the apparatus described, is very simple. The bus speed is controlled by varying the engine speed. To start the bus, the controller or selector switch is moved to the desired operating position, the throttle is operated, and the bus accelerates to the desired speed.

To stop the bus, the accelerator is released and the brakes applied. The operation of the accelerator coordinates the generator battery or teaser field, with the engine throttle. As the engine and generator speed increases, the generator voltage builds up and, as a result of the increasing voltage applied to the driving motors, the bus increases in speed. The releasing of the accelerator causes the engine speed and, therefore, the generator voltage, to decrease.

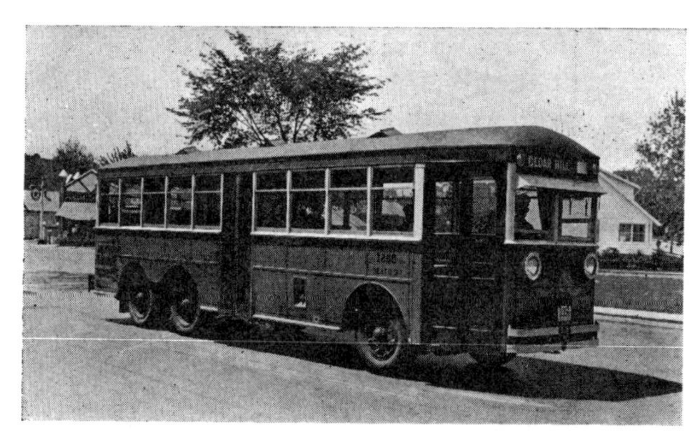

A Modern Six-Wheel Cincinnati-Westinghouse Gas Electric Bus

It is not necessary to move the controller or selector switch to the "off" position when stopping the bus. Whenever the controller or selector switch is shifted to another position, the accelerator first should be released. This insures smooth operation of the bus.

Operators of Westinghouse-Equipped Gas Electric Busses

Public Service Coordinated Transport
Boston Elevated Railway Company
Surface Transportation Corporation
New England Transportation Company
Montreal Tramways Company
The Chicago & Alton R. R. Company
Cleveland Railway Company
City of Detroit, Dept. of Street Railways

A Cincinnati-Westinghouse Electric Coach of the Utah Light & Traction Company, Salt Lake City

ELECTRIC TROLLEY COACHES

THE electric trolley coach is being seriously considered by many operators as a useful and economical means of solving certain urban transportation problems, where the density of passenger traffic is not sufficient to warrant the investment in new track necessary for the continuance of street car operation. This situation exists in many cities of moderate size where patronage has decreased due to the prevalence of the family automobile. On many properties there have been always lean cross town and feeder lines which never have returned a profit. It is undesirable to discontinue service on many of these lines and it becomes a problem of furnishing adequate and economical service. The application of the electric coach seems to offer the most service at the least cost.

Other features favorable to the application of the trolley coach are:

There are still many communities where electric railway companies contribute heavily to paving charges. This constitutes a financial burden which almost prohibits operation of lean transportation lines. The electric coach in many instances, in common with automotive vehicles, may be operated without paving expense.

The electric coach operates on cheap electric power and the existing substation and distribution equipment may be used also with certain modifications in the trolley wires.

High schedule speeds are possible due to the concentration of power and the relatively unlimited electric energy supply. The increased horsepower per ton permits all the speed that can be used safely in city streets and gives the passenger the feeling that he is actually "on his way".

The electric coach has adequate maneuvering ability to pass around other vehicles that would delay street car operation.

The electric coach is designed for one-man operation. Costs on this account are comparable with any other type of vehicle that will perform the service.

The maintenance costs per coach mile on present installations compare favorably with other railway and highway vehicles.

Improved transportation service may be afforded with the modern electric coach. The operation is quiet due to pneumatic tires. There is an atmosphere of comfort and ease due to the smooth acceleration, leather seats, improved riding qualities of the chassis, and low floor, that pleases the daily rider.

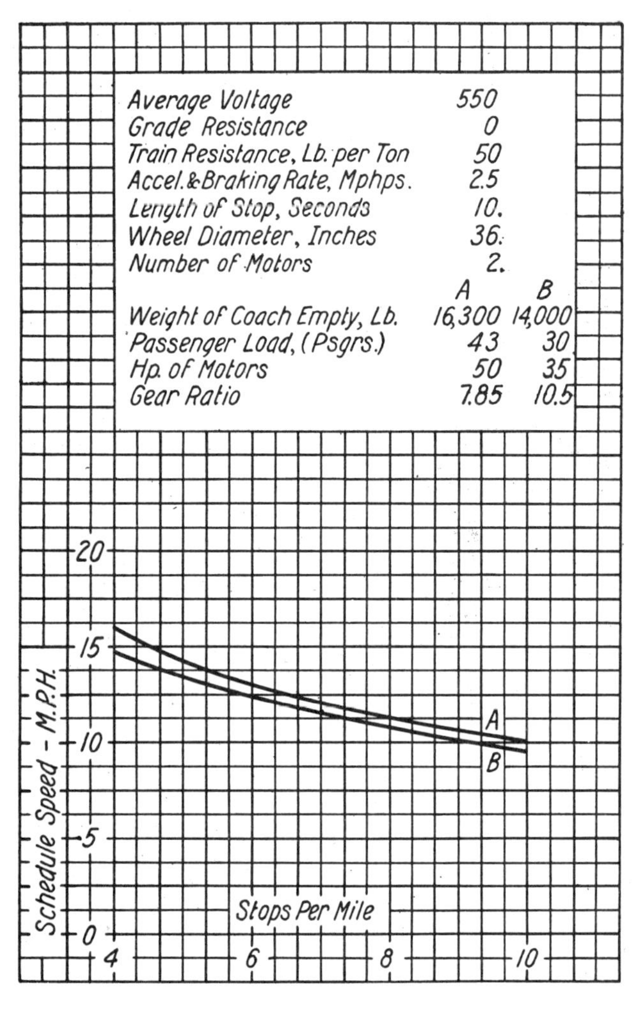

Schedule Data

Electrical Equipment

Electric coaches are available through standard bus and car manufacturers in sizes and arrangement to fulfill a variety of operating requirements. Westinghouse engineers have been active in the development and perfection of this vehicle since its earliest inception. Many of the equipment features of the latest model coaches are made possible by the development of motors and control for high-speed street car drives and gas electric busses. The Westinghouse Electric and Manufacturing Company has available a complete line of equipment for any type of electric coach that may be required to meet local conditions.

A brief description of standard 600-volt electric motive power apparatus available is as follows:

Two—35-hp., type 1427-AT-6 motors geared 10.5:1 on 36-in. wheels and single-end foot-operated, automatic acceleration control. This equipment may be applied to a 14,000-lb., 4-wheel vehicle, seating approximately 30 passengers, in normal city service without dynamic braking.

Two—50-hp. type 1426-AT-6 motors geared 7.85:1 on 36-in. wheels and single-end foot-operated, automatic acceleration contactor control. This equipment may be applied to a 16,300-lb., 4-wheel or 6-wheel vehicle, seating approximately 43 passengers, in normal city service without dynamic braking.

Two—50-hp., type 1426-AT-6 motors geared 10.5:1 on 36-in. wheels and single-end foot-operated, automatic acceleration contactor control. This equipment may be applied to a 16,300-lb., 4-wheel or 6-wheel vehicle, seating approximately 43 passengers, in normal city service with dynamic braking.

Where a coach weighing 13,000 lb. or less is applied in service with relative ease, a single 50-hp. motor, type 1426-AT-6, and either hand or automatic acceleration control may be used.

A Light-Weight Structural Steel Trolley Base

Performance

Schedule making possibilities of trolley coaches without service dynamic braking are shown in the accompanying curve. Maximum accelerating rates up to 5.5 miles per hour per second may be obtained. Balancing speeds at 600 volts on tangent level roadway are 31 mph. and 34 mph. for the double 35-hp. equipment on a 14,000-lb. coach and double 50-hp. equipment on a 16,300-lb. coach, respectively.

Service dynamic braking may be applied to the double 50-hp. equipment resulting in average retardation rates of 3 miles per hour per second and improved operation where the operating profile has many grades.

Operation

Westinghouse type VA control equipment for electric trolley coaches is designed for simple and convenient foot operation. In the front end are the accelerating controller, reverser, and control switches. The controller resembles a standard automobile brake pedal and provides eleven accelerating notches with positive stops in series and parallel positions. A momentary pause is required on the first notch to permit the line switch to close, preventing a too rapid start. The eleven notches assure smooth acceleration. The control is automatic, spring return, and is located for operation by the right foot. Line switches and contactors making and braking the main currents are arranged for cabinet or group mounting depending upon the body design.

The automatic dynamic braking feature is operated by an additional foot pedal located at the left of the accelerating foot controller and requires the addition of contactors, sequence switch, limit relay, resistance, etc.

Type XM Foot Controller

Coaches may be equipped with heaters utilizing the accelerating and braking currents.

Great improvements have been made in the design and manufacture of propulsion motors for application to electric coaches. Due to the perfection of the high-speed type motor, the weights per rated horsepower have been reduced over 50 per cent and at the same time improvements in all round efficiency accomplished. The types 1427-AT-6 and 1426-AT-6 are rated at 35 hp. and 50 hp., respectively, on the A.I.E.E. basis. These motors are sturdily designed to furnish high rates of acceleration and operate reliably in frequent-stop city service. Four brushholders are provided on each motor. Roller armature bearings are standard.

An Electric Coach of the New Orleans Public Service—Built by Twin Coach with Westinghouse Electrical Equipment

Electric trolley coach electrical equipment is designed for easy inspection and reliable operation and to operate with low maintenance.

Typical Bill of Material for Foot-Operated, Electro-Pneumatic Control, Automatic Acceleration, Electric Trolley Coach with Two 50-Hp. Motors.

Item	Req.	Apparatus	Wt.—Lb.
1	2	W-810 Trolleys Complete	250
2	1	Type 220-A-3 Fuse Box	11
3	1	Type UP-21 Line Switch	87
4	1	Type 371-M-19 Overload Relay	14
5	8	Type UP-21 Electro-magnetic Contactors	136
6	1	Type XD-524 Reverser	40
7	1	Type TK-752 Main and Motor Cutout Switch	8
8	1	Set of Type M Resistors	70
9	1	Type XM-14 Foot Controller	30
10	1	Type XS-116 Sequence Switch	90
11	1	Type 379-M Limit Relay	6
12	1	Set Pneumatic Details	35
13	1	Set Insulating Details	10
14	1	Set All Rubber-Covered Main and Control Cable	120
15	1	Set Knuckle Joint Connectors	8
16	2	Type TC-2 Control Switches	6
17	2	Type 1426-AT-6, 50-Hp. Motors	1650
		Total	2571

Costs for Service with Trolley Coaches

Expense Item	Cents per Mile
Way and Structure	1.64
Maintenance	7.26
Power	2.32
Conducting Transportation	10.18
General	3.55
Total	24.96

A Light-Weight Interurban Car of the Cincinnati, Hamilton & Dayton Railway Company

LIGHT-WEIGHT INTERURBAN CARS

INTERURBAN electric railway properties are meeting successfully the competition of paved highways and private automobiles by operating the economic type of rolling stock and equipment particularly adapted to the locality served. Such cars, regardless of size, are designed for the maximum of passenger comfort with the highest permissible safe operating speeds. For lines of moderate traffic possibilities, remarkable and rapid developments by manufacturers cooperating with operators have made available the light-weight interurban car.

Several years ago it was common practice to build interurban cars weighing from 70,000 to 90,000 lb. and equipped with 4 motors varying in size from 85 to 125-hp. each. Wheels as large as 37 inches were used; the distance from the rail to the car floor was correspondingly great and the term "battleship" applied to this type of car was a very appropriate one.

The interurban car of today is much different, as indicated by the data in an accompanying table.

A notable advance has been made in reducing the weight. There are few interurban roads which have made recent purchases of new cars weighing over 60,000 lb. Car weights of light-weight interurbans purchased within the past five years may be said to vary between 35,000 and 60,000 lb.

One property recently replaced cars weighing 72,000 lb. by a modern type weighing 49,500 lb. complete; another property is replacing 56,000-lb. cars with 35,000-lb. cars. These decreases in weight have

Type 535 Motor

been accomplished in several ways. A better knowledge of car design and the use of better materials have enabled the car builder to contribute his share; the electrical manufacturer has been able to put more horsepower into the given space and this has reduced the size and weight of the motor accordingly.

Thus, smaller wheels have been made possible, contributing to a further reduction in weight. Light-

Modern Light-Weight Interurban Car Data

Property	Car Wt.—Lb.	Seating Capacity	Wheel Dia. In.	Motors	Total Hp.	Con-trol	Stops Per Mile	Schedule Speed Mph.	Balancing Speed Mph.
Monongahela West Penn Pub. Ser. Co.	34,000	48	22	4–1425	140	K-75	2.47	15.98	40.0
Monongahela West Penn Pub. Ser. Co.	35,000	44	26	4–516	200	K-75	1.40	25.00	55.0
Missouri & Kansas Rwy. Co.	35,050	52	28	4–510	140	K-35	0.89	22.90	44.5
Chicago & Joliet Elec. Rwy. Co.	38,000	48	26	4–510	140	K-35	1.05	24.20	45.0
Chicago, North Shore & Milwaukee R. R. Co.	40,100	52	26	4–514	160	HL	1.00	22.50	36.0
Steubenville, E. Liverpool & Beaver Valley Trac. Co.	45,140	51	26	4–532	200	K-35	1.20	19.00	43.0
Pittsburgh Railways Co.	49,500	53	28	4–535	240	HL	1.40	23.20	55.0
Cincinnati, Hamilton & Dayton Rwy. Co.	58,780	54	28	4–535	240	HL	1.30	23.20	54.0
Philadelphia & Western Rwy. Co.	60,206	51	30	4–535	240	AL	0.96	26.10	44.5

weight interurban cars rarely have wheels exceeding 28 inches in diameter. This reduces the distance from rail to floor, cuts down the stop time, and increases the schedule speed. The control has been improved by decreasing the weight per horsepower handled as well as by making the maintenance problem much more simple. The latter has been accomplished by building the control on the unit switch plan; that is, each part of the control is a unit which may be mounted in the most convenient location and maintained with a minimum of effort. As an outgrowth of the unit switch idea, cabinet control has been evolved. This permits practically all the control to be mounted in a cabinet where it is given the maximum of protection and where it may be maintained most easily.

A Light-Weight Interurban Car of the Chicago & Joliet Electric Railway Company

A Light-Weight Interurban Car of the Pittsburgh Railways Company

Passenger comfort and car appearance by no means have been forgotten in the design of the car. Seats are much more comfortable and luxurious than they used to be; the interior of the car is of a more pleasing appearance and the lighting has been improved greatly; in many instances the passengers may get an unobstructed view through the front of the car. In older designs this often was impossible due to the baggage compartment or to a bulkhead. Electric heating in most instances has superseded the coal stove and hot water system of former cars. This has resulted in a more even temperature and the absence of stove fumes.

Westinghouse engineers are prepared to make studies and to recommend equipment and cars for light-weight interurban service. Powerful motors geared for high-speed operation and Westinghouse electro-pneumatic unit switch control for cabinet or under floor installation, permits the operation of one or more cars in trains, when desirable. The functioning of the electrical equipment is positive and reliable; it is designed for easy inspection and to operate with low maintenance.

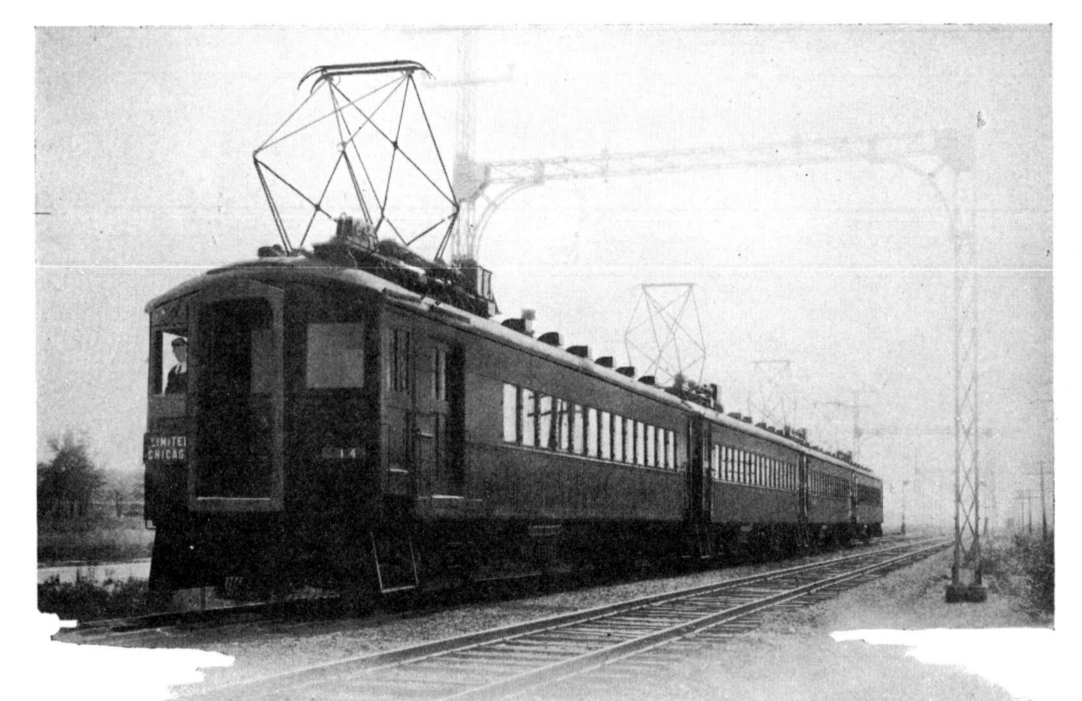

A Heavy Interurban Train on the Chicago, South Shore & South Bend Railroad

HEAVY INTERURBAN CARS

ORIGINALLY built to provide passenger service more frequently, yet comparable as to comfort and speed with the steam roads, the heavy electric inter-urban railway now encounters competition from the bus and private automobile. This situation is being met in some instances by coordinating service with the bus and using it as a feeder. In other instances, on railways operating out of the larger centers of population, the purchase of new luxurious higher speed steel rolling stock has been the solution. Tracks have been rebuilt for fast operation, block signals installed so that speedy and comfortable service can be furnished with almost absolute safety. Because of many accidents on the highway, the passenger has a feeling of security and safety when riding on the heavy interurban.

The accompanying table outlines briefly schedule speeds and service available on interurban roads where receipts from passenger patronage have shown an increase since modernization.

The diversification of its service to the community through the development and expansion of an electric freight haulage service has also added greatly to heavy interurban system receipts. Baldwin-Westinghouse electric locomotives and baggage car type locomotives, successful in this service, are treated in the following chapter.

Type 567 Motor

For heavy high-speed interurban service, Westinghouse has proven electrical equipment for operation on 600, 750, 600-1200, or 750-1500-volt direct-current power from either third rail or trolley.

Heavy Interurban Schedule Speed Data

Property	Car Wt.—Lb.	Seating Capacity Per Car	Wheel Dia. In.	Motors	Total Hp.	Control	Run	Distance	Mph. Sch. Speed	Mph. Bal. Speed
Chicago, South Shore & South Bend R.R.	133,000	44	36	4-567	840	HBF	South Bend to Chicago	89.80	44.90	72
Chicago, North Shore & Milwaukee R.R. Co.	102,000	52	36	4-557	560	HLF	Milwaukee to Chicago	87.19	43.59	67
Indiana Service Corporation	96,000	50	37	4-333	500	HL	Ft. Wayne to Peru	58.70	39.10	60
Union Traction Co. (of Indiana)	90,000	56	37	4-333	500	HL	Muncie to Indianapolis	63.60	36.30	60
Columbus, Delaware & Marion Electric Co.	100,000	37	34	4-557	560	HL	Columbus to Marion	49.30	29.60	66
Lake Shore Electric Ry. Co.	84,500	64	38	4-557	560	AL	Cleveland to Toledo	120.00	28.80	67

High-Speed Multiple-Unit Operation on the Chicago, North Shore & Milwaukee Railroad

Westinghouse electro-pneumatic unit switch control may be adapted to meet any requirements of modern electric railway operation with reliable and efficient performance, maximum safety, and low maintenance. The HL control for this application includes the same desirable features that have made it the standard of the world for subway, elevated, and rapid transit train operation.

The Westinghouse Electric and Manufacturing Company has a large range of motors for interurban application. Data on standard motors are given in the accompanying table.

Engineers are available upon application to the nearest Westinghouse Office to assist railway companies in the selection and application of equipment for interurban operation.

Heavy Interurban Car Motors

Type	Hp.	Rpm.	Wheel Dia. In.	Gear Ratio	Wt. Lb.	Axle Dia. In.
600-1200 VOLTS						
333-VV-7	125	808	33	16:61	3850	6
333-VV-8	125	725	33	16:61	3850	6
557-A-7	145	1010	33	16:61	4050	6
557-A-8	140	920	33	16:61	4050	6
567-A-5	170	705	33	16:61	4900	6
567-A-6	165	645	33	16:61	4900	6
750-1500 VOLTS						
545-C-7	85	915	33	16:73	3175	6
545-C-8	85	810	33	16:73	3175	6
334-V-7	115	925	33	16:61	3850	6
334-V-8	115	760	33	16:61	3850	6
567-A-5	210	895	33	16:61	4900	6
567-A-6	200	830	33	16:61	4900	6

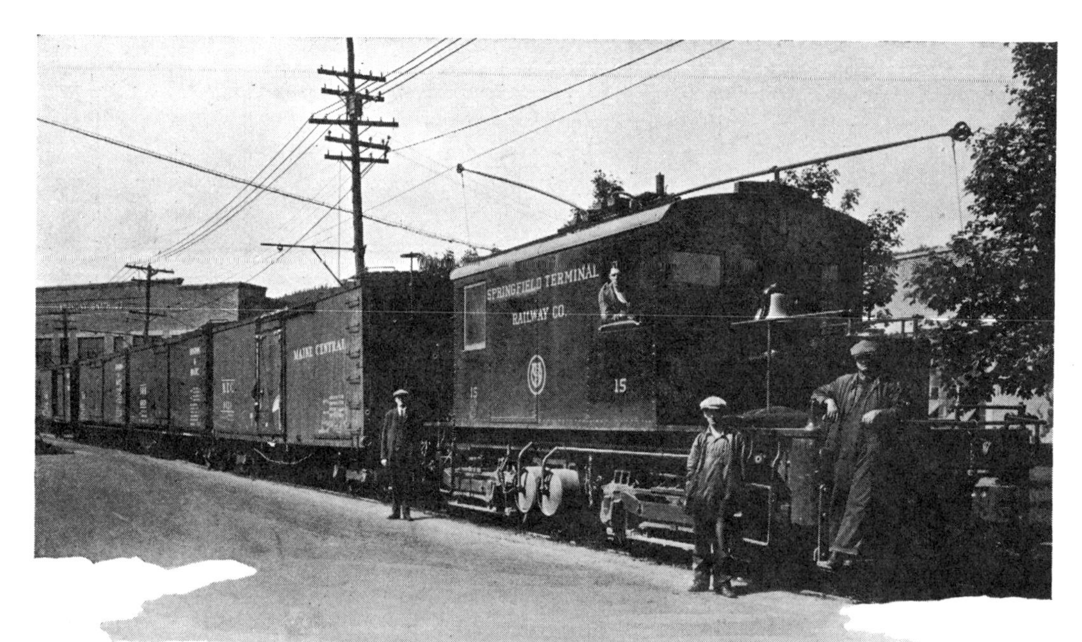

Class B-1, 50-Ton Locomotive in Service on the Springfield Terminal Railway

FREIGHT HAULAGE

THE ELECTRIC LOCOMOTIVE

CITY and interurban railways have invested billions of dollars in rights-of-way, tracks, overhead lines, substations, transmission lines, rolling stock, and shops. In addition, many of them own land and buildings strategically located for use as freight terminals and real property in suburban areas admirably situated for transfer yard purposes jointly with trunk line railroads. Competition among the great railroad systems for freight is extremely keen and golden opportunities exist in almost every center of population for electric railway properties to increase their profitable revenue through the utilization of present assets.

That electric freight haulage service pays is not an idle boast is proven by statistics recently gathered showing the increase in gross freight business between the years 1920 and 1928 on 13 large and small American railways. These data are given in an accompanying table and indicate an increase of 50 per cent in the freight business for the eight year period. While not startling, this increase is substantial and helps many of the properties to maintain their total gross earnings in the face of passenger business lost to the family automobile.

For many railways the only requirements to enter this profitable freight haulage business are a traffic department to solicit the shipments and electric

Electric Railway Freight Haulage Growth
1920 to 1928

Operator	No. Locos.	Type Freight	Total Gross Freight 1920	Total Gross Freight 1928	Per Cent Increase
1	4	Oil—Glass—Sand	$ 375,080	$ 445,625	19
2	2	Potatoes—Flour	82,625	145,867	76
3	7	General	759,252	958,970	26
4	17	Coal—Autos	1,026,734	2,185,110	117
5	5	Sugar—Coal	519,695	544,139	5
6	6	General	369,220	417,760	13
7	5	Interchange	112,000	370,700	230
8	3	Interchange	53,750	80,800	50
9	7	Interchange	246,853	400,100	62
10	5	Interchange	59,400	143,500	142
11	2	Rock—Lumber	119,975	146,132	23
12	53	General	4,065,843	6,004,833	50
13	1	Fruit—Autos	2,221	12,890	575
	117		$7,792,648	$11,710,244	50

locomotives and equipment with which to haul the tonnage.

Baldwin-Westinghouse locomotives operated throughout the world have demonstrated conclusively the following advantages of freight movement with electric motive power:

Reliability, due to the application of apparatus designed especially for locomotive service.

Low maintenance cost is indicated by an actual average figure of 6.24 cents per locomotive-mile, based on 56 Baldwin-Westinghouse locomotives varying in size from 34 to 72 tons each, with total operations of 891,100 locomotive miles.

High degree of availability, can be retained in service 1000 to 3000 locomotive miles with only minor inspections.

Low energy requirement, due to high-torque low-speed field control motors especially designed for minimum power consumption.

Practically any desired speed may be provided on electric locomotives for a given train under given conditions.

Flexibility is insured by double-end operation and multiple-unit features permitting operation of two or more machines by one crew when desirable.

Simple to operate and inspect.

A 55-Ton Storage Battery Locomotive of the Twin Branch Railroad Company

Classification of Baldwin-Westinghouse Standard Electric Locomotives

Class	A	B-1	B	D	E
Nominal Wt.—Tons	35	50.0	50	60	75
Minimum Wt.—Tons	30	41.5	45	59	67
Maximum Wt.—Tons	35	50.0	55	61	84
Standard Application Total Hp. of Motors	300	400	400	560-800	1000

An 80-Ton Locomotive on the Chicago, South Shore & South Bend Railroad

Baldwin-Westinghouse electric locomotives have been classified and standardized as to weight and horsepower to meet the requirements of the railway industry for freight haulage service.

All standard electric locomotives are built ready for operation with the following conservatively designed and well known equipment:

Baldwin Locomotive Works

Steel cab of steeple design with structural steel under-frame

Standard gauge swivel type trucks with wheels, axles, and journals of either M.C.B. or A.E. R.A. Standards

Westinghouse Electric & Manufacturing Company

High-torque modern railway motors

Electro-pneumatic unit switch control

Westinghouse Airbrake Company

Straight and automatic air brakes

Air compressors

The Class B-1 locomotive as illustrated is typical of seventy 50-ton Baldwin-Westinghouse locomotives now successfully operating in more than 22 states as well as in a number of foreign countries. This locomotive is so built that it fulfills the Interstate Commerce Commission's requirements as to electric motive power and may be arranged for operation on 600, 750, 600-1200 or 750-1500-volt power. It is equipped with four type 562-D-5, 100-hp. commutating-pole motors and electro-pneumatic unit switch field control of hand acceleration type with resistance designed for switching service.

***Tonnage Data for Slow-Speed Class B or B-1 50-Ton Locomotive**

Total Hp.—400

Speed—One hour rating 11.5 mph.
Continuous rating 13 mph.

% Grade	Tons, Trailing Load Starting 30% Adhesion	Running 1 Hr. Rating	Running Cont. Rating
0.0	1045		
0.5	750	680	455
1.0	580	410	270
1.5	470	285	180
2.0	390	215	135
3.0	290	135	80
4.0	225	95	50

***Tonnage Data for High-Speed Class B or B-1 50-Ton Locomotive**

Total Hp.—680

Speed—One hour rating 19.3 mph.
Continuous rating 22.1 mph.

% Grade	Tons, Trailing Load Starting 30% Adhesion	Running 1 Hr. Rating	Running Cont. Rating
0.0	1045		
0.5	750	750	444
1.0	580	454	261
1.5	470	318	177
2.0	390	239	129
3.0	290	153	75
4.0	225	106	47

Locomotives are delivered complete with all accessories and may be placed in haulage service without delay. Complete specifications will be furnished upon application, with a statement of service conditions, to the nearest Westinghouse office.

Typical performance as to what standard 50-ton locomotives will haul in gross trailing tonnage is given in accompanying tables. Starting tonnages are within the short time limit of the electrical equipment and running tonnages are given for the one-hour and continuous ratings of the four motors operating with forced ventilation and on the short field notches of control.

Approximately 60 Baldwin-Westinghouse electric locomotives of the 60 to 85-ton class are operating in freight haulage service.

These locomotives resemble the design of the standard 50-ton unit but are built and powered for heavier duty and faster service.

BAGGAGE CAR EQUIPMENT

Many electric interurban railways are conducting successfully a fast freight service with the high-speed baggage car type of locomotive. This type of equipment lends itself very well to rapid freight movement where fast passenger service cannot be impaired by slow-speed freight trains.

A speed of 25 to 35 miles per hour will allow this equipment to operate without interfering with passenger schedules. Furthermore, in some localities franchises do not permit the operation of electric locomotives through city streets but baggage car operation is not prohibited.

The baggage car locomotive may be used for less than carload freight and at the same time to haul carload freight. This type of locomotive can be arranged for double-end operation, and multiple-unit features can be furnished permitting the operation of two or more motored units from a single station.

A baggage car weighing 40 tons light may carry a maximum load of 20 additional tons and the total weight of the car loaded would vary for tractive effort purposes from 40 tons to a maximum of 60 tons.

Performance

Typical performance as to what 40-ton baggage cars will haul in gross trailing tonnage is given in an accompanying table. Starting tonnages are within the short time limit of the electrical equipment and the running tonnages are given for the one-hour and continuous ratings of the four motors operating with forced ventilation and on the short field notches of control. For the high-speed baggage car the equipment consists of quadruple Westinghouse type 333-V-8 motors geared 16:61 to 36-inch wheels.

***The following factors were used in computing these data:**

Locomotive train resistance—15 pounds per ton.
Trailing train resistance—7 pounds per ton.
Straight track.
Starting data based upon clean dry rails—30 per cent adhesion.
Train accelerating rate—0.2 mphps.
Grade resistance—20 pounds per ton.

***Tonnage Data for High-Speed, 40-Ton, 600-Volt Baggage Car Locomotive**

Total Hp.—500

Speed—One hour rating—20.9 mph.

Continuous rating—26.7 mph.

	Tons, Trailing Load		
% Grade	Starting 30% Adhesion	Running 1 Hr. Rating	Running Cont. Rating
0.0	840	540	
0.5	600	200	460
1.0	465	110	275
1.5	375	70	190
2.0	315	45	140
3.0	230		85
4.0	180		55

Construction

The cabs for baggage car locomotives may be built by car builders or by the operating companies themselves. This causes the general design of the cars to vary somewhat, but the cars are usually arranged with an operating station at one or both ends with the remaining space available for baggage. The control apparatus, consisting of switch groups, reversers, grids, and also the air brake details, is mounted under the car floor.

The two trucks upon which the car body is mounted are of two-axle equalizing type, with standard brake rigging, while the cars themselves have the M.C.B. standard type of coupling. Straight and automatic air brake equipment is used, having all details except the master valve and gauges mounted under the floor as is the practice on most interurban electric cars.

Rapid and Efficient Interurban Freight Service on the Western Ohio Railway

Electrical Equipment

The Westinghouse Electric & Manufacturing Company has standard proven motive power apparatus for this application which includes type HL electo-pneumatic control and amply powered motors of proper speed characteristics for the service desired. Equipment may be arranged to operate from 600, 750, 600-1200, or 750-1500 volts direct-current power from trolley or third rail.

A Typical Bill of Material of a Double-end Electrical Equipment for a 40-Ton, 600-Volt Baggage Car Locomotive

No.	Item	Wt., Lb.
2	Trolleys Complete	321
1	Lightning Arrester	8
1	Main Switch and Main Fuse	57
2	Type TC-2 Control and Reset Switches	8
2	8 Switch Unit Switch Groups, Which Include Cutout Switch, Terminal Board, Relays, and Details Mounted in Sheet Steel Boxes	1,500
1	Reverser	220
1	Set of Starting Resistors	700
2	Master Controllers Giving the Necessary Series and Parallel Starting Steps and also Reversing Operation	275
3	Junction Boxes	24
1	Set of Insulating Details	20
1	Set of Pneumatic Details	115
1	Set of Main and Control Cable	320
20	Knuckle Joint Connectors	10
1	Set of Lighting Details	30
4	Type 333, 125-hp. Motors Complete	15,400
	Total Weight	19,008

The main switching operation is obtained by the use of unit switches mounted in two switch groups under the car. These switches are of the Westinghouse electro-pneumatic type. The switch groups also include the necessary relays, terminal board, and motor cutout switches; the reverser itself is mounted in a separate box.

The motorman's compartment is arranged so that the motorman can control the starting, stopping, and reversing of the car easily and positively by means of the master controller and the air brake valves. Such other details of the control as are necessary for ease in handling the car are also mounted in the motorman's compartment, such as the TC-2 control and reset switch.

The Westinghouse Electric & Manufacturing Company has furnished electrical equipment for a large number of high-speed baggage cars operating successfully in conjunction with interurban service in the United States,

A 6,000-Ton Virginian Railway Train on Two Percent Up Grade

A History of the Development *of the* Single-Phase System

By

J. V. Dobson *and* *F. C. Hanker*

Reprint 357, March, 1929

Westinghouse Electric & Manufacturing Company
East Pittsburgh, Pa.

A History of the Development

of the

Single-Phase System

THE history of the development of the electric railway both for light and heavy traction is so closely interwoven with the engineering activities of Mr. B. G. Lamme, that a review based on memoranda left by him is of inestimable value. The design and development of the earliest types of railway motors, the forerunner of the modern motors, was one of his outstanding achievements. Subsequently he was the most important factor in the development of the synchronous converter or, as he almost invariably called it, the rotary converter. Later he devoted much of his engineering and inventive genius to the development of the single-phase system. These represent only a small part of the contribution of Mr. Lamme to the development of the railway art, as is fully appreciated by those who had the good fortune to be associated with him.

Almost since the early days of light electric traction, the problem of heavy electric traction has been given much thought. In the earlier considerations, the principal difficulty was supposed to lie in the ability of the manufacturers to produce sufficiently powerful equipment. However, it was not long before more serious difficulties presented themselves and some of these are with us even to the present day.

In the first place, with direct-current generation and distribution at 550 to 600 volts, which was the accepted standard, the electric railway people were confronted with the problem of transmitting large units of power for considerable distances. This meant closely spaced power supply stations and, with direct-current generation, this implied subdivision of the power supply into a number of separate units, which was in opposition to the growing tendency to concentrate the power generation into large central stations. In other words, this was in the direction of subdivision instead of centralization.

The second serious difficulty was in supplying this power to a moving locomotive or train. With low voltage and large power the collection and handling of the current was then, and still is, a serious problem.

With these two conditions encountered, heavy railway electrification was practically at a standstill, and, in the minds of those who knew, there seemed but little promise for future railway electrification on a large scale. It is true that a number of attempts were made to build electric locomotives in the early days, but these were mostly by those who were familiar with the traction problem only and who had but little or no experience with the generation and transmission problems.

Development of Rotary Converter and Third Rail

Confronted with the conditions mentioned it is no wonder that the coming of rotary converters in large capacities was recognized as a great boon to the heavy electric railway. Here at least was a solution of one of the difficulties, for by means of polyphase alternating-current generation and distribution at high voltage, and conversion to direct-current through distributed substations, the problem of centralization of the power stations was accomplished. However, the problem of collection and handling of the low voltage current was still to be met, so that the rotary converter only partially solved the heavy railway problem.

Another step, however, represented a very material advance, namely, the development of the third rail. This permitted relatively high amounts of current to be transmitted from the substation to the moving vehicles, and thus it appeared, for awhile, to solve the second part of the heavy railway problem. In consequence, after the development of the rotary converter and the advent of the third rail, the heavier work went forward by leaps and bounds. However, it may be noted that this was only where traffic was very heavy, such as for elevated and subway service in the larger cities, for city terminals, suburban railway service, etc.

As soon as application of the third rail and the rotary converter system began to be considered in connection with longer lines, with infrequent service, the limitations of this arrangement began to be apparent, these being largely of a financial nature. In other words, while it would be possible, from an engineering standpoint to build and operate successfully a long main line railway by means of rotary converters and the 600-volt third rail system, yet the figures showed most positively that the first cost and the operating expense would render such an arrangement unprofitable.

Thus, the rotary converter and third rail solution of the problem was not a general one and applied only to isolated cases where the traffic was very heavy and the service very frequent, or where there was some serious limitation to the use of steam, necessitating either the abandonment of the steam locomotive

An Early Rotary Converter

altogether, or an excessive expense in some form due to its retention.

The smoke nuisance and danger in the Baltimore & Ohio Tunnel, the St. Clair Tunnel, and the Pennsylvania and New York Central terminals at New York, may be cited as illustrations where it was considered necessary to adopt electric operation regardless of its possible economy compared with steam operation.

Up until about 1900, the situation stood as described. The engineers throughout the world looked longingly at the possibilities of the alternating-current system for solving the heavy railway problem; not because there were any particular merits in alternating-current apparatus in regard to traction itself, but because here was a high voltage, flexible system, which if it could only be used on the trains or locomotives themselves, would solve at once the problems of generation, transmission, collection and handling of large units of power on a moving vehicle.

The well known Zossen experiments in Germany with a three-phase, high-voltage trolley system, with induction motors, attracted a great deal of attention, all out of proportion to the true value of the results. This was a very spectacular undertaking and indirectly had a bearing on the future railway work.

High Voltage Necessary for Heavy Electric Traction

It was beginning to be recognized that high trolley voltage was a necessary condition in the general solution of the railway problem. With alternating-current it was easy enough to meet the high voltage, but there were other limitations. In the three-phase traction system, as brought out by the Ganz Company in Europe, three-phase motors of large power could be used but there was the handicap of two overhead wires at different potentials, thus involving a double collection of current. Moreover, this system was apparently limited to about 3000 or 4000 volts and if one was to use alternating-current at all, there should be no such limitation in the voltage. Furthermore, for lighter service, involving relatively small motors, the polyphase induction motor did not seem to furnish a very satisfactory solution of the traction problem.

Direct-current was recognized, even at this time, as a possible solution, provided very materially higher voltages than 600 could be used, but almost everybody had doubts as to the practicability of sufficiently high voltage, either on the generators or on the motor equipment. Thus much thought was given to the possibilities of single-phase alternating-current, for here one could use the single overhead trolley with the voltage limitations very largely removed. However, engineers were faced by the fact that there was as yet no suitable single-phase motor available.

A Series Commutator Type Motor Developed

Mr. B. J. Arnold made a noteworthy attempt toward single-phase operation, by trying to use a single-phase induction motor to drive a car through a special variable speed gear. This apparently was the first published attempt to solve the traction problem by single-phase. However, fired by the enthusiasm of Mr. Westinghouse for the alternating-current system, Mr. Lamme and his associates had already been working on the same problem but along radically different lines, namely, the development of a series type single-phase motor with commutator, resembling in characteristics the series type direct-current motor.

One of the First Westinghouse Series Commutator Type Single-Phase Motors—The No. 132

It had been recognized for years that the variable speed characteristics of the series type motor were ideal for traction service and thus in attempting this new solution of the railway problem, the Westinghouse engineers attempted to retain the fundamental characteristics of the existing direct-current system. To accomplish this meant the commutation of alternating-current on a relatively large scale, something which was believed as totally impracticable at that time. However, the company had had sufficient experimental experience with the commutation of alternating-current commutating motors to indicate that the problem was entirely possible, especially if the frequency used was quite low.

As early as 1892, a pair of 10-hp. commutator type railway motors had been built for an experimental car for Mr. Westinghouse. The frequency used was too low, however, to handle the car satisfactorily, and the motors used were also too small, so that the tests were soon abandoned. About 1896, the problem was taken up again, but was not carried very far. About 1899, some 60-cycle series motors of about 40 hp. were built and operated for about six months, and while not

An Experimental Single-Phase Car Operated at the Home of Mr. Westinghouse about 1894

entirely satisfactory, yet the evidence was such that it looked very much as if, with frequencies as low as 25 cycles or less, railway motors of the single-phase commutator type could be built for quite large capacities.

First Commercial Application of Single-Phase in 1902

The result was that in 1901 and 1902, the engineers of the Westinghouse Company took up the question of building single-phase railway motors and in 1902 when a contract was taken to equip a high-speed electric line with single-phase series type railway motors, Mr. Lamme announced to the engineering profession the first commercial application of the single-phase commutator motor to railway service. This was the true practical beginning of the present single-phase railway system.

It was recognized then, and always has been recognized, that the single-phase commutator type railway motor is not in itself as economical or efficient as its direct-current competitor, but against this it was figured that the simplification, and economy, of the transmission system together with the more economical speed control would offset the decreased economy of the motor itself. From the speed control standpoint, the single-phase system was far ahead of the direct-current, for the flexibility of the alternating-current system allowed voltage variations for controlling the motor speed, without the use of regulating rheostats for absorbing the extra voltage and power. Here was considered one of the major advantages, especially for locomotive work.

Adoption of Single-Phase Spreads Rapidly

The announcement of the series single-phase commutator traction motor was a great spur to railroad electrification activities. Nor was this activity confined to the United States. Much contemporary work had been going quietly along and Mr. Lamme's motor brought others forward in quick succession in European countries. The Government of Switzerland had taken up the problem of railroad electrification on account of their abundant water power and on account of their fuel shortage. Germany also redoubled her efforts and the single-phase motor development paralleled the tests of Zossen.

Like all new things, the single-phase system, when first brought out, was misapplied in many ways. In a number of cases where the direct-current system did not seem applicable, the single-phase system was applied and was also found inapplicable, the fault, however, not lying directly in the system of electrification. In a number of cases there was an attempt to use the alternating-current system in connection with large direct-current systems already established, thus involving much complexity in equipment. In fact, within a few years, it developed that the single-phase system was not a satisfactory alternative to the existing direct-current system in general, but that it had its own field, and this field was where the special characteristics and advantages of high trolley potentials would apply. In other words, the single-phase system really began where the existing direct-current system was handicapped by trolley limitations.

Two of the 66-Ton, 3300-Volt, Single-Phase Baldwin-Westinghouse Motive Power Units Placed in Service on the Grand Trunk Railway in 1908

While there were at first some misapplications of the single-phase system, due largely to over-enthusiasm, yet within a very few years it began to be applied to heavy service and in all such installations it has persisted, and not only persisted, but has enlarged its field. One of the first systems to adopt this was the St. Clair Tunnel under the Detroit River, which was initiated about 1906. This electrification was necessitated on account of the smoke problem. The equipment consisted entirely of slow-speed locomotives and the first locomotives installed are still in use with a most remarkable record of reliability and low maintenance cost.

The First Baldwin-Westinghouse Electric Locomotive Hauling a Train at East Pittsburgh in 1895

New Haven Electrification brings Many Problems

A second large electrification, initiated practically at the same time as the St. Clair Tunnel, was the well-known New Haven electrification. This attracted real attention,—some of it favorable and some of it otherwise. After this contract was taken for electrification at 11,000 volts, 25 cycles, single-phase, many engineers, undoubtedly with all sincerity, insisted both privately and publicly that the thing was a physical impossibility, and that large passenger and express trains could not be handled by single-phase equipment.

Two of the Original 108. 8-Ton, Single-Phase Motive Power Baldwin-Westinghouse Units Placed in Service on the New York, New Haven & Hartford in 1908 Which Have Now Traveled over a 1,250,000 Miles

Mr. Westinghouse did not worry about the opinions of others in this matter, and was always willing to take up the cudgel in favor of alternating-current traction. In fact, he was such a believer in the alternating-current as a general solution of all problems that he felt that in the end alternating-current traction systems must dominate all others. With this in view, he obtained the American rights to the Ganz three-phase system, not necessarily because of any need for such a system, but partly because he desired to have more than one solution of any problem. When the New Haven contract was taken he said to some of the Westinghouse engineers—"Now I have dropped you into the middle of the pond and it is up to you to swim out".

The history of the New Haven development and experience need not be given here. It may be said, however, that while much trouble developed which had to be cleared away piecemeal, yet the real troubles were not where they were anticipated. The first forty locomotives built were of the gearless type, that is, with the armatures surrounding the axles but driving through flexible connections. Many wise men shook their heads over these motors, as a gearless single-phase commutator type motor for 300 hp. was something which had never been attempted before,—which might be said of everything else in this system. However, it was not the motors which developed trouble. In fact, these motors have made about the best record of any of the elements which made up this great system.

Some of the greatest troubles developed in connection with the overhead trolley system and its protective devices. Short circuits, and voltage and current surges, had been encountered in all alternating-current systems, in connection with alternating-current power distribution in general, but these were only semi-occasional. In the New Haven system, at first, these were not only of daily occurrence, but sometimes very many times a day, and apparatus which might stand a few surges during the year and still have reasonably long life, was found to last only a few weeks on the New Haven system.

New circuit breakers, new selective arrangements, new protective devices, new methods of insulation, new problems of trolley suspension, new problems of under-running trolleys, all had to be handled. For the first two years some lively work had to be done, but it was seen quite early that most of the difficulties to be overcome were not fundamental in character and the solutions were not prohibitive in cost or otherwise.

The New Haven Proved to be a Huge Success

Behind all this, Mr. Westinghouse still had full confidence in the system and the engineers on the New Haven Railroad also had confidence. With these powers behind it, the system soon began to loom up as a success instead of the failure which many predicted. Whatever else might be said of the Westinghouse Company in connection with this great installation and development, one is obliged to admit that in no great undertaking of any kind has there been shown greater persistency, stamina and resourceful engineering, than in this New Haven electrification.

As stated before, the single-phase system was brought out to meet a need, this need being for a higher voltage on the trolley system. However, the advent of the single-phase railway spurred on the advocates of direct-current to greater activities in the direction of higher voltages, so that soon the 1200-volt trolley system came out and proved to be a marked success. This eventually proved to be the most satisfactory solution for the interurban traction systems where the 600-volt limit was prohibitive. Also, the 1200-volt system, operated in connection with the existing 600-volt lines, proved a simpler and

One of the More Recent 178-Ton, Single-Phase Baldwin-Westinghouse Locomotives Placed in Service on the New Haven Railroad in 1924

easier problem than the mixture of single-phase and direct-current. The 1500-volt was an offshoot of the 1200 and is usually classed with it.

Recognizing that the 1200 and 1500-volt systems were still insufficient for heavy main line service; the

2400-volt and the 3000-volt trolley systems were brought out. A notable first instance of the latter system is the Chicago, Milwaukee, St. Paul and Pacific electrified section, comprising many hundreds of miles through the mountains.

Regenerative Control an Important Feature

An important adjunct of three-phase operation is regenerative control on the locomotive by which the energy delivered by a train going down grade will return power to the line and thus relieve the brakes of the major part of their usual service.

This system was used on the three-phase traction systems installed by the Ganz Company, and by the Italian Westinghouse Company in Italy. With three-phase operation, regeneration is a relatively simple matter. When the three-phase induction motor runs above its synchronous speed, it automatically begins to generate power.

On the Norfolk & Western electrification, initiated some years ago, using a single-phase trolley system, with phase splitters for developing three-phase on the locomotive for use with induction motors, regeneration has been used from the first. Regenerative control is also used on 3000-volt direct-current electrifications, but in this case auxiliary apparatus is necessary in

One of the 195 3-Phase, Westinghouse-Equipped Locomotives and Train on the Italian State Railway

order to produce, more or less automatically, the regenerative characteristics. The usual series type direct-current motor is not in itself capable of feeding power back to the line in a stable manner. Stability in practice is obtained by field excitation derived from a separate source, and the regenerative devices used on direct-current equipments represent very interesting solutions of this problem.

The Norfolk and Western Uses Three-Phase Motors

About fifteen years ago in taking up the electrification of the Norfolk & Western system on the mountain grades in West Virginia, the problem appeared to be almost entirely one of carrying heavy loads up a steep grade and letting them down on the other side. In other words, to some extent, it represented a hoisting problem. There were certain conditions presented from the operating standpoint which seemed to be met better by the induction motor than by any other type. However, the induction motor would not operate from the single-phase contact line and the polyphase system was not applicable on account of the complication of trolleys and the difficulties of high voltage. Consequently, a modification of the single-phase system was used in which single-phase current was supplied to the locomotive and was there converted to polyphase current by means of a "phase-splitter" or "phase-converter". The rest of the locomotive was practically of the three-phase type.

Two of the 151-Ton, Single-Phase, Westinghouse-Equipped Motive Power Units on the Norfolk & Western Railway

One of the principal requirements of this system was that any locomotive must come up against its load with full load, or even overload torque for a period as long as five minutes. At that time, this was a prohibitive condition for the single-phase commutator-type motor, and, therefore, the three-phase seemed to be the most practicable solution. The Norfolk & Western installation, therefore, simply represents one of the modifications of the single-phase system in general.

Subsequent orders for locomotives for the Norfolk & Western were filled with locomotive equipments more powerful than the original and with improved auxiliary equipment. This has resulted in an improvement in the line power factor and in consequence an improved regulation of trolley voltage is obtained.

The Virginian Also Uses Three-Phase Motors

Heavy traction electrification has been attended with big and difficult problems. Many of these undertakings were filled with heart-breaking experiences taxing the ingenuity, and stick-to-itiveness of their backers. How well most of the problems have been surmounted may be demonstrated by the successful

In 1925 the Virginian Railway Placed in Service Thirty-Six 215-Ton, Single-Phase Westinghouse-Equipped Motive Power Units

installation and performance of the electrification of the Virginian Railway. This railroad, like the Norfolk & Western, which it closely parallels, includes heavy grades against loaded movement.

The remarkable success attained on the Norfolk & Western with the split-phase system resulted in the adoption and extensive use of this system on the Virginian Railway and with splendid success.

It is interesting also to note that the three-phase electric locomotives on the Italian State Railways were built by the Italian Westinghouse Company.

Pennsylvania Electrification Uses Single Phase

Somewhat later the Pennsylvania Railroad, in electrifying its Philadelphia suburban service, adopted the single-phase system for multiple-unit trains, with power purchased from the Philadelphia Electric Company. Here is an instance of a main line railway purchasing power from a central generating station, for operating its equipment. The New Haven company has also done the same thing, but to a lesser

Westinghouse Equipped Single-Phase, Multiple-Unit Train Entering Broad Street Station, Philadelphia

extent, for within recent years the New York Edison Company has furnished considerable power to the New Haven. But, in general, this has been supplied from special single-phase generators in one of the New York Edison stations. In the Philadelphia installation on the other hand, the two main branches of the electrification, namely, the Paoli and the Chestnut Hill lines, are fed from the three-phase generators of the Philadelphia Electric Company.

The use of phase balancers on this system provides a thoroughly practical means of taking off a relatively heavy single-phase load with equalization of load on the three-phase supply. Extensions to 60-cycle power system resulted in the development of the frequency changer sets having a specially designed 25-cycle generator to care for the single-phase loadings. Experience had taught that it was impracticable to use the frequency changer sets of usual mechanical design on account of vibrations set up by the single-phase pulsating torque. Damper windings were a recognized necessity in the generator pole faces.

80-Ton, Single-Phase Baldwin-Westinghouse Locomotives Handle all the Switching Operations in the Oak Point Yards of the New York, New Haven & Hartford Railroad

More recently a simple spring mounting has been interposed between the generator frame and the foundation which has effectively eliminated the vibrations from substation and switchboards.

Along with the development of high-voltage direct-current, as represented by the Chicago, Milwaukee, St. Paul & Pacific Railway, the Chilean State Railway and the Paulista Railway, the Westinghouse Company has continued to develop the single-phase system with commutator motors so that the latter has become capable of meeting all the requirements of freight and passenger service under extremely heavy conditions. Moreover, it can handle multiple-unit, and small car service with equal facility and it is particularly well adapted for electrification of freight yards, as exemplified by the Oak Point yards of the New Haven Railroad, probably the finest illustration of electrified freight yards that can be found anywhere in the world.

The Mercury Arc Rectifier

Recognizing that the mercury rectifier might some day become an important factor in the electrification program, extensive developments were carried out by the Westinghouse Company. Sufficient work was done to demonstrate that the rectifier on the car or locomotive was a complete engineering possibility and that its use was purely an economic question. Also an experimental line was operated in 1914 and 1915 at Grass Lake on the Michigan Railways lines having 5000 volts direct-current on the trolley and

The Most Successful of Early Power Arc Rectifiers Built by Westinghouse in 1910

fed through mercury rectifiers in the substation from an alternating-current power line. The experiment being entirely satisfactory, the voltage was later reduced to make it conform to that of the rest of the system.

Westinghouse Equipped First Motor-Generator Locomotive in America

Much attention today is directed to the motor-generator type of locomotive. This is one whose motive power is from a single-phase trolley through a motor-generator set consisting of a synchronous motor and direct-current generator and driving direct-current axle mounted motors. The control is of the variable voltage system. A locomotive of this type was built by Oerlikon Works in Switzerland about 1904. It evidently did not compare favorably from the standpoint of cost and weight. However, as designs were improved and higher weight efficiencies were obtained in rotating machinery this type of locomotive came to the foreground.

In 1924 the Ford Motor Company designed an electric locomotive which incorporated this system. The Westinghouse Electric & Manufacturing Company furnished the electrical apparatus for this novel locomotive. This is the first of the motor-generator type locomotive in America. Other locomotives utilizing this system have since been built by Westinghouse for the Great Northern Railway.

Thus it may be seen that in the heavy railway field, Mr. Westinghouse and the Westinghouse Electric & Manufacturing Company have been at the forefront in development and progress. The only radically new system brought out, namely, the single-phase, originated with the Westinghouse Company.

The First Motor-Generator Type Locomotives in America—Two 186.4-Ton, Single-Phase Motive Power Units, Equipped by Westinghouse for the Detroit, Toledo & Ironton Railway

The split-phase system which is really a branch of the single-phase, was experimented with as early as 1896 or 1897, in connection with the possibilities for electrifying the Manhattan Elevated in New York, and a phase splitter and induction motor were so operated on experimental test. Credit for the commercial application of the split-phase system, therefore, lies with the Westinghouse Company.

In consequence, it may be said that the Westinghouse Electric & Manufacturing Company has had practical operating experience with all systems which have been seriously proposed for railway electrification, and has carried each through to a successful operating stand-point. From this experience it knows the advantages and disadvantages of each system and is familiar with the limitations of each.

Although the single-phase system is by far the most flexible for extensive heavy traction electrification, as proven by past experiences, there are certain instances where high-voltage direct-current electrifications may be used to good advantage and in any such cases the Westinghouse Company is ready with an open mind to embrace the use of such a system.

Two of the 178.75-Ton, Single-Phase Motor-Generator Type Baldwin-Westinghouse Motive Power Units which were Placed in Service on the Great Northern Railway in 1927

The Application & Equipment of the Safety Car

By G. M. Woods
Westinghouse Electric & Manufacturing Co.

THE AUTHOR DISCUSSES THE CLASSES OF SERVICE IN WHICH SAFETY CARS HAVE BEEN AND CAN BE USED, THE POSSIBILITIES AND THE LIMITING FEATURES OF OPERATION. THE PROBLEM OF APPLICATION TO A SPECIFIC PROPERTY IS EXPLAINED AND VARIOUS PARTS OF THE ELECTRICAL EQUIPMENT ARE DESCRIBED AND ILLUSTRATED

Merchandising Transportation

By F. G. Buffe
General Manager The Kansas City Railways

Safety Car in Combined Residential and Business Service, Baltimore, Md.

CONTENTS

THE APPLICATION AND EQUIPMENT OF THE SAFETY CAR

FOREWORD

TRACTION SERVICE

VERY phase of the life of a community is affected by the success or failure of its traction lines in providing a service which is indispensable to the public. Careful regulation, combined with concern for the business stability and the success of traction lines, should prevail in each city of the country. Adequacy of service at the lowest rate compatible with continued efficiency is the paramount consideration from the point of view of the public, and neither factor can be sacrificed to the other without public detriment. Each community is urged to consider the situation of its traction service from these two angles, in order that it may be ascertained whether the increased costs common to all business have been unaccompanied by added revenues sufficient to maintain the service requisite for the industrial and commercial efficiency of the community.

CHAMBER OF COMMERCE OF THE UNITED STATES

APRIL, 1920

General Application of the Safety Car

A DISCUSSION OF THE CLASSES OF SERVICE IN WHICH SAFETY CARS HAVE BEEN AND CAN BE USED WITH AN OUTLINE OF THE POSSIBILITIES OF SUCH SERVICE AND THE LIMITING FEATURES OF THE CAR

URING the last ten years there has been a general tendency toward lighter weight in street car construction and during the last five or six years street railway operators and manufacturers have directed their attention especially to the development of light weight cars to be operated by one man. Five years ago few individuals could conceive of a car of this type operating in any service other than in small towns or on lines of extremely light traffic in towns of moderate size. The development and general use of the now well known "Safety Devices", along with the shortage of labor and the unfortunate financial condition of most of the railway companies have resulted in the adoption of the Safety Car for a variety of traffic conditions. The Safety Car has been successful because of its own merits, but its widespread adoption has been expedited by the peculiar situation which has existed since its development, and which is largely a result of the war. At the present time there are a few railway operators who feel that the Safety Car is strictly a "small town" vehicle. At the other extreme are those who feel that the Safety Car is the suitable and economically correct car for all surface lines of any city.

The Safety Car is the most nearly universal type of car operating today and actual use has demonstrated its adaptability to a large percentage of the surface lines of the country. However, necessarily it has certain limitations and it is only through a proper recognition of these that the car can be most successfully operated. The average street railway company today recognizes the economies and increased receipts incident to Safety Car operation, and the question is not, "Can we use Safety Cars?", but "How can we use Safety Cars to the best advantage?" Local conditions vary so widely in different cities that no definite rules can be applied. Each problem must be analyzed and the solution derived by a careful study of the chief characteristics of the various lines and the application of the principles demonstrated by Safety Car operation under similar conditions in other cities.

Operation Under a Variety of Traffic and Climatic Conditions

Safety Cars are now operating in all parts of the United States from the smallest town boasting a traction system to the largest cities of the country. The problems arising from the large negro population of the South and those arising from the extreme cold and heavy snows of the North have been solved with equal facility. A study of a number of lines which have been changed to Safety Car operation shows them to be divided into the following general classes:

1. Lightly traveled lines on which the substitution was made on a car for car basis.
2. Lightly traveled lines where materially increased service was furnished.
3. Heavy service lines radiating from rapid transit or main trunk surface lines.
4. Lines paralleling rapid transit lines.
5. Main surface lines.

In the first class are those lines which usually do not pay when cars are operated with two men and which sometimes do not pay even when Safety Cars are substituted. Many of these lines are operated merely to hold a franchise and many Safety Car shuttle lines running through a thinly populated district act as feeders for the main lines, and serve to develop the territory through which they operate. It is usualy undesirable to substitute Safety Cars on a car for car basis especially where large double truck cars were previously operated. The patrons of the railway at first feel that the Safety Car is only a scheme for saving the wages of one man. Its resemblance to the old single-truck cars with the objectionable "teetering" and usually with longitudinal seats, makes the car-riding public feel that such substitution is a step backward, and eventually political capital may be made of the elimination of one of the platform men.

For the preceding reasons it is advisable, when starting Safety Car operation on any property, to choose for the installation lines where the traffic warrants an appreciable increase in service. Just as the Safety Car operators are entitled to a share in the profits in the form of increased wages so are the car

riders entitled to increased service. There are, however, many lines now operating with two men on the cars where a substitution of Safety Cars for an equal number of the old cars is the only proper course, and meets with public approval if the railway management

Virginia Railway & Power Co., Petersburg, Va.

has shown its willingness to improve the service on other lines where an increase in service is justified.

Many cases can be cited where railway companies began Safety Car operation on one or more of the busiest lines, furnishing vastly improved service and at the same time or shortly thereafter substituted Safety Cars for old single truck cars on unimportant lines on a car for car basis. This plan has met with universal success.

There have been various instances where car for car substitution was made on practically every line of a property. It cannot be said that the patrons were particularly enthusiastic about the change but they continued to ride the cars and although in some cases the receipts decreased slightly the saving in operating expense was sufficient to change a losing business to an earning one and to enable the company to pay an attractive return on the investment in the Safety Cars. On some lines receipts have been increased from 10 to 20 per cent as the result of a car for car substitution. The reason for the increased patronage is often found in the fact that any new car attracts traffic. The Safety Car in many instances is more comfortable and pleasing than the cars displaced and pilfering of fares by the platform men in very nearly eliminated.

Frequent, Reliable Safety Car Service Creates Traffic

One of the most gratifying features of Safety Car operation is the greatly increased revenue which has resulted from short headways on lines previously regarded as unprofitable. The increase in receipts resulting from more frequent service has always been recognized, but for many years the problem of handling the greatest number of passengers per platform man has resulted in the use of extremely large units running at infrequent intervals. When shorter headways are operated with the large car the increased revenue is counterbalanced by the increased operating expense. When the Safety Car was first used to replace large double truck cars it was necessary almost in every instance to operate at least 50 per cent more mileage in order to handle the traffic properly. While some increase in car riding generally was expected the decrease in operating expense was the chief reason for the adoption of Safety Cars. Actual results of operation showed that in many cases the increase in revenue was greater than the saving in operating expense.

The appreciation of the traffic building qualities of the Safety Car operated on short headways resulted in the second class of application, namely; on lines of light traffic where increased service was furnished. On many lines operating from 15 to 30 minutes headway one could walk from the outskirts of the town to the business district during the interval between cars. The private automobile and the jitney bus flourished under these conditions. When Safety Cars were used to give a 50 per cent increase in service the percentage increase in receipts was in many cases equal to or greater than the percentage increase in car mileage. On one property the service on a line, whose receipts were less than 13 cents per car mile, was increased 47 per cent and the receipts increased 147 per cent or to approximately 21 cents per mile.

The Safety Car makes it more convenient to ride to the business district on the cars than to use an automobile, especially where parking restrictions are enforced. The pedestrian and the jitney rider are both converted by frequent service to car-riders. In addition people who stay at home and shop by 'phone when the car service is poor will ride to their shopping when a superior service is provided.

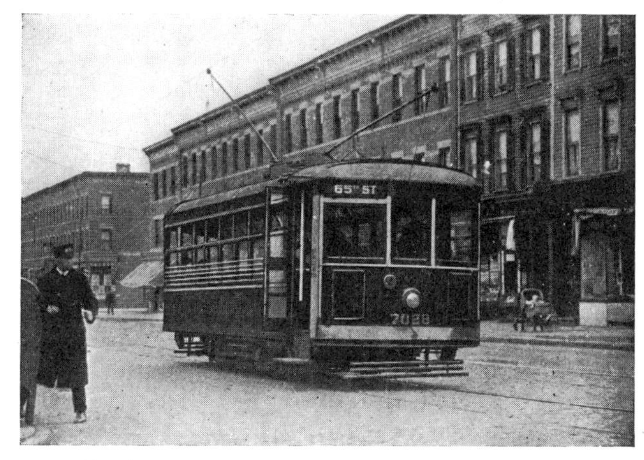

Brooklyn (N. Y.) City R. R. Service in the Large City

It must not be inferred that an increase in car riding will result whenever and wherever the headway is decreased. A line must first have traffic possibilities or the increased service idea may be carried too far for

favorable financial results. For instance, because the headway on any particular line was profitably reduced from 20 minutes to 10 minutes does not indicate that the receipts could be expected to show a similar increase if the headway were further decreased from 10 minutes to 5 minutes. Each line has a limit beyond which it is uneconomical to increase service.

In starting Safety Car operation on lines of light traffic as well as on heavy traffic lines it is better to err on the side of too much service than too little. The attitude of the public toward a railway company is becoming more and more important. The feeling of the public about an innovation is largely a matter of first impression. If the service is actually improved and the public is thoroughly informed of the improvement, the Safety Cars are almost certain to meet with approval. On the other hand if there is no improvement in service, first impressions are apt to be unfavorable and the latitude of further Safety Car application restricted. A certain property substituted Safety Cars for an equal number of large double-truck cars on a light traffic line. During the rush hours the cars were crowded and there was an appreciable decrease in receipts. It became necessary to add more cars to the line, but today the attitude of the public toward the cars in that town is tolerant where in a similar neighboring town where the headway was greatly decreased at the time of installation the car riders are most ardent Safety Car boosters.

Safety Cars Admirably Adapted to Service on Lines Radiating from Rapid Transit Lines

The application of Safety Cars to lines radiating from rapid transit lines or main trunk surface lines has taken place where the cars have been used in

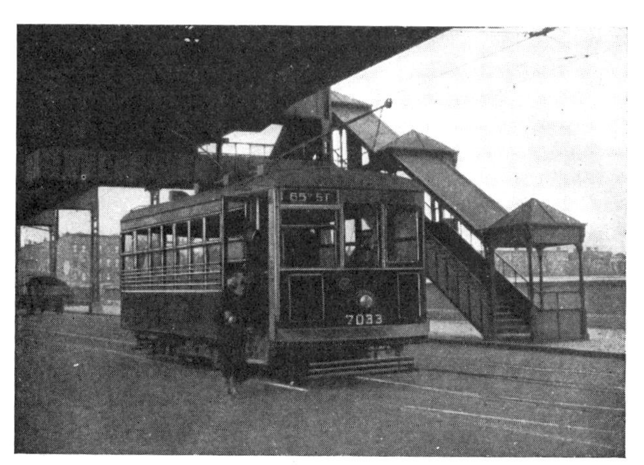

Heavy Traffic Feeder Service in Brooklyn

larger cities. The Safety Cars are admirably adapted to service of this character, especially when they connect with rapid transit lines. The rapid service which can be furnished by subway and elevated lines is rendered doubly desirable when supplemented by frequent, convenient Safety Car operation radiating from the terminal. Heavy loads are picked up at the rapid transit terminals and distributed through subur-

The Connecticut Company, Bridgeport, Conn.

ban districts or to outlying sections where the development does not justify the extension of the rapid transit lines. The speed of rapid transit lines and the "at the door" service of the surface lines are thus combined.

Under these conditions the morning loads are picked up gradually and discharged at one point. In the evening the majority of passengers board the car at one point and leave in small groups. The collection of fares can be simplified by having pre-payment and post-payment areas at the rapid transit terminals. All passengers boarding the car at other points can pay as they leave on inbound trips and as they enter on outbound trips. Street collectors are sometimes used at the heavy loading points. These conditions also lend themselves readily to the recently popular scheme of "Pay As You Enter" on inbound trips and "Pay As You Leave" on outbound trips.

The use of Safety Cars to distribute passengers from main surface lines is a matter regarding which opinion differs widely. In a majority of cities there is a marked dislike to transferring from one car to another. In cases where a transfer will save from 5 to 10 minutes out of 45 the vast majority of passengers prefer to use the through cars rather than suffer the inconvenience of changing from one car to another with the possibility of having a long wait at the transfer point and the probability of being unable to obtain a seat in the car to which the transfer is made. For this reason it is held that where a line operates several miles from A to B and has various branches at B running to C, D, E, etc., that the logical service consists of Safety Cars operating from A through to C, D, and E. The ideal service for each of the branch lines can thus be provided and the overlapping of the lines between A and B with, perhaps, certain additional service will adequately take care of that section.

On the other hand, it is urged that in the larger cities the use of subway or elevated lines in conjunction with the surface lines is resulting in a decrease of the aversion for the transfer and that the patrons of the railway lines are becoming more inclined to take

The Kansas City Rys. Co. Service in Congested Districts

the route which consumes the minimum of time between place of business and home even at the expense of a transfer. It is further maintained that the operation of large cars in trains is the most economical way of transporting passengers over the main section of the line and that Safety Car service should be limited to the sections radiating from the main line. In the majority of cases the all—Safety Car service appears to be more desirable. Certain limiting conditions sometimes exist which seriously handicap the Safety Car operation and render the combination of large cars most easily and economically operated.

The Fresno (Cal.) Traction Co. Service in Residential Districts

Another class of Safety Car service which will be limited to large cities possessing subway or elevated lines is that where surface lines parallel the rapid transit lines. In the case of elevated roads these lines are usually located under the elevated structure. All through passengers use the rapid transit lines and the receipts of the surface lines which formerly were lines of extremely heavy traffic are reduced to the point where operation of two-man cars is unnecessary and unprofitable. To discontinue operation entirely would result in a material loss of revenue because the possible short-haul passengers, rather than use the less convenient rapid transit line, will walk short distances. The use of Safety Cars on short headways generally results not only in the holding of all the old short-haul patronage, but also in creating additional short-haul patronage.

Safety Car Capable of Handling Heavy Traffic

The final class of service is that met on the main surface lines of large cities. The methods of applying Safety Cars to these lines vary in diffrent sections of the country, and there is an even greater variation in opinions regarding the practicability of their application to this service. The most common use of Safety Cars on main surface lines is where the car mileage is increased from 50 to 100 per cent. The rapid acceleration and retardation of the Safety Car, the smaller number of passengers per car and hence fewer stops, the low floor design, the elimination of signals between conductor and motorman and the better view traffic officers have of the door combine to speed up the movement of individual cars.

Serving Both Colored and White Patrons in Birmingham

The safety provisions of the door and step control and the collection of fares by the car operator tend to increase the duration of stops. As previously mentioned this effect is counteracted in some cases by the use of street collectors and by pre-payment areas. Where the management of the street railway company honestly endeavors to improve the service and the publicity describing and pointing out the advantages of Safety Car service is wisely handled, the people will cooperate by having the exact change ready for fare.

Unfortunately it seems that the larger the city the less willingness there is to assist in this respect. The sale of tickets at a reduced rate, however, often decreases the delay due to making change, In one city 95 per cent of the passengers use tickets.

The congestion arising from the greater number of units also tends to slow up traffic. The congestion is not proportionate to the number of units because of the features which tend to accelerate the movement of the Safety Car, and because of their smaller size. It is certain that a number of Safety Cars can pass over a congested section of city street more rapidly than an equal number of large double-truck cars and still more readily than an equal number of two-car trains. Observations made of congested city service when Safety Cars and double-truck cars operate on the same track shows that the Safety Cars are frequently waiting for a double-truck car to get out of the way and that in the neighborhood of 50 per cent more Safety Cars than double-truck cars can be operated over a given section of track in a given time.

Winter Time Operation in Waltham, Mass.

Before the Safety Car demonstrated its ability to handle city traffic quickly, one-man operation was regarded as inherently slower than two-man operation. However, actual use of the Safety Car showed the factors which tend to increase its speed in their true light.

Track Capacity the Limiting Feature to Safety Car Operation

Track capacity will not be reached over a given line as a whole except in rare instances. In practically every city various car lines overlap in the business district on a few of the principal streets. In other cities the topography is such that there are only a few routes over which every line of the city has to enter the business district and a number of lines operate for a considerable distance on each of these routes. In general, a headway of 30 to 40 seconds for distances up to one mile on surface lines is about the limiting frequency even for Safety Car operation. In this connection it is well to remember that the width of the street, the number of other vehicles using the street and the attitude of traffic police toward facilitating

The End of the Line in Brooklyn Residence District

street car movement are becoming more important factors in track capacity. The number of cars per hour is not the only consideration.

On those lines which are so ideally located as to possess no "neck of the bottle" the track capacity of the line as a whole depends also to a certain extent on condition of streets, vehicle traffic, etc. Experience indicates that, with two-man cars of the same seating capacity as Safety Cars and of equally rapid accelerating and braking characteristics, it is difficult to maintain the car spacing and schedule with minimum rush hour headway of one minute. One operator of

Interior Arrangement of the Standard Safety Car

wide experience has expressed the opinion that where the traffic density is equivalent to ten passengers per car mile, when the non-rush headway with double-truck cars is four minutes or lower, it is impracticable to handle the entire service with Safety Cars.

Congestion and the difficulty of handling large numbers of people in small units has given rise to various schemes for using Safety Cars in conjunction with double-truck cars in such a way that the rush hour loads would be divided according to size of unit. Obviously it is incorrect to provide cars of different seating capacities at equal intervals, all other conditions being the same. Either the large car is larger than necessary or the Safety Car is too small. Theoretically an unequal spacing could be arranged to equalize the loads but it is difficult to maintain the correct spacing in actual practice. The most general scheme is to operate Safety Cars over the entire line during non-rush hours and to add double-truck cars during rush hours, short-routing the Safety Cars. This arrangement results in a suitable division of the load and in desirable short headways over the portion of the line on which traffic is heaviest.

Under present limitations of "swing runs" it is almost as expensive to use a man a few hours as it is to use him the entire day. The present tendency on lines of heaviest traffic is toward train operation. Several large properties operating a number of Safety Cars under a variety of conditions also operate two-car trains on lines of extremely dense traffic. Another large property is operating three-car trains on a four-minute headway over a ten-mile section of line.

Bearing in mind that no material increase in riding can be anticipated by decreasing headways under three or four minutes the use of Safety Cars for the entire day does not seem advisable on the heaviest lines even where their operation is physically possible. On the majority of heavy surface lines Safety Cars used in combination with double-truck cars, short-routing the former during rush hours, meets with the most general favor.

Oil City (Pa.) is Enthusiastic Over Safety Cars

Seattle, One of the Pioneers in Safety Car Operation

An interesting development of the short-routing principle is found in a large Western city. The shopping district is at present about a mile from the business district terminal of the line. The persons boarding the cars in the shopping district were usually unable to obtain seats and criticism of the service resulted. The operation of Safety Cars between the large cars, the former looping in the shopping district and running only over the densely settled portion of the line was tried with marked success. The increase in receipts and better service resulting from this method of operation were so pronounced that the same general plan was followed on other lines. At present 15 per cent of the mileage in this city is Safety Car mileage and a large percentage of these cars is used in conjunction with double-truck cars.

Another plan suggested is the use of Safety Cars for the non-rush hours and the substitution of double-truck cars during the rush hours. The objections to this plan are the additional car barn and shop space required, the additional investment and the great increase in number of men during the rush hours.

One reason for operating Safety Cars on lines of heavy service when the cars are first purchased is to settle at once the opinion likely to exist in the public mind that the cars are suited only for lines of light traffic. If the first cars are placed in the outskirts of the city the opinion is strengthened and each time their use is extended to other lines there is opposition to the change.

From the standpoint of the public, street railway service must have the desirable characteristics of frequency, reliability, speed, safety and comfort. This service must also be produced at the minimum expense, for in the long run the car-riding public pays for what it receives. On the vast majority of city surface lines the Safety Car can furnish the service required more economically than any other vehicle. In city transportation, there is a field for subway, elevated and surface trains, for double-truck cars and for Safety Cars. Each must be used in the class of service to which it is best adapted. It is only through the proper co-ordination of these various agencies that the maximum of service and efficiency can be reached.

The Eastern Massachusetts Street Railway
Has 200 Safety Cars in Operation

Safety Car Operation by Virginia Railway & Power Co.
Has Resulted in Repeat Orders

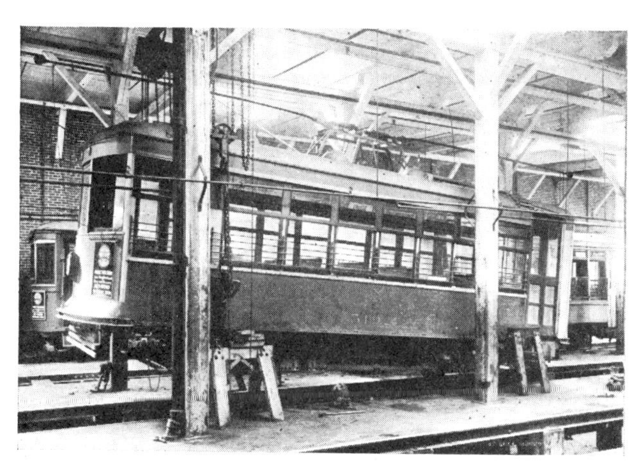

The Safety Car Occasionally Has to be Shopped but is
Easily Handled

Serving One of the Best Residential Districts
in New Haven

Safety Car Operation in a Small California City

Congested Service on the Eastern Massachusetts

The Safety Car Solves Traffic Problems in Many New England Cities, Eastern Massachusetts Street Railway

Application of the Safety Car to Specific Properties

CAREFUL AND THOROUGH STUDY OF TRAFFIC CONDITIONS, TRACK SYSTEM, PRESENT EQUIPMENT AND POSSIBLE ECONOMIES SHOULD BE MADE BEFORE SAFETY CARS ARE INSTALLED

N the majority of the large railway properties the organization is such that reports on proposed Safety Car operation can be prepared by engineers on the property who are thoroughly familiar with the local operating conditions. Many small properties, however, do not have the organization to take care of matters of this kind, regular duties requiring the entire time of those who are fitted for this work. For this reason engineers outside of the organization are frequently called upon to investigate the economies of such innovations. While it is impossible to equal the detail and exactness which results from an every-day association with a railway property, a complete knowledge of the important factors, coupled with frequent conferences with the railway operators, to a large extent overcomes the handicap of lack of intimate personal association with local conditions.

In making a study of a railway property as a whole, a considerable amount of general data is required to give a "bird's-eye view" of the entire property. Bearing in mind that the increase in gross revenue to be anticipated from increased service is of prime importance and that the chief economies are effected in platform labor, power and maintenance, the basic data required for showing the results of Safety Car operation on any particular line are obvious.

Transportation Data Essential

The first thing required in the investigation is a map of the track system indicating all double track and single track sections, and all sidings. The routing of all lines should then be shown on the map. An indication of the kind of territory around the various lines, whether business, industrial, residential, etc., and approximate data on the distribution of population and the riding characteristics are desirable.

The information to be obtained on each line includes: round trip distance; headway, basic and rush hour; running time, basic and rush hour; number of cars, basic and rush hour; regular car hours per day; regular car miles per day; type of car, designated by class number; revenue passengers per year; non-revenue passengers per year; gross income per year; tripper car hours per day; tripper car miles per day; trailer car hours per day; trailer car miles per day.

The transportation department always has very good ideas as to the best lines for Safety Car operation. From the data previously tabulated, and from general observation, a mutual agreement can be reached with the operating officials. The number of cars per hour over certain sections of track can be shown on the map for both rush hour and non-rush hour conditions. Certain lines may require re-routing in order to avoid a particularly crowded section of track, and in most cases where there are single track sections, re-located sidings, or new sidings are required.

It is not necessary to study every line on the system for the initial report, but enough lines should be considered to indicate the results of Safety Car operation in various classes of service. The change to this service usually will be made on only a few lines at first because most companies desire to proceed slowly on any new practice. After the results of operation of the first lines are available, it is relatively easy to determine along what line to proceed with the further use of the car in order to obtain the most satisfactory results.

At least one of the lines picked out for the initial operation should have sufficiently heavy traffic to show the ability of the Safety Car to handle such service and to warrant a material decrease in the headway. At least one of the lines should run through the business district. Lines with extremely heavy peak load riding and lines operating over sections where the track is worked close to its capacity should be avoided. In the South the first lines changed to Safety Car operation should not have an unusually high percentage of colored passengers. In some cases the residence of an important stockholder of the railway company, a local politician, or a labor leader on a certain line may be the final factor in determining the choice of one line over several others. In short, the

Safety Cars should be so used that the most influential citizens of the town are convinced of their real worth and are enthusiastically in favor of their extended operation.

Rush-Hour Service in New England

Equipment Data the Next Consideration

A complete list of cars is necessary. This list should include the car number, motor type, number of motors, gear ratio, wheel diameter, number of seats, length, and total weight of car and type of brakes.

From the seating capacity, maximum capacity, receipts per car mile, headways and general service observations, the amount of Safety Car service to handle properly any of the lines can be determined. The amount of additional service justified from the

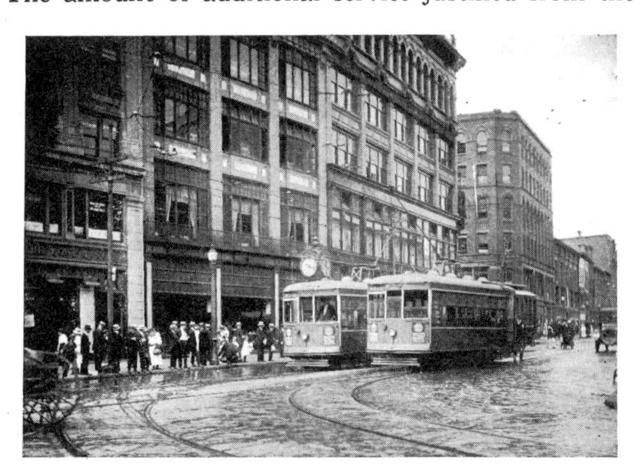

In the Best Shopping Districts are Safety Cars

standpoint of pleasing the public and building traffic cannot be determined by any mathematical calculation, but can be approximated after a study of the results obtained in other cities under similar operating conditions. It is advantageous to increase the service materially and then if the increased riding does not come up to expectations, quietly reduce the car mileage to the proper amount.

PLATFORM EXPENSE.—In order to calculate

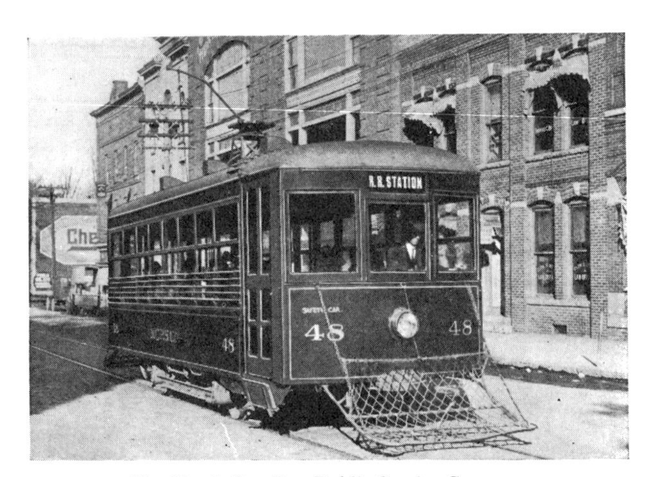

The North Carolina Public Service Company

the saving in platform expense the only data required are the car hours with two-man operation and the proposed car hours with Safety Car operation; the average present wage and proposed Safety Car wage. In general, an increase in wages of 10 per cent for Safety Car operators is recommended. The actual car hours paid for should be checked against the scheduled car hours.

ENERGY CONSUMPTION.—While on most properties the energy consumption and the maintenance cannot be determined for any particular class of car or for any particular line, the regular reports of the company are an indication of the correct amount

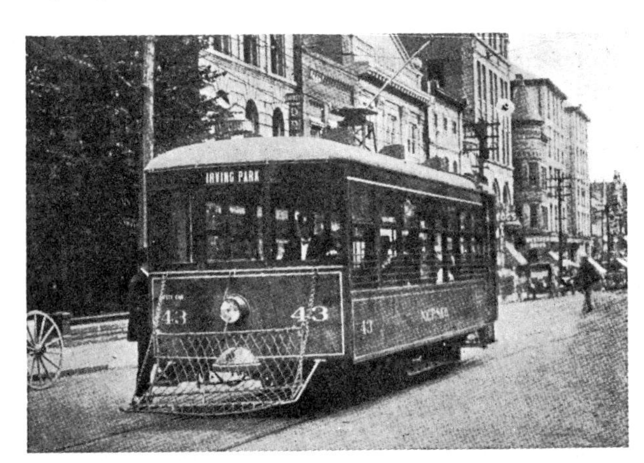

The Safety Car in Greensboro's Business District

when the car weight and equipment and the car mileage for each particular class of cars are taken into account.

The energy consumption should be obtained for both Winter and Summer months where heaters are

used, for a large part of the additional power used in Winter will be due to the heaters. In the Winter, frictional resistance is higher and the snow on the track and wet rails also increase the energy consumption. From data pertaining to grades, stops and

On the Bleak and Barren Coast of Maine

schedule speeds, the energy required for the propulsion of the various cars can be calculated. This will be approximately proportional to the weight of the cars, for while the Safety Car will make fewer stops, due to the greater number of cars and fewer passengers per car, the train resistance in pounds per ton will be higher for the lighter car. From calculations on the energy consumed in propelling the car and the known amount for air compressors, lights and heaters, a very close estimate of the saving due to Safety Cars can be made.

MAINTENANCE.—Many operators believe that

Citizens Traction Company, Oil City, Pa.

Safety Cars have not been in service long enough to obtain reliable maintenance data. Operating results indicate that in Southern climates the cars can be maintained for from 1.0 to 1.25 cents per car mile. Under the more severe conditions of the North the cars can be maintained for not more than 2.0 cents per car mile. Local conditions and maintenance methods vary so widely on different properties that the best plan is to base the Safety Car maintenance on that of the cars in operation on the particular

Safety Cars Ready for Rush-Hour Service

property. The maintenance of the Saftey Car will be approximately one-half that of modern double-truck cars with quadruple equipments. The expense of inspection of Safety Cars will be more than one-half that of the double-truck cars, but that will be counterbalanced by the cost of repairs which will be less than one-half the cost of repairs on the double-truck cars. This estimate taken in conjunction with the maintenance expense of Safety Car operation under similar conditions will result in a reasonably accurate estimate of Safety Car maintenance.

MISCELLANEOUS SAVINGS.—Some decrease

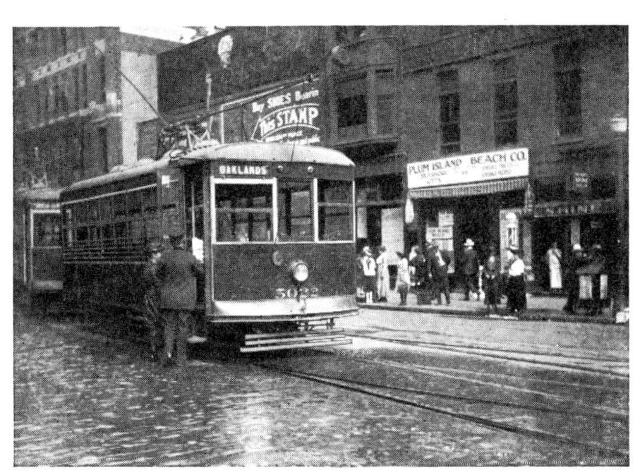

Service for Light Traffic as well as Heavy

in maintenance of way can be expected but sufficient reliable data are not available to justify including this saving in a conservative estimate of the results of Safety Car operation. Some saving in accident expense also will result, but any estimate of this should

be omitted. Superintendence and general expenses should remain approximately the same for any given property and need not be included. Since the car miles are increased these expenses in terms of cents per car mile will be decreased in inverse proportion to the car mileage.

INCREASE IN RECEIPTS.—The increased riding to be expected cannot be calculated. The experience of railways operating Safety Cars under a variety of conditions shows that the percentage increase in receipts is equal to at least one-half the percentage increase in car mileage where the cars are properly applied.

The fundamental principles are practically the same regardless of the size of the city. If a certain number of persons ride on the street cars when a certain headway is operated, a shorter headway will not only attract many of those who formerly walked, but will also stimulate additional travel. The percentage increase in receipts for a given increase in service is affected by the former interval between cars, the length of the line, the density of the population and the extent to which short haul traffic was formerly developed.

An example of an analysis of a property and a tabulation of the saving in operating expenses is given on pages 31 and 32 of Westinghouse Electric and Manufacturing Company Special Publication 1614-A.

Safety Car Installations in Connecticut are Increasing

The Safety Car in 1920

More Than 4000 Safety Cars in Use or on Order Oct. 1, 1920

Number of
Safety Cars Purchased per year
Since Jan. 1, 1916

1916 - - - -	190 Safety Cars
1917 - - - -	280 " "
1918 - - - -	650 " "
1919 - - - -	1620 " "
1920 (Jan. 1 to Sept. 1)	1500 " "

Regular Inspection of Safety Car Equipment in Brooklyn

Electrical Equipment of the Safety Car

A DESCRIPTION OF THE IMPORTANT FEATURES OF THE ELECTRICAL EQUIPMENT ESPECIALLY DESIGNED FOR THE SAFETY CAR

THE entire electrical equipment of the Safety Car has been designed with a view to producing the most reliable and efficient apparatus of the minimum weight consistent with conservative design. The most important part of any car equipment is the motor. The Westinghouse No. 508-A motor has been especially designed for Safety Car service after an analytical study of Safety Car operating conditions and requirements. All of the design details which have contributed to the success of Westinghouse railway motors are embodied in its construction. Among the special features of the No. 508-A are peculiarly effective ventilation, all through-bolts for axle caps, improved protection of commutator and bearing from dust and great mechanical strength of all parts.

Particular attention is called to the special details of construction which combine to produce a motor of low maintenance and reliability in service.

The split between the *Axle Caps* and the frame is at such an angle that the weight of the motor is taken off the axle cap bolts. Each axle cap is held by four through-bolts. *Oil and Waste Lubrication* is provided for both armature and axle bearings. Separate chambers permit of gauging the oil in the bearings. Large waste pockets are provided.

The *Armature Coils* are of insulated copper ribbon wound with no sharp cross-overs. Each armature coil is hot pressed, dipped and thoroughly baked. The coils are protected at the ends of the slots by U-shaped pieces of insulation. A strip of tin is placed around the periphery of the completely wound armature and over this the steel wire band is wound, while the armature is hot, and then soldered together, forming a solid hoop.

The *Brush Holders* are supported by insulated studs. Heavy, flat coil springs provide the tension while braided shunts carry the current The spring tension is adjusted by means of a pin passing through one of a number of holes in the casting. The right and left hand brush holders are interchangeable.

No. 508-A Standard Safety Car Motor

Soft steel punchings riveted together between end plates are used for the cores of the *Main Field Coils*. All sharp corners are removed and cushion springs hold the coils firmly in place. Vibration and chafing of the insulation are thus eliminated. The field coils are wound with square copper wire and are thoroughly impregnated.

Small steel forgings securely bolted to the frame between the main poles form the cores of the *Commutating Coils*. A sheet brass punching is pressed

Sectional View of Bearing Housing

into a groove at the end of the commutating pole, forming a support and protection for the coil. The commutating field coil construction is similar to that of the main field coil.

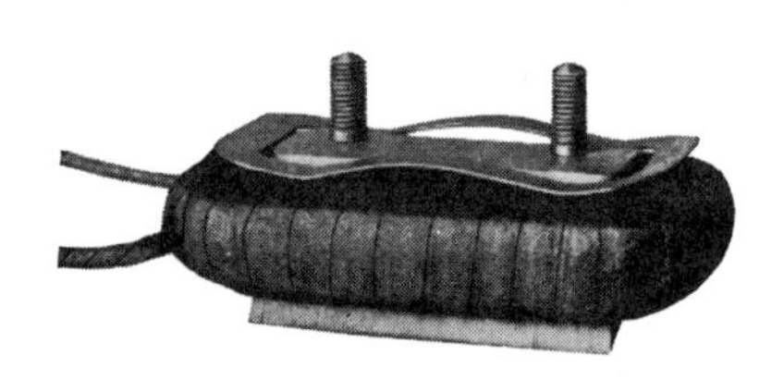

Main Field Coil and Pole Piece

The primary function of the *Gear Case* is to retain the gear lubricant and to keep out foreign material. The pressed steel gear case illustrated serves these ends admirably. It is made of heavy sheet steel of

Complete Armature Showing Ventilating Fan

high ductility, bent into shape and amply reinforced at the suspension points. It is provided with a lapped joint along the split to avoid leakage. The two-point suspension gives rigid support and is entirely free from lateral strains.

The improved malleable iron *Axle Collar* and dust

Partially Wound Armature

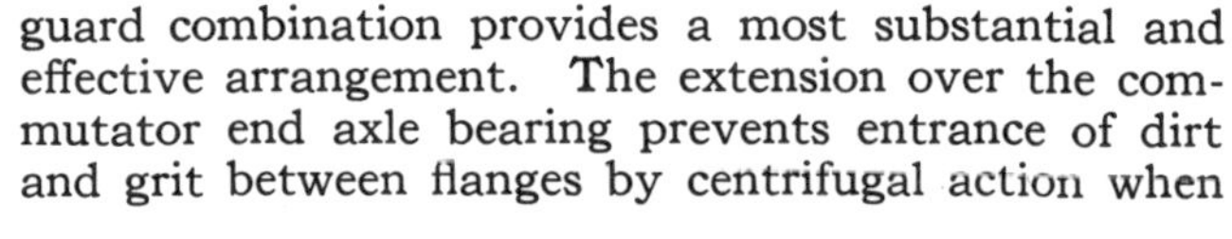

guard combination provides a most substantial and effective arrangement. The extension over the commutator end axle bearing prevents entrance of dirt and grit between flanges by centrifugal action when

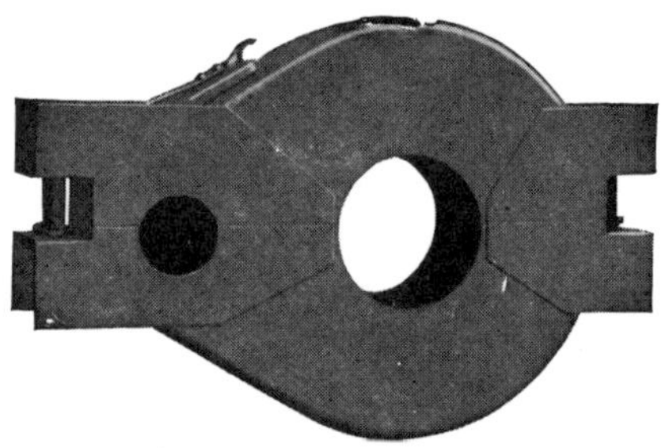

Pressed Steel Gear Case

rotating, resulting in increased life of bearings. An *Axle Dust Shield* made of sheet-steel completely encases the axle between the bearings; two windows permit inspection of axle bearings without removing

Combined Axle Collar and Dust Guard

the casing. The *Commutator Cover* is made of pressed steel. It is light in weight, unbreakable and easily handled. The cover is held securely in place

by an effective, yet easily operated latch.

The K-63-BR *Controller* is the result of years of development of platform control. Its light weight,

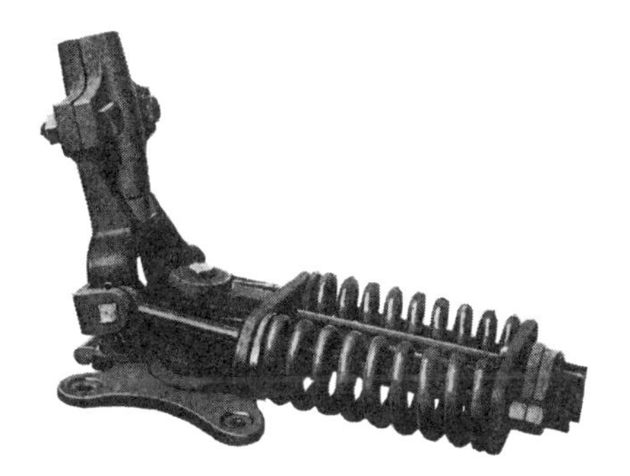

No. 15-C U. S. Trolley Base

small space requirements and ease of manipulation are especially desirable on Safety Cars. The 611-type

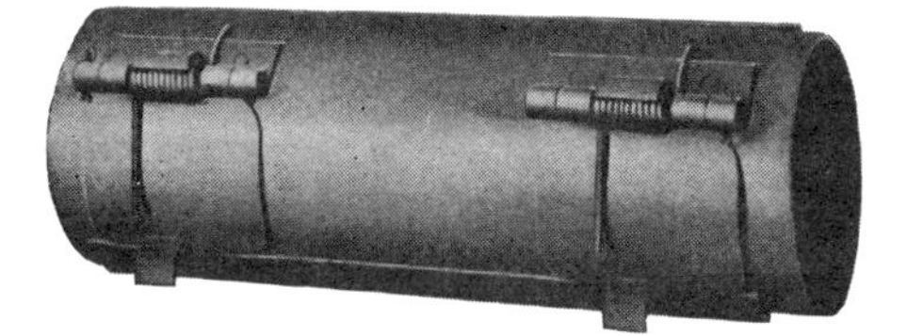

Axle Dust Shield

Circuit Breaker is so arranged that it is tripped and reset by the same handle and is thus readily adapted

Type 611 Circuit Breaker

to use with the *Standard Safety Devices*. It is light and compact and is designed with a particularly effective magnetic blowout.

A detail of car equipment that is seldom given the prominence it deserves is the *Trolley Base*. On cars operated by one man, it is particularly desirable to

Grid Resistor, 5-in., 3-point

have the trolley base and overhead construction so designed and maintained that instances of the trolley

Commutator Cover

leaving the wire are reduced to a minimum. With a 13-foot pole the No. 15-C base easily maintains 20-lb.

Brush Holder

pressure on a trolley wire eight feet above the base. The No. 15-C trolley base complete with 13-foot pole, harp and wheel, weighs only 110 pounds.

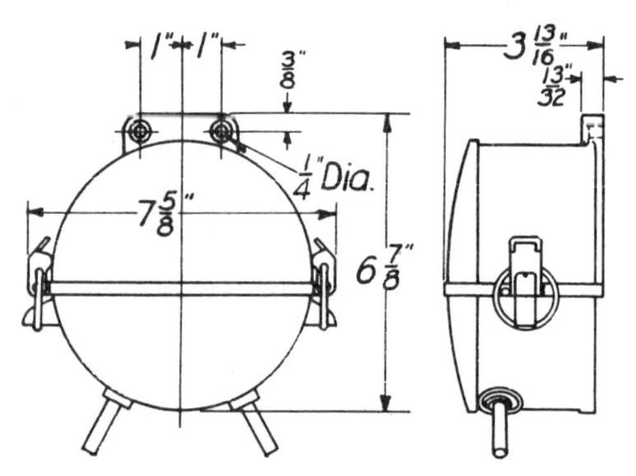

Outline of Type MP Lightning Arrester

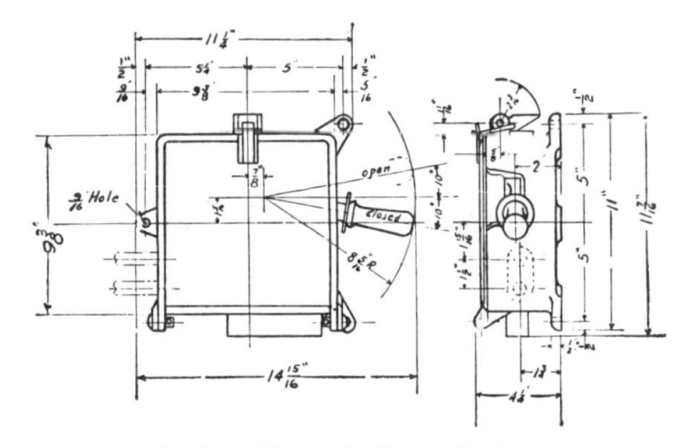

Outline of Type 611 Circuit Breaker

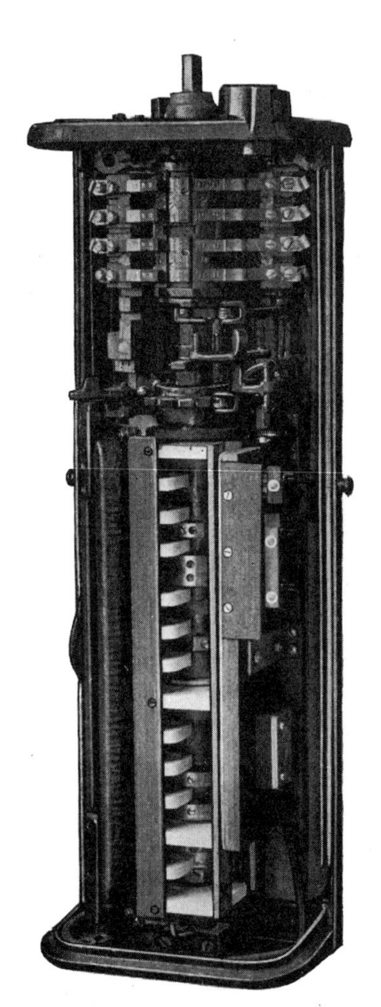

Arc Deflector Closed

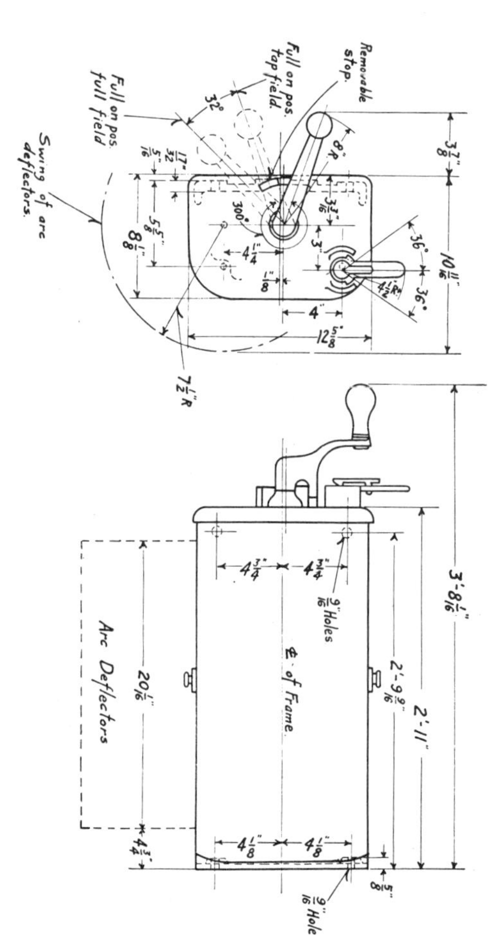

Outline

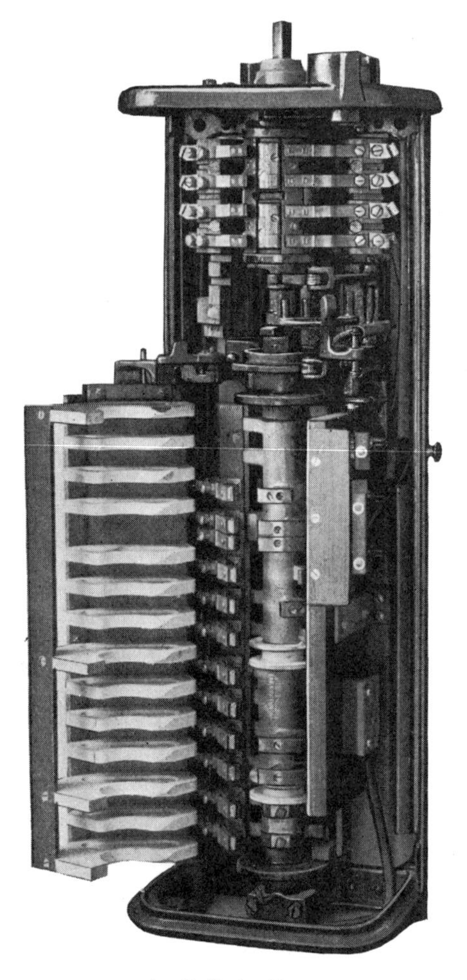

Arc Deflector Open

TYPE K-63 BR CONTROLLER

The Grid Resistor is made up of one frame in which are mounted two rows of 5-inch 3-point grids. Space and weight economy are effected by the single frame arrangement. The individual grids are composed of a cast alloy material resulting in a mechanically strong element of relatively light weight. The grids are assembled on mica-insulated tie rods and are clamped between galvanized sheet steel end frames. The steps of resistance are so proportioned that rapid acceleration is obtained without the jerking, uneven motion which results in wear and tear on the motors, gears and in fact,upon the entire car and equipment.

A complete list of apparatus (including weights) comprising the electrical equipment for Standard double end Safety Cars follows:

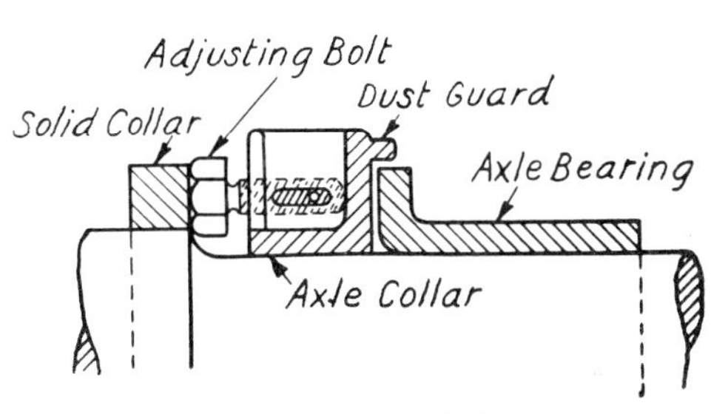

Detail of Axle Collar

Equipment for Standard Double End Safety Car

Apparatus	No. Req'd	Type	Weight
MOTOR ITEMS			
Motors	2	508-A	1700 lb.
Gear cases	2	Sheet steel	
Axle bearings	2	P. E.	
Axle bearings	2	C. E.	
Axle collars	2	M. I.	300 lb.
Gears, Solid Tr. St.	2	74-Tooth	
Pinions, Forged St.	2	13-Tooth	
MAIN CIRCUIT CONTROL ITEMS			
Trolley base with 14 ft. pole	2	No. 15	
Trolley harps	2	No. 25	220 lb.
Trolley wheels	2	No. 40-A	
Lightning arrester	1	M. P.	8 lb.
Circuit breakers	2	611	48 lb.
Controllers	2	K-63-BR	268 lb.
Reverse handle	1	K-63-BR	
Grid resistors	1 set	5 in.-3-pt.	75 lb.
EQUIPMENT DETAILS			
Main cable	695 ft.	7 x 0.0545 in.	90 lb.
Knuckle joint connectors	8	Pivot	1 lb.
LIGHTING DETAILS			
Keyless wall receptacles	20		8 lb.
Mazda lamps	20		1½ lb.
Snap switch	1		½ lb.
Transfer switch	1		½ lb.
Snap switch (D.P.D.T.)	1		½ lb.
Cable	400 ft.	19 x 0.0142 in.	18 lb.
			2748 lb.

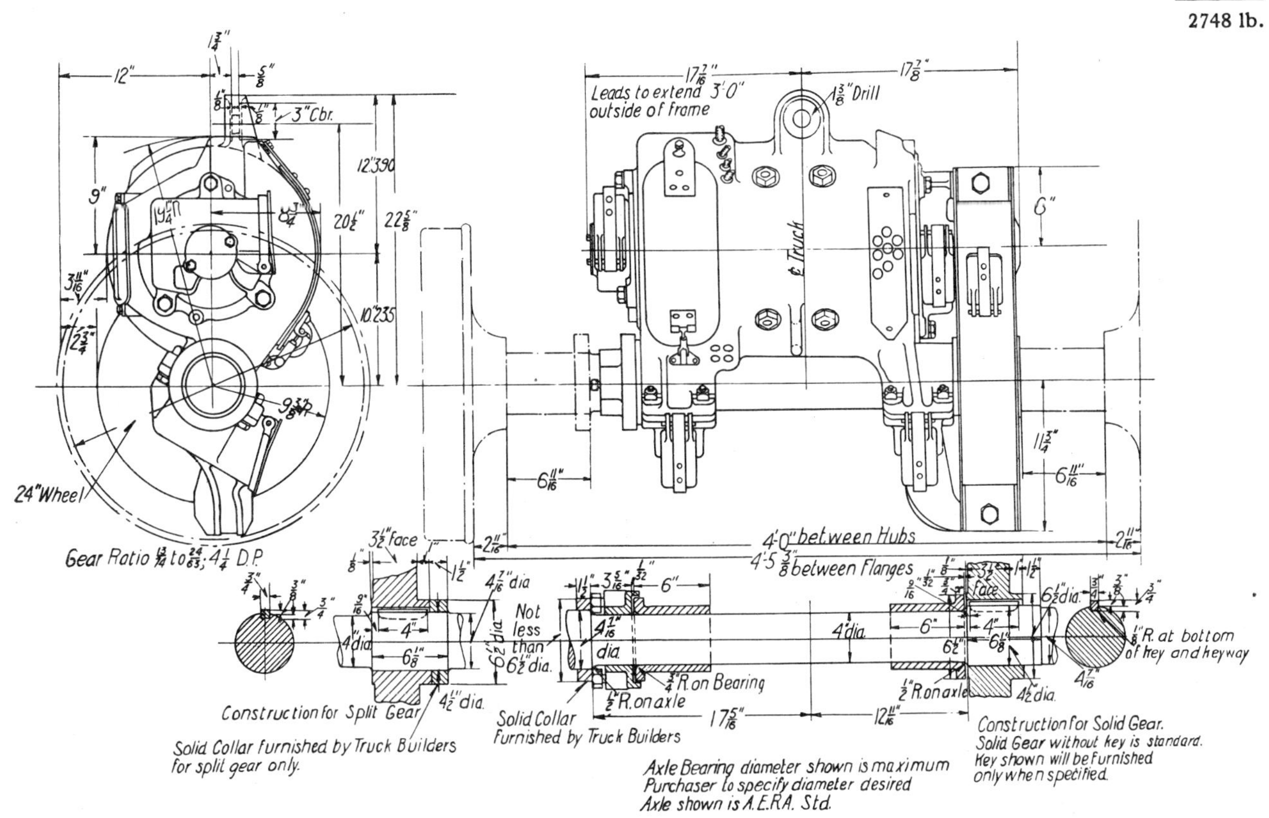

Outline Dimensions of 508-A Motor

Safety Cars Prove Successful on Some of the Heaviest Traffic Routes in Kansas City

Merchandising Transportation*

BY F. G. BUFFE
General Manager The Kansas City Railways Company

PRESENT CONDITIONS HAVE EMPHASIZED THE NECESSITY OF STUDY ALONG THESE LINES—VARIOUS METHODS ARE DISCUSSED, INCLUDING MEANS FOR IMPROVING THE SERVICE, PUBLICITY OF IMPROVEMENTS AND SECURING CO-OPERATION OF TRAINMEN.

One of the subjects assigned by the American Electric Railway Transportation and Traffic Association for committee investigation and discussion at the October convention is "Merchandising Transportation". No doubt if some of the operators who have passed to a region where transportation is unnecessary could return to this convention they would be surprised that the association was seriously discussing this subject. In the old days street railway transportation was provided and the public could take it or walk. Such an idea as attempting to dispose of street car rides by merchandising methods never occurred. As a matter of fact, the necessity for department store methods in transportation business had not arisen. However, "necessity is the mother of invention," and under conditions today in the street railway business, even stronger than necessity, it is a case of self-preservation.

Those of us who in the past three or four years have gone through one nightmare after another, including every trouble to which an industry can be subjected, who have seen surpluses turned into deficits, who have watched receipts vanish in the smoke of gasoline, look back with envy and amazement at the halcyon days when labor was plentiful and wages low; when there were no coal strikes or "flu" epidemics; when our nearest competitor was the "one-horse shay" and when 5 cents could be divided into operating costs, taxes, interest and still leave something for dividends.

The war hastened the industry's arrival at the point where it had to be up early in the morning and stay up late at night to sell what it had to offer. Regardless of the war, however, we had already reached the place where salesmanship methods were necessary. The business is in some respects no longer a monopoly, and while "competition may be the life of trade," it has come mighty close to being the undertaker in the street railway field. As an example, in 1915 and 1916 Kansas City pointed with pride to its ownership of 10,000 or 12,000 automobiles. Today approximately 35,000 of them are in use. Figured at the conservative estimate of two passengers per automobile operated, this means some 70,000 people who are no longer buying their transportation at the old stand. In some cities, to a greater or less extent, the jitneys have hung out their shingle, and where they have done so they are hurting legitimate transportation.

In addition, increased street railway fares have developed another very likely competitor, which is no more or less than the sidewalk. Not that there has been any tightness in money matters evidenced by the people, nor has it appeared that any one desired to economize on street railways rides, or on anything else, yet we have had to face the antagonism and mental stubbornness brought about in many localities by fare increases. Our riders, who very willingly pay double for food, clothing, rent, doctor bills and entertainment, have for so long coupled 5 cents and street car rides together in their minds that our raising the ante called forth a stormy protest which in many cases turned short-haul riders into pedestrians. It is some relief to know that this condition has changed. The 5-cent fetich seems to have passed. The public

---

MR. KEALY SAYS:

SOME Safety Car operating figures taken from our experiences in Kansas City may be of interest to street railway men elsewhere. In round numbers these figures show that on the more important lines service in car miles has been increased from 26 to 31 per cent, headways have decreased from 30 to 37 per cent, and sixty-six Safety Cars have worked out an actual saving of $425 expenses per day. In other words, through the introduction of Safety Cars service has been very materially increased, headways reduced and the cars have saved in power and platform costs alone approximately $2,000 per year, per car.

---

*Abstract of paper presented at annual meeting of Missouri Association of Public Utilities, Jefferson City, Mo., June 3, and reprinted from *Electric Railway Journal*, June 19, 1920.

now is beginning to think of rides in terms of service and cost. As a result practically every city in the country with increased fares reports increased riding. Even Boston, where antagonism to a 10-cent fare almost amounted to a boycott, has overcome this feeling. There the riding habit has returned.

The above are some of the very pressing reasons why transportation will have to be merchandised if we are going to sell. The subject offers rather unlimited scope for discussion and ideas. It covers too much territory to do more than suggest some outstanding features in an article of this nature.

The subject naturally divided itself as follows: First, direct methods of selling, which include service, advertising, education of employees, methods of handling and collecting ticket fares, methods of meeting jitney competition and the safety car. Second, indirect methods, such as those dealing with opportunities of awakening public service commissions, chambers of commerce, councils and other civic bodies to the necessity of lending their aid to increase business for the traction property.

Adequate Service the Basis for Success

In the very nature of things, service, in a selling campaign, must come first. The service we give is not only the display of our goods in the show windows, but it is the measure of the quality of the article we sell. We cannot rightfully go to the public for adequate support unless by the same token we give the public adequate service in every sense of the word. The absolute necessity of this is too obvious to require explanation to any operating man. It is true there was a time when some operators thought increasing operating costs must be met by decreased service. That this is a fallacy that will inevitably lead to disaster has been proven time and again. Decreased service spells decreased business in more than a direct ratio. It not only loses business but it loses, too, public confidence. It results in fattening the purses of our competitors at the expense of our own. Our business is such in its very nature that expense cannot be reduced as prices increase. Our whole structure is built upon an adequate, necessary public service, and relief from increased costs must come from increased fares, which principle is of course economically sound and morally right.

There is a tendency for operation to fall into a rut. There is such a thing as obsolete methods establishing themselves by prescription. Examples of this can be seen in the tenacity with which certain routes are maintained and obsolete stop systems continued. Many of us have gone on the theory that because a certain route has become established by usage it must be continued indefinitely regardless of changed conditions. The old theory of a stop at every city block irrespective of the interval between stops is based upon custom and usage. On some lines stops are so spaced that when the time for acceleration and braking has been taken out the car runs at full speed for a few seconds only. Any one driving an automobile knows what this condition means to gears and engine, to say nothing of the loss in time. Therefore, our wares must in any case be better displayed and better adapted to the needs of the public by revision in routes and stops. Especially are changes necessary in congested districts.

Scientific Traffic Study Made

For example, in Kansas City we have for some months enjoyed the services of John A. Beeler, consulting engineer of New York City. Mr. Beeler has acquired a most enviable reputation in the United States in straightening out traffic tangles. His work in Washington during the war, in Boston and in Philadelphia speaks for itself. Very recently Mr. Beeler has been retained by the Public Service Corporation of New Jersey and the Chicago Surface Lines. Through a rearrangement of stops in the downtown district, by the use of the double berthing system and loading platforms, Mr. Beeler has been able to secure on many lines an increase in speed of more than 100 per cent. On one important street, in a block where formerly sixty cars an hour passed in one direction, we are now able to put through ninety-three. The effect of this is of course most significant. It means that those cars scheduled to hit the downtown district at the beginning of the rush hour are there on time when they will do the most good. Due to Mr. Beeler's rearrangement, we now find it possible to maintain our system speed at more than 9 m.p.h., and of course increased schedule speed and regular headways mean just as large an increase in service as the addition of more cars. In fact, it is more, because additional cars, if not at the right place when needed, serve no useful purpose.

I think therefore every one will admit that in merchandising transportation the first essential is to see that service is all modern operating methods can make it. Service, after all, starts in the carhouses. Clean, well maintained, well painted cars are our biggest advertisement. Very often an entire system is judged by the apperance of its cars on the street, and too much attention cannot be paid to this phase of operation. A policy of retrenchment that starts with the equipment will end in disaster if continued.

Let the People Know

Service being the first step in selling car rides, it very naturally follows that keeping the public informed of this service should be the next. There is no more practical reason for a street railway company "hiding its light under a bushel" than there is for a department store. The public is very appreciative of the printed word. The repeated suggestion that your city has the best street railway service in the country,

if in any way at all backed up by facts, will very shortly meet a receptive mood in the public mind. Local pride in one's city will help bring about this mental condition. People can be educated to point proudly to their street railway service the same as to their public buildings and parks.

Advertising from the standpoint of selling transportation is a different problem than that presented by the good will and public policy advertising campaigns which have been carried on so extensively in the past five years by public service companies. A most excellent medium is the space provided by the car itself. Perhaps there is no more effective advertising than dash cards. A hanging frame on the inside from the car roof is also most desirable space. A notable example of this is furnished in Philadelphia, and each week sees a different message on Philadelphia's service in the frames. The New York Interborough uses the space on the two front windows of each car, and under the title of the "Subway Sun" communicates new data to the car riding public each week.

This direct advertising should enlarge on service changes, and copy for it should be worked out with the sole purpose of directing the attention of the rider to facilities offered by the street railway system and service given. One good feature which can be emphasized is the cost of operating an automobile. It has been demonstrated that no type of car, not even the smallest, can be operated for less than 10 cents a mile. In the larger cities to drive an automobile downtown to business and back costs from 75 cents to $1.50 a day. Another strong feature which can be utilized is the fact that the street railway system in any community is essential, that upon it the community's growth has been predicated and that system and service have both been outlined on the basis of handling all the people in the community all the time. This being the case, it is to the interest of every citizen to see that the local car line is supported, that jitney competition is not allowed and that a full measure of co-operation be given by the public. Under our present service-at-cost franchises (and in those states which have utility commissions practically all franchises resolve themselves into this type) the rate of fare depends very directly upon the amount of riding. There can be no hope for decreased fares unless the volume of riding increases and service is utilized to the fullest extent.

The moving picture theater offers a very productive field for the advertising man. Few people who ride the cars have any conception of the machinery back of their daily ride. They have but a limited knowledge of shops, power plant, carhouses, regular inspection force, track department and other branches of the organization that produces city transportation. These things can be filmed and will serve to educate the public in the rudimentary elements of its transportation.

Education of Trainmen Important

Another essential point in selling transportation is proper education of the salesman himself. Trainmen are the company's representatives, and too often their training involves only operation of cars and not stimulation of business. I believe the greatest field ahead of street railway operators today lies in the training of transportation forces, and at the wages today being paid there is no good reason why we cannot secure the service of men's heads as well as their hands.

Such training involves courtesy, politeness, careful operation, and goes even deeper. We must first make trainmen realize that in a sense they are salesmen, that they are responsible for much more than the ordinary operation of cars. We must bring them to a realization of the importance of their own position. We must awaken in them a higher sense of their responsibility. This means education, and, like most things worth having, will take time and unceasing patience to secure. It starts at the employment office, where greater endeavor must be made to secure a higher type of employee. It means more intensive schooling and instruction. Frequent meetings of employees and talks by officials of the company help. A company publication is another good method. In Kansas City, through published articles and talks to the men, we have endeavored to get them all interested in the company's financial statement, which shows the result of their own operation. These statements, in easily understood terms, are published in the employee's magazine every month. Every wage increase we have made in the past sixteen months has been based upon such financial statements. We have succeeded in a large measure in this policy, and I believe today employees of The Kansas City Railways are better informed as to the company's financial position and its policies than is general throughout the country in traction systems.

Just last week more than 3,100 employees signed petitions addressed to the City Council demanding that jitneys be driven off the streets. In several divisions there were not to exceed a dozen employees who refused to sign the petitions. The employees themselves requested that these men be discharged.

The active interest and co-operation of the employees of any company is a most vital factor in the sale of street car rides. Although there are many other reasons why Philadelphia has been able to succeed without fare increases, the biggest factor in the success of that company has been the co-operation of its 10,000 men. This co-operation has not been the result of a month's work or a year's work, but has been secured by constant hammering along the same lines for the past seven or eight years.

Other Means of Merchandising Transportation

For many companies a new element has been introduced from the merchandising standpoint in the necessity for ticket fares, brought about by fractional rates, which have been put into effect in many cities. It has meant in some cases the organization of a complete ticket distribution and handling system. St. Louis, according to the latest reports, has some 700 ticket agencies. There are about 450 in Kansas City. These agencies must be supplied, their interest in the work stimulated and the people induced to save money by the purchase of tickets.

There are two good merchandising reasons why ticket sales should be pushed: One arises from the fact that we have to meet daily the competition of the automobile. A purchaser of a book of tickets, having paid his fare for some days in advance, will be much less inclined to forsake the street car for its competitor. Another feature of the ticket business comes from the psychological fact that the same respect is not ordinarily paid tickets or tokens that is commanded by money itself. In other words, a stack of white chips never did as much, seemingly, as a stack of silver discs for the tickets have been purchased and the money for them paid out. After they have passed from the custody of the owner they become cash for rides and their possession undoubtedly induces short haul business.

The rapid evolution of the safety car has brought about other possibilities from the merchandising standpoint. The principle upon which safety car operation depends is that of more frequent service. Its use allows reduced headways, and the less time prospective patrons are forced to waste at street corners the better chance we have of securing their business. Especially is this true where we have jitney competition to fight. Practically all companies who have adopted safety cars and who have followed this principle in their use report increased patronage. As one customer expressed himself some time ago in reference to safety car operation on the line on which he lives: "While the little fellows are not as comfortable in many ways as the double-truck cars, yet it seems to us that there is a car in sight on our line all the time." While not literally true, there was some merit to his statement, inasmuch as a double-truck five-minute headway had been replaced by a safety car every three minutes.

Some safety car operating figures taken from our experience in Kansas City may be of interest to street railway men elsewhere. In round numbers these figures show that on the more important lines service in car miles has been increased from 26 to 31 per cent, headways have decreased from 30 to 37 per cent, and sixty-six safety cars have worked out an actual saving of $425 expenses per day. In other words, through the introduction of safety cars service has been very materially increased, headways reduced and the cars have saved in power and platform costs alone approximately $2,000 per year, per car.

The foregoing ideas are merely scattered attempts to present some of the more outstanding features of the subject to your attention. The details very naturally will be different on every property. The main fact, however, it seems to me, is very plain. Our business can be increased by direct selling methods. We have competition to meet, and we must meet it upon a new basis. The new order of things very naturally involves some radical changes, and in this connection what could have been more radical than the introduction of the safety car? All of us know the opposition brought to bear against the little unit five years ago, which was if anything stronger within the ranks of the industry than from the public. The car was finally taken up and adopted because it offered one of few possibilities of meeting the situation which arose with the beginning of the war. Some of us actually had to be clubbed into its adoption. We are daily seeing other innovations. The work Mr. Beeler is at present engaged in is in many respects a radical departure from old methods. It is not uncommon now to read of the appointment on many systems of traffic agents, whose duties are to stimulate and encourage travel. Without question the field covered by this subject offers many opportunities as yet hardly touched. The business has been going through a fight for its very existence, and it has existed against reverses which would have overthrown and swamped any other industry, merely because it was essential and had to "carry on". The growth and well being of every community in the United States is predicated more or less directly upon its means of urban transportation. The withdrawal of street railway service is a catastrophe which could not be contemplated in any city. For this reason therefore the industry has survived and every indication points to a more optimistic future. As operators we have had to do things we never contemplated before and we have all had to go through a mental shakeup that has been good for us individually. Ideas which formerly were regarded as fantastic have had to be taken up in dead seriousness, on the principle of a drowning man grasping at the first straw that offered itself. Many of us were surprised to find these so-called "straws" were in reality good, substantial logs upon which to swim out of some of our troubles. The struggle has sharpened the wits of the entire industry, and without doubt we have learned and are learning to sell the product which our manufacturing plant produces.

The STORY of the CEDAR VALLEY ROAD

Electric Freight Haulage

Westinghouse Electric & Manufacturing Company
East Pittsburgh, Pa.

Special Publication No. 1575 December 1917

A Heavy Tonnage Main-Line Freight Train

Freight Haulage

HE movement of the enormous wheat, corn and cotton crops and the mining and industrial products during the past strenuous year, has conclusively demonstrated the inadequacy of America's Transportation System. Increased shipping facilities are imperative, if we are to keep abreast of our steadily increasing agricultural, industrial and commercial growth.

Branch Line L. C. L. Freight

*THIS is an opportune time for **Electric Traction** to come into its own as an important factor in the solution of this economic problem. There are innumerable opportunities for introducing **Electric Freight Service,** and a study of your field may prove a pleasant surprise.*

With the view of helping other Electric Railways to increase their revenues by building up freight business, the operation of the "Cedar Valley Road" is herein described.

Freight haulage, the great revenue builder of the Electric Railway, is well worked out in regard to methods of securing and handling traffic, on the Waterloo, Cedar Falls and Northern Railway. This is one of the roads which recently changed over its system from 650 volts to 1300 volts direct current as standard.

*The "Cedar Valley Road" conceived by master minds, and built in the center of a highly competitive Steam Railroad Territory is an excellent example of what can be accomplished by **Electric Railway Freight Haulage.***

"A steam railroad with a trolley wire over it," completely conveys the idea of the substantial manner in which the Waterloo, Cedar Falls & Northern Railway has been constructed. This road, built for service, admirably renders it to its patrons. The operation of the "Cedar Valley Road" demonstrates conclusively how an Electric Railway desiring to enter this much neglected, but lucrative field may perfect its organization, and successfully increase its revenue.

A 60-Ton Westinghouse Equipped Locomotive and Main-Line Freight Train

The Cedar Valley Road

Organization

Pioneers

IN 1895 the Waterloo & Cedar Falls Rapid Transit Company was formed to give local street car service in the city of Waterloo and town of Cedar Falls with a connecting interurban line (8 miles long) between these two points. This organization bought out the horse-car line in Waterloo and a gasoline line in Cedar Falls.

The rolling stock of this earlier rapid transit road was equipped with Westinghouse No. 12-A motors in city service, and No. 38-B for interurban lines between Waterloo and Cedar Falls. The original equipment for this early road consisted of a 75-kw. Westinghouse generator connected to a Westinghouse single-acting engine.

About 1901 the interurban line from Waterloo to Denver, Ia. (14 miles) was built, and a new Westinghouse 250-kw. double-current generator installed with three single-phase, 50-kva., 376 to 22,000-volt, 25-cycle transformers. This supplied a 22,000-volt transmission line which ran to a substation, then called Glasgow, halfway between Denver and Waterloo. This was the first high tension line in the state of Iowa. At Glasgow there were step-down transformers and a 150-kw., Westinghouse rotary converter.

This line (1903) was later extended to Denver Junction (connecting with the Chicago Great Western R. R.). At the same time steam service was inaugurated between Waterloo and Sumner, Iowa, over the electrified section to Denver Junction, and then over leased tracks of the Chicago Great Western, 44 miles to Sumner, via Waverly. With the opening of this service both passenger and freight business was handled.

THE completion of the Waverly line in 1910 marked the beginning of a new epoch in the history of this road. It was strengthened financially and was the real beginning of the present Waterloo, Cedar Falls & Northern Railway. In connection with the Waverly extension it was necessary to supply more power, and therefore the present power house was started and equipped with a 1500-kw. turbo-generator unit. Two 500-kw., rotary converter substations were built, one at Denver and the other at Cedar Falls.

In 1910 operation over the Chicago Great Western was discontinued and a line built from near Denver Junction to Waverly (7 miles). In 1912 the line from Waterloo to Cedar Rapids was started and completed in September 1914 as a 650-volt direct-current line, but finally placed in operation as a 1300-volt direct-current line in March 1915.

Early Freight Haulage

The first freight work was done in 1899 when brick from a plant near Cedar Falls was transported to Waterloo and delivered on the city streets on flat cars for building purposes. This was done between midnight and morning. One flat car at a time was handled by an extra interurban motor car.

The first electric freight engine was built in 1900 and the second in 1901, the latter weighing 26 tons and equipped with four Westinghouse 12-A motors. This locomotive handled one car of coal at a time to the normal school in Cedar Falls, and in doing so it was required to operate up a two per cent grade.

East Waterloo

Main-Line Freight Train

General Description

Rock Crusher

THE Waterloo, Cedar Falls & Northern Railway Company, known as "The Cedar Valley Road," operates a system of electric railway lines through the Cedar River valley from Waverly on the north to Cedar Rapids on the south. The country served is a rich agricultural community ranking with the best in Iowa. The cities connected are prosperous manufacturing centers.

Waterloo, which is the hub of the system, is a progressive city of 35,000 population, having over one hundred factories. In this city the Cedar Valley Road operates a ***freight belt-line*** which connects a large number of the factories with its own interurban lines and with the steam railroads.

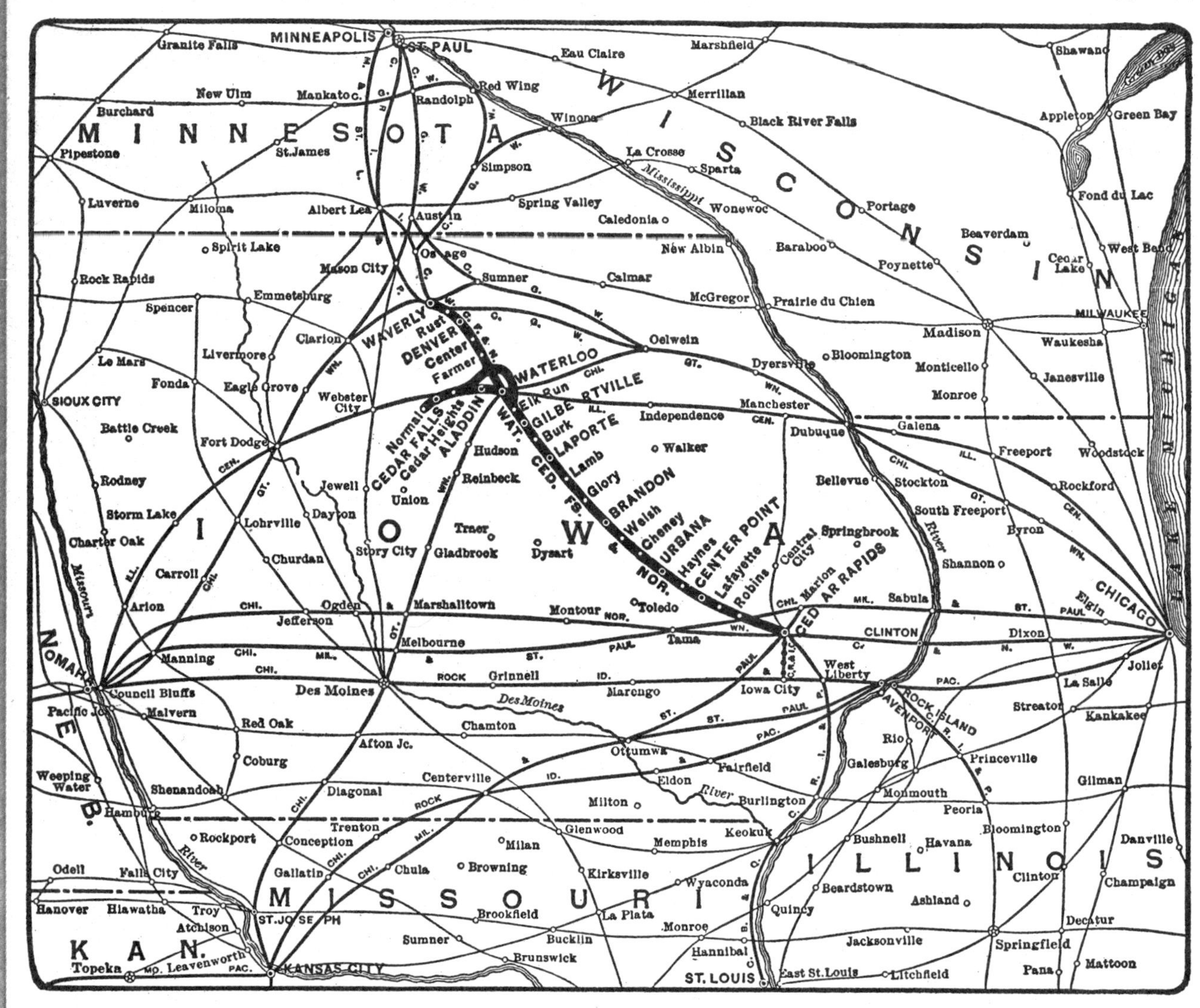

Map—Waterloo, Cedar Falls & Northern Railway

THE Waterloo local street car service is also operated by this company, including local lines radiating from the business centers to the factory districts, suburbs and parks.

Cedar Falls, a city of 10,000 people, 8 miles west of Waterloo, is the terminus of the first interurban line built by the "Cedar Valley Road," and this has been in operation about twenty years. The railway company operates local service on a loop connecting the business section with the Iowa State Teacher's College, all interurban and local cars passing a terminal depot located in the center of the business district.

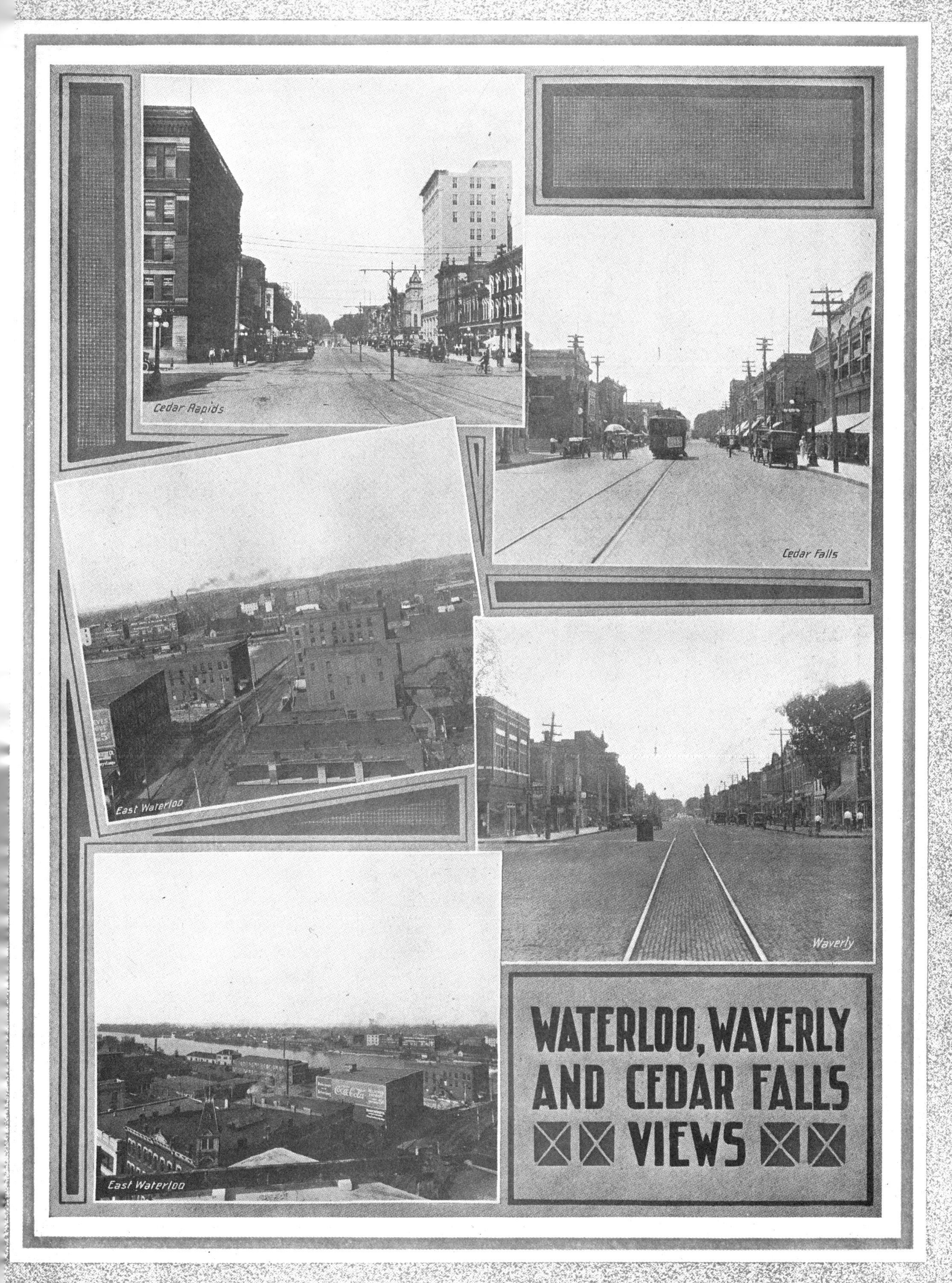

WATERLOO, WAVERLY AND CEDAR FALLS VIEWS

NATIONAL OATS COMPANY

DOUGLASS STARCH WORKS

QUAKER OATS COMPANY

CEDAR RAPIDS INDUSTRIES

A SEPARATE freight depot, where car-load freight is handled, is located near the business section. Forty-eight passenger and four freight trains are operated daily between Waterloo and Cedar Falls.

Waverly, 22 miles from Waterloo, the terminal of the northern division, is also a manufacturing city of about 6000 people. This line serves the towns of Denver and Farmer which are trading centers for prosperous farmers. Intermediate sidings are provided where car-load freight is unloaded by the farmers and where grain and stock is loaded.

Typical Station Scene Showing Grain Elevators, Coal and Stock Yards Located on the Industry Track and Off the Main Line Entirely

WHEN this interurban line was built, though paralleling steam lines most of the way, it branched out into exclusive territory striking several progressive communities which were not served by steam railroad facilities. The country traversed by the "Cedar Valley Road" consists principally of rich agricultural lands and these have increased in value easily 25 per cent on account of the excellent transportation facilities afforded by the service of the Waterloo, Cedar Falls and Northern Railway.

Cedar Rapids, a city of 50,000 people, 60 miles southwest of Waterloo, is the terminal of the Southern Division. This line serves Gilbertville, La Porte City, Brandon, Center Point, Lafayette and Robins.

OME of the towns along this route, after the "Cedar Valley Road" went through, doubled in population as, for instance, the towns of Urbana and Brandon. These were on the old "Prairie Schooner" Route, and although the country around them was of the best agricultural nature, for over fifty years they failed to secure any steam railroad service, thus leaving them entirely isolated from the rest of the country as far as railroad facilities were concerned.

On account of the excellent freight service, stock and grain has been routed from ***competitive steam road territory*** via the "Cedar Valley Road" for shipment. At all stations, there are one or more grain elevators, and stock yards are also to be found.

In going through this territory, when the road was first projected considerable foresight was shown in that land adjacent to the route was purchased in all towns and in plots large enough so that all industries could generally be placed adjacent to the railroad property. In other words, all industries located with track facilities along the route at the different towns are on leased land owned by the "Cedar Valley Road." This secures advantageous location of industries, from the standpoint of interesting the people, and also operation.

One of Five 60-Ton, 1300-Volt Locomotives Equipped With Westinghouse Motors and HL Control

Center Point—A Stucco Station Building With Substation

Terminals and Stations

La Porte City Station

THE purchases for terminal and station sites were based on the requirements for future developments, and in very few cases did the management allow the price of property required for its purposes to alter its decision as to future demands. As a result of this foresightedness the company now owns its belt lines and a 35-acre tract for freight terminals within six blocks of the wholesale district of Waterloo and on a paved street. On part of this site a freight station has been erected.

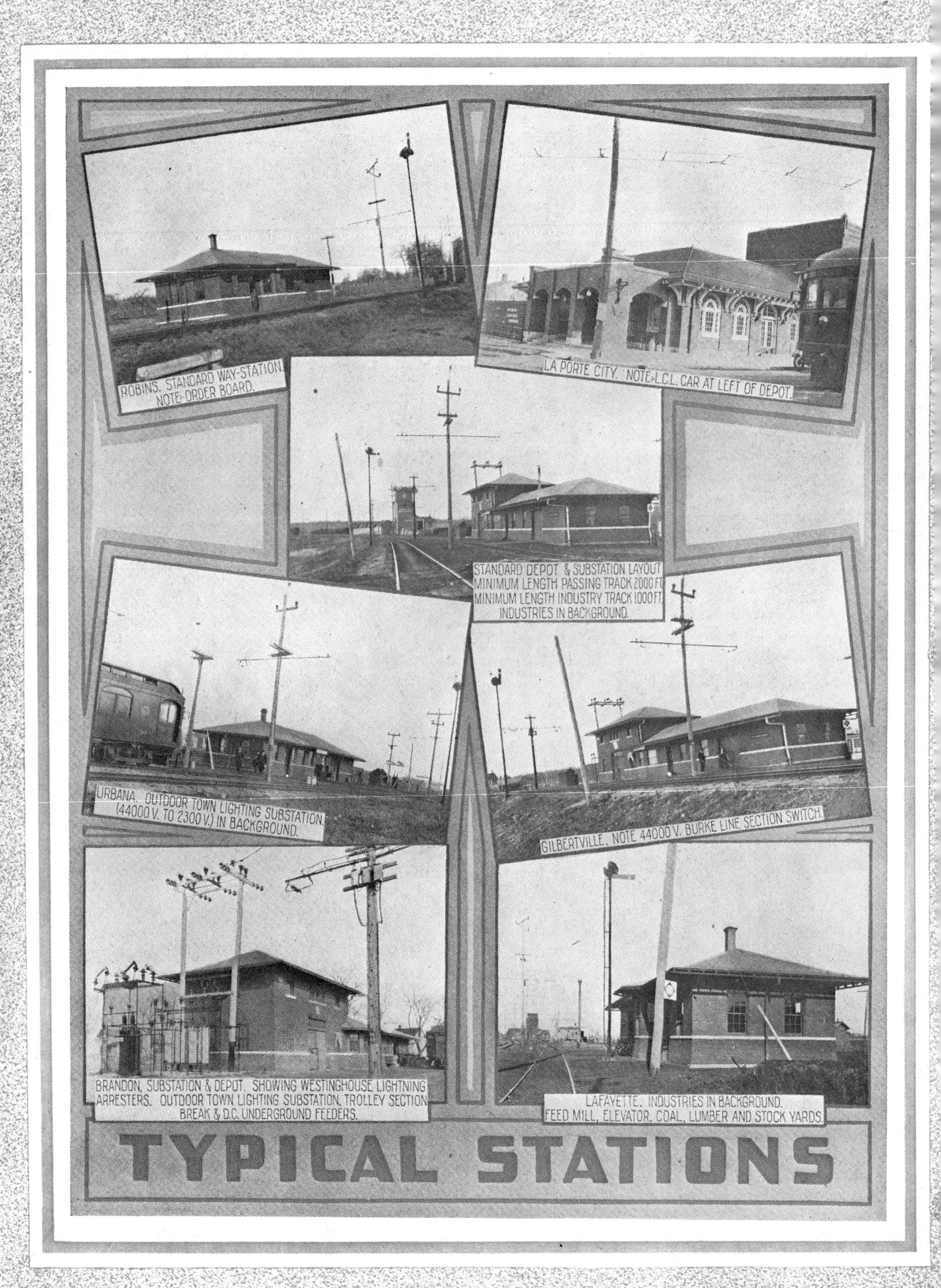
ROBINS. STANDARD WAY-STATION.
NOTE-ORDER BOARD.

LA PORTE CITY. NOTE-L.C.L. CAR AT LEFT OF DEPOT.

STANDARD DEPOT & SUBSTATION LAYOUT
MINIMUM LENGTH PASSING TRACK 2000 FT.
MINIMUM LENGTH INDUSTRY TRACK 1000 FT.
INDUSTRIES IN BACKGROUND.

URBANA. OUTDOOR TOWN LIGHTING SUBSTATION.
(44000 V. TO 2300 V.) IN BACKGROUND.

GILBERTVILLE. NOTE 44000 V. BURKE LINE SECTION SWITCH.

BRANDON, SUBSTATION & DEPOT. SHOWING WESTINGHOUSE LIGHTNING ARRESTERS. OUTDOOR TOWN LIGHTING SUBSTATION, TROLLEY SECTION BREAK & D.C. UNDERGROUND FEEDERS.

LAFAYETTE. INDUSTRIES IN BACKGROUND.
FEED MILL, ELEVATOR, COAL, LUMBER AND STOCK YARDS.

TYPICAL STATIONS

New Cedar Valley Terminal, Waterloo, Iowa—For Interurban and City Lines

Station Facilities

Main Line

THE mileage between the important way stations on this line made it possible to build a combination passenger, freight and substation at nearly every point. The standard design includes a building of brick, concrete and steel, 101 feet in length, and 22 feet in width. The substation occupies 30 feet, 6 inches of length at one end of the building. A ticket office extends 12 feet across the width of the building, and a waiting room 16 feet in width and a freight room 36 feet, 6 inches in length, occupies the other end of the building.

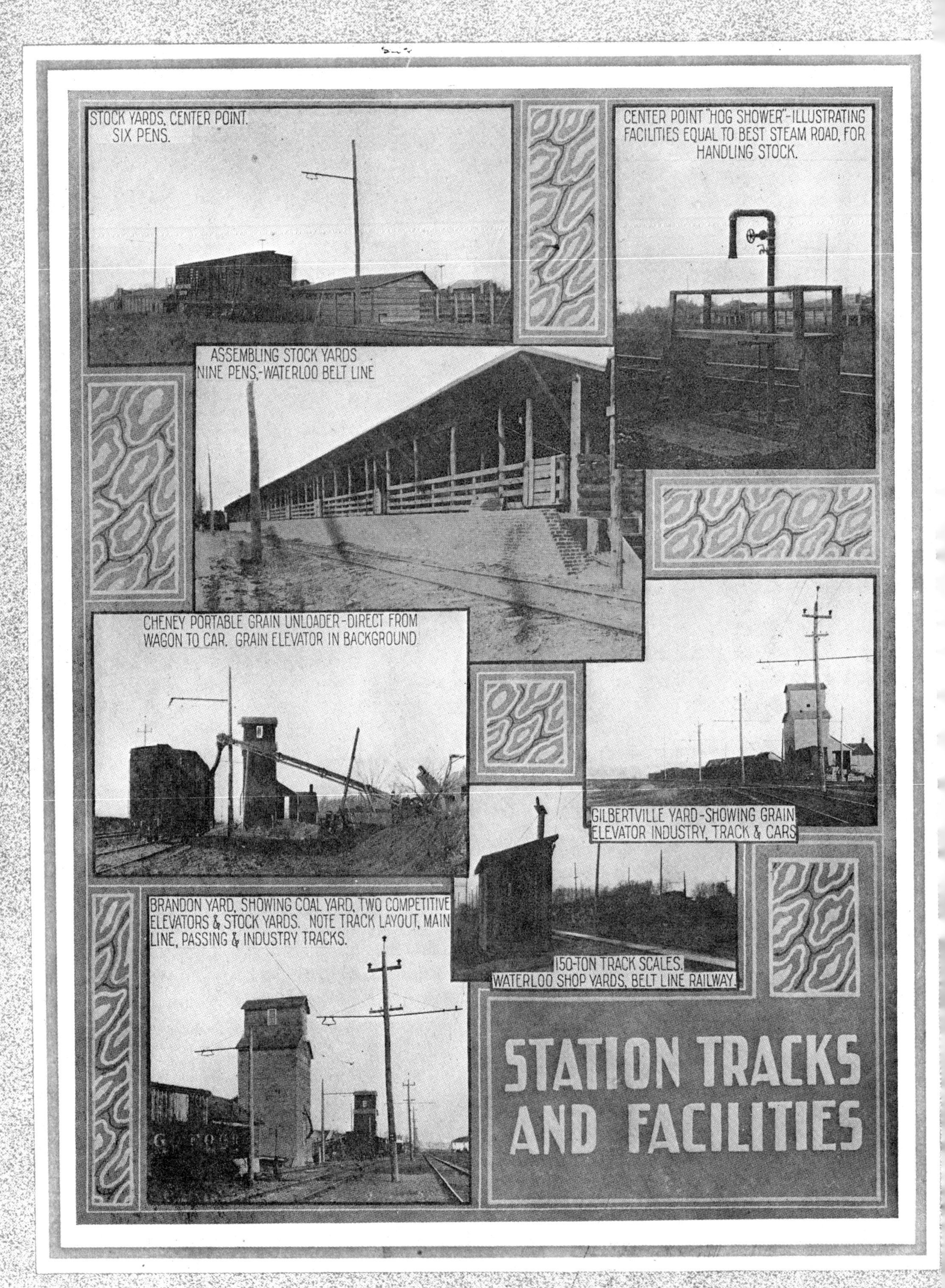
STOCK YARDS, CENTER POINT.
SIX PENS.
CENTER POINT "HOG SHOWER"-ILLUSTRATING FACILITIES EQUAL TO BEST STEAM ROAD, FOR HANDLING STOCK.
ASSEMBLING STOCK YARDS
NINE PENS,-WATERLOO BELT LINE
CHENEY PORTABLE GRAIN UNLOADER-DIRECT FROM WAGON TO CAR. GRAIN ELEVATOR IN BACKGROUND
GILBERTVILLE YARD-SHOWING GRAIN ELEVATOR INDUSTRY, TRACK & CARS
BRANDON YARD, SHOWING COAL YARD, TWO COMPETITIVE ELEVATORS & STOCK YARDS. NOTE TRACK LAYOUT, MAIN LINE, PASSING & INDUSTRY TRACKS.
150-TON TRACK SCALES.
WATERLOO SHOP YARDS, BELT LINE RAILWAY.
STATION TRACKS
AND FACILITIES

Station Scene Showing Freight Train on Passing Siding. All Switching to Industries is Done From This Track. Keeping Main Line Clear

S*TATION Facilities* include a team track, well-built stock pens, grain elevators, and loading chutes. The station proper, wherever possible, includes a tract of land approximately 300 feet by 2000 feet in length, permitting the railway to offer attractive long term leases for grain elevators, milling facilities, etc. Each station is provided with a Union Switch & Signal Company manually-operated twin train order semiphore board, with blades displayed in the upper left and right hand quadrants controlling train movements in both directions.

Valuable track facilities are owned in Aladdin where the William Galloway Company has an extensive plant. ***Other freight producing industries are situated and served by the company's belt line.*** A tract of land 200 feet by 500 feet, just outside the business district of Waterloo, serves as a site for the four-bay fire-proof carhouse herein shown, which is 70 feet by 200 feet in plan. The design and location of this carhouse on the property are such that three more sections may be added when more storage space is required.

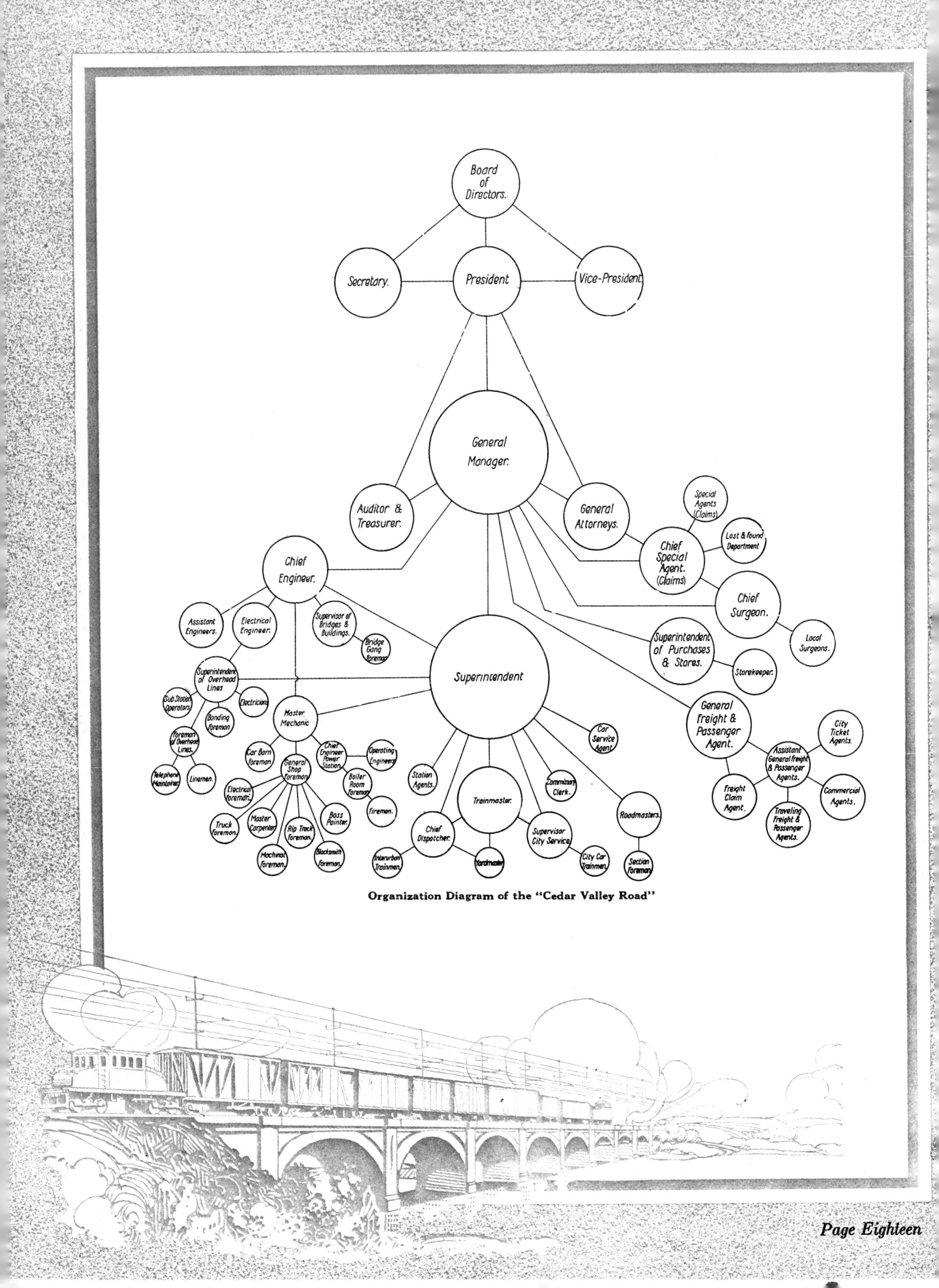

Organization Diagram of the "Cedar Valley Road"

Special Movement—An 800-Ton Train of Buick Automobiles Shipped From Flint, Mich., to Waterloo, via the "Cedar Valley Road"

Traffic

60-Ton Freight Locomotive

ONE of the secrets of the rapid and extensive growth of the Waterloo, Cedar Falls and Northern Railway has been its complete traffic arrangements, and well organized methods of securing business.

The Traffic Department is headed by a ***General Freight and Passenger Agent*** whose title indicates complete charge of all traffic. In addition, there are the following officers directly under him: Assistant General Freight and Passenger Agent at Waterloo, and another similar officer at Cedar Rapids; Commercial Agent at Waterloo, and a Travelling Freight Agent making Cedar Rapids his headquarters.

Freight Cars and Tonnage Moved
A Typical Month

Date	CEDAR RAPIDS DIVISION						WAVERLY DIVISION						CEDAR FALLS DIV.						SYSTEM TOTAL		
	EAST			WEST			EAST			WEST			EAST			WEST					
	Loads	Empties	Tons	Loads	Empties	Tons	Loads	Empties	Tons	Loads	Empties	Tons	Loads	Empties	Tons	Loads	Empties	Tons	Loads	Empties	Tons
1	9	8	556	28	2	1518	1	1	50	3	—	118	2	2	125	10	—	548	53	13	2915
2	14	7	715	28	1	1884	3	—	110	3	—	137	2	2	100	7	—	439	57	10	3385
3	16	17	897	38	2	1938	3	2	170	6	3	310	3	1	100	6	2	315	72	27	3730
4	12	4	735	25	6	1784	2	7	219	2	5	190	5	1	195	10	—	552	56	23	3675
*5	8	8	482	36	1	1634	3	—	77	—	—	—	—	—	—	—	—	—	47	9	2193
6	13	14	864	31	1	1522	5	3	262	9	—	435	—	2	40	7	—	353	65	20	3476
7	11	20	1031	26	8	1356	—	1	20	5	—	226	1	1	45	6	—	331	49	30	3009
8	11	9	659	35	2	1852	1	3	148	5	1	198	—	1	20	4	—	170	56	16	3047
9	13	13	769	26	1	1347	2	4	160	3	—	97	—	1	20	5	—	255	49	19	2648
10	26	9	1225	24	2	1207	2	2	90	3	5	170	1	4	105	4	—	255	60	22	3052
11	7	5	475	15	2	844	4	2	184	3	2	133	1	3	85	3	—	105	33	14	1826
*12	21	4	698	29	1	1527	4	—	120	—	—	—	—	—	—	—	—	—	54	5	2345
13	12	19	825	23	2	1126	1	—	25	1	—	25	3	—	85	5	1	285	45	22	2371
14	18	1	630	24	—	1138	2	—	89	4	—	202	—	5	100	3	—	160	51	6	2319
15	16	18	1006	17	11	893	1	1	50	4	2	266	3	1	115	5	1	305	46	34	2635
16	20	7	797	29	—	1695	3	2	115	4	1	172	2	2	125	5	2	266	63	14	3170
17	15	15	974	34	—	1453	1	1	50	4	—	152	—	2	40	5	—	188	59	18	2857
18	8	1	305	10	3	597	1	7	165	6	5	326	1	2	80	3	—	85	29	18	1558
*19	22	13	973	34	—	1745	4	—	105	—	—	—	—	—	—	—	—	—	60	13	2823
20	14	4	654	33	4	1687	—	3	60	8	2	427	1	8	185	4	—	181	60	21	3194
21	25	14	1356	30	—	1578	2	1	86	1	1	45	2	1	70	6	1	312	66	18	3447
22	18	20	1182	27	—	1212	4	1	180	5	3	214	1	3	95	4	1	226	59	28	3109
23	26	14	1099	19	11	1332	7	1	206	5	1	220	2	5	180	3	1	147	62	33	3184
24	21	12	1182	28	1	1454	1	3	85	3	—	150	2	1	80	2	—	53	57	17	3004
25	4	6	289	13	1	621	4	3	148	3	6	251	1	5	125	1	—	25	26	21	1459
*26	22	27	1372	29	—	1248	2	—	55	—	—	—	—	—	—	—	—	—	53	27	2675
27	18	2	867	29	1	1149	1	—	25	1	1	43	1	—	25	3	1	181	53	5	2636
28	14	8	860	29	3	1514	—	1	18	5	1	250	4	3	235	4	—	180	53	16	3057
29	30	3	1327	27	4	1415	2	1	125	4	—	175	3	—	122	1	1	45	67	9	3209
†30	7	6	364	15	—	750	2	—	72	—	2	40	—	—	—	—	—	—	24	8	1226
Total	471	308	25168	788	70	41366	68	50	3269	100	41	4972	41	56	2497	116	11	5962	1584	536	83234

Switching Service:—Waterloo Belt-Line Yard Engine handled 1717 loads and 1504 empties. *Sundays. †Thanksgiving day.

F53. 10M—7-25-15—C

Cedar Valley Road

WATERLOO, CEDAR FALLS & NORTHERN RAILWAY COMPANY

DAILY SITUATION REPORT

INSTRUCTIONS: The Chief Dispatcher will make the necessary number of copies of this report and mail them the first thing in the morning of each day. Information called for on this form must be furnished in every report.

Waterloo, Iowa, Nov. 30, 1916. ALL LINES—From 6 A. M. Nov. 29 to 6 A. M. Nov. 30, 1916.

WEATHER AT 6 A. M. Nov. 30, 1916 — **PAST 24 HOURS**

TEMPERATURE	SKY	WIND	PRECIPITATION	TEMPERATURE	WIND	PRECIPITATION
A 25° Above / B Below	C Clear	D Velocity Calm	FA Rain — / FB Snow —	GA Min. 25° / GB Max. 40°	HA Velocity — / HB Direct'n —	JA Rain — / JB Snow —

REGULAR PASSENGER TRAINS RUN

District		Trains	Percent on Time
1st CEDAR RAPIDS DISTRICT	XA EAST	11	100
	XB WEST	11	100
2nd WAVERLY DISTRICT	XA EAST	7	100
	XB WEST	7	86
3rd CEDAR FALLS DISTRICT	XA EAST	24	100
	XB WEST	24	100

SPECIAL PASSENGER TRAINS RUN

FA	FB From	FC To	Miles	FD Service Required Acct. of
Eng. 30	Shaver 9:05A	Waterloo 4:55P	60	O.S. Lamb and Party.
Eng.				
Eng.				
Eng.		Total		

FREIGHT TRAINS RUN

		Regular	Extra	Loads	Emts.	Tons
K CEDAR RAPIDS DISTRICT	EAST (KA/KB)	2	0	MA 30	MB 3	NC 1327
	WEST (KC/KD)	2	0	NA 27	NB 4	OC 1415
Q WAVERLY DISTRICT	EAST (QA/QB)	2	0	OA 2	OB 1	RC 125
	WEST (QC/QD)	2	0	RA 4	RB 0	UC 175
S CEDAR FALLS DISTRICT	EAST (SA/SB)	2	0	UA 3	UB 0	WC 122
	WEST (SC/SD)	2	0	WA 1	WB 1	MC 45

PASSENGER TRAINS ANNULLED

	Train No.	Acct.
4th. Cedar Rapids Dist.	None	
5th. Waverly Dist.	None	
6th. Cedar Falls Dist.	None	

FAST FREIGHT TRAINS

AX Train	BX Left	CX At	DX Arrived	EX At	GX Stock Chicago	HX. Stock C. Rapids	JX Stock Waterloo	KX B. & E.	MX Other	NX Remarks
402	Shops	5:35A	Shaver	3:41P		1			11-0-	632 Tons
498	"	7:05P	"	11:42P		1			17-3-	695 "
401	Shaver	9:30A	Shops	7:10P					12-3-	613 "
499	"	2:05A	"	6:45A					15-1-	802 "

PASSENGER TRAIN DELAYS

GE No.	A Mtg. Trains	B Connection at Waterloo	C Connection at C. Rapids	D Connection at Waverly	F Waiting Orders	G Hot Boxes	H Motor Failure	I Derailment Motors	J Derailment Cars	K Bad Weather	M AC and CD Trouble	N Phone and Wire Failure	O Track Obstruction	Q Other Causes	R Misc.
43														15	

FAST FREIGHT TRAIN DELAYS

BC Train Nos.	A Conn. and Trains	B Mtg. 1st Class Train	C Mtg. Other Trains	D Getting Orders	F Hot Boxes on Cars	G Eng. Failure	H Slow Track	I Picking Up Stock	J Holding to Load Stock	K Switching Out & Pkg. Up Cars	M AC and DC Trouble	N Track Obstruction	O Derailments	R By Acc'd's. To Other Trains	S Bad Weather	U Other Causes and Remarks
499	35									10						30 Double Belt Line

BD Switch Engines Working

BP Waterloo 2
Cars Handled 69-34

BG C. Rapids —
Cars Handled —

BI Work Trains 183
Galloway Trap Car #1 — 1 — 21

BJ Service Chargeable to: Roadmaster; " ; Over Head Lines

BM Coal on Track at Power House, 6:00 A. M.
Iowa 13
Illinois 0

BN Freight Cars on Our Rails at 6:00 A. M.
Foreign 190
Ours 12

BO

DISPATCHERS	HOURS ON DUTY
H.A. Gee	9 A.M. - 3 P.M.
E.J. Miller	3 P.M. - 12 A.M.
F.H. Schrader	12 A.M. - 9 A.M.

H. A. Gee
Chief Dispatcher.

A Two-Car Westinghouse Equipped Multiple-Unit Train Crossing Seven-Span Bridge

THE ***Assistant General Freight and Passenger Agent at Waterloo*** is in charge of certain assigned duties. The ***Assistant General Freight and Passenger Agent at Cedar Rapids*** has charge of solicitation of business and traffic matters. ***Travelling Freight Agent out*** of Cedar Rapids makes all competitive towns on the system. ***General Agent*** at Cedar Falls has charge of local traffic solicitation there.

All freight claims also come under this department, and the road is a member of the Freight Claims Association.

Typical Industry Scene

USINESS to or from any point in the United States is fostered and secured by a very comprehensive method in that the General Freight and Passenger Agent spends about one-half of his time visiting the various trunk-line points throughout the country. His work is followed up by Special Agents across the United States who call upon large manufacturers who do not allow trunk-line solicitation, and thus makes personal contracts for routing shipments via the Waterloo, Cedar Falls and Northern Railway. Also, agents and representatives of foreign lines, with whom this road has business, are visited.

Statistical Records, compiled from information secured by the ***Traffic Department,*** are of considerable value in the way of keeping track of the volume of business and its origin. Records are kept of business by firms, by towns, by comparison reports, and information secured by personal contact with brokers.

In addition to the extensive facilities, such as grain elevator locations and stock yards, which will be mentioned later, there are several excellent schemes pursued which are conducive to increasing the traffic that might be easily overlooked by an ***electric railway in the freight business.***

Competitive Points on the road are visited by the Travelling Freight Agent, and he endeavors, for instance, to educate elevator men to ship their grain over the Waterloo, Cedar Falls & Northern Railway, and also in this way other commodities which have their origin on this road are, whenever possible, ***routed via the electric line.***

Reciprocal Switching Arrangements exist between the Waterloo, Cedar Falls & Northern Railway, and all trunk-lines making connections in cities served by the ***Electric Line,*** which enables business from industries located on other roads to be handled the same as if they were on the ***Electric Line.*** Thus, by this arrangement an industry can be reached by the ***"Cedar Valley Road"*** in any of the cities served by it, no matter whether the industry is located on it or not.

LSO, by such an agreement all classes of business can be solicited regardless of where they are located. An example of such traffic handled is vividly shown and well illustrated by the freight train of Buick automobiles shipped from Flint, Mich., making a train of 800 tons. The automobile shipments in the last year and a half have amounted to over 1000 machines.

Milling in Transit Arrangement permits grain, coming from points on the ***Electric Line,*** its trunk-line connections, or other points in Iowa and surrounding states, to be milled in transit, and then shipped to destination as a completed product. For example: a car load of corn is shipped from La Porte City, Ia., to Chicago; while in transit it is "set out" at Cedar Rapids and the corn is milled into starch, and then sent on to Chicago, via the steam line, on the original billing. Thus, by this method of shipment there is every advantage that could be offered by a steam road.

General Traffic—The diversity of business is remarkable and in every way is the same as found on trunk-lines; raw material comes in on the ***Electric Line*** and goes out as manufactured products. In this connection it may be of interest to note that the inbound business exceeds the outbound.

In Waterloo, there are 155 different manufacturing plants producing gasoline engines, farm tractors, cement mixers, separators, wagons and many other products. Of the 3000 articles manufactured in Waterloo factories, gasoline engines lead the production, supplying 23 per cent of the total output of the United States. One factory does a mail-order business of about one carload per day. Waterloo also has a large packing and cold storage plant specializing in pork products.

Passenger Traffic—Interchange arrangements enable the Waterloo, Cedar Falls & Northern Railway to sell tickets to all points in the United States and Canada in connection with all roads. Another special service to the people in the territory served is the handling of complete trains consisting of sleepers and baggage cars from connecting steam lines; and, as an example of this service, on October 18, 1916, a special dairy train went from Waterloo via the "Cedar Valley Road" to Cedar Rapids, thence via foreign lines to Boston as a solid train. This consisted of three sleepers, one business and twelve palace stock cars.

Shaver Yard, Cedar Rapids—Large Freight Yard at Southern End of System
Heavy Tonnage Trains are Made Up Here

Steam Road Interchange

Main-Line Freight Locomotive

THE Waterloo, Cedar Falls & Northern Railway was the ***pioneer electric line*** to compel steam railroads to interchange freight with electric lines. As a result, over 70 per cent of the switching from steam roads entering Waterloo is performed by the ***Electric Line*** since many of the 155 factories of Waterloo are located on the ***Electric Belt Line,*** belonging to the "Cedar Valley Road." Also, a belt-line service is furnished by the Cedar Rapids Terminal and Transfer Company which receives freight from the Waterloo, Cedar Falls & Northern Railway and from all of the steam and interurban roads entering Cedar Rapids, distributing it, and doing local switching.

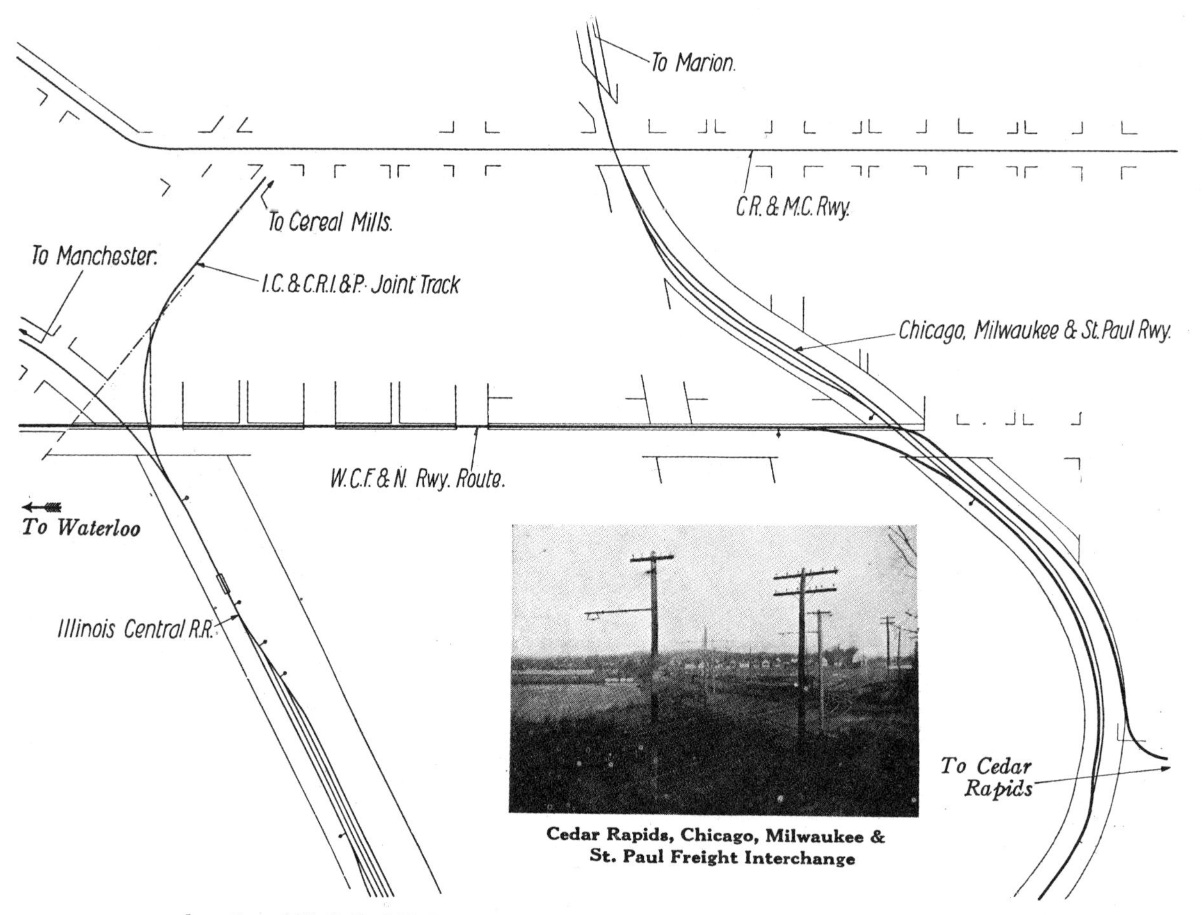

Cedar Rapids, Chicago, Milwaukee & St. Paul Freight Interchange

Location of W. C. F. & N. Ry. With Respect to Steam Railroads Entering Cedar Rapids

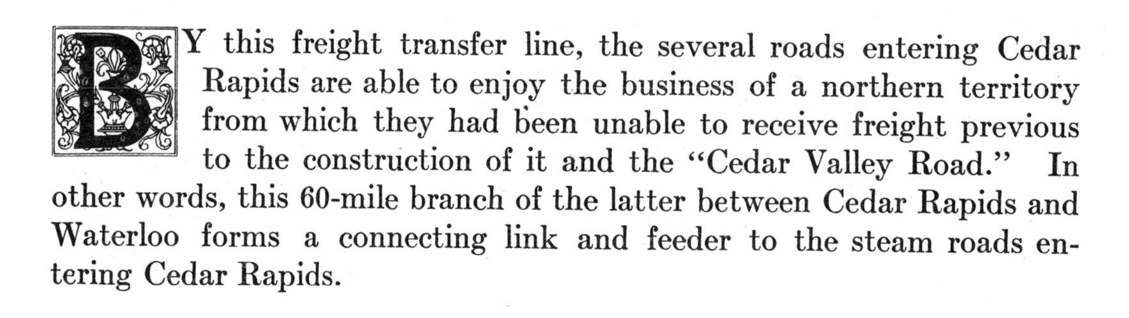

BY this freight transfer line, the several roads entering Cedar Rapids are able to enjoy the business of a northern territory from which they had been unable to receive freight previous to the construction of it and the "Cedar Valley Road." In other words, this 60-mile branch of the latter between Cedar Rapids and Waterloo forms a connecting link and feeder to the steam roads entering Cedar Rapids.

Steam Road Interchange

AT *Waterloo* interchange connections are made with the Illinois Central, Chicago, Rock Island and Pacific, and Chicago Great Western Railroads. These are made at the following points: East Waterloo—Chicago Great Western (Chicago-Kansas City Division), known as Highland Transfer, and is at one end of the Cedar Valley Belt-Line, also the Illinois Central R. R.

West Waterloo—Chicago, Rock Island and Pacific Railway (Minn. Division). Chicago Great Western Railroad (Chicago-Kansas City Div.)

Waverly Division

At Waverly—Omaha Division, Chicago Great Western.

Cedar Falls Division

At Cedar Falls—Chicago Great Western Railroad.

Waterloo-Cedar Rapids Division, Main Line

At *Cedar Rapids*—Chicago, Milwaukee & St. Paul Railway, handles all cars to and from the Chicago and North Western, Rock Island, and Illinois Central Railroads from points served from Cedar Rapids, as well as industries in Cedar Rapids.

At *Center Point*—Chicago, Rock Island and Pacific Railway (Decorah Division).

At *La Porte City*—Chicago, Rock Island and Pacific Railway (Minneapolis and St. Paul to Chicago and Kansas City), Minnesota Division.

La Porte City Interchange—Just north of La Porte City a comprehensive interchange track layout is located between the main lines of the Waterloo, Cedar Falls and Northern Railway and the Chicago, Rock Island and Pacific Railway on the adjoining right-of-way. It was built to relieve the interchange track facilities between these roads in West Waterloo and to serve at the same time for the reception and delivery of freight coming from the South. This interchange layout includes two parallel storage tracks sufficient in length to properly clear 15 cars on each, one serving as a delivery track and the other as a receiving track. "Leads" from both steam and electric roads approach these two storage tracks at both ends.

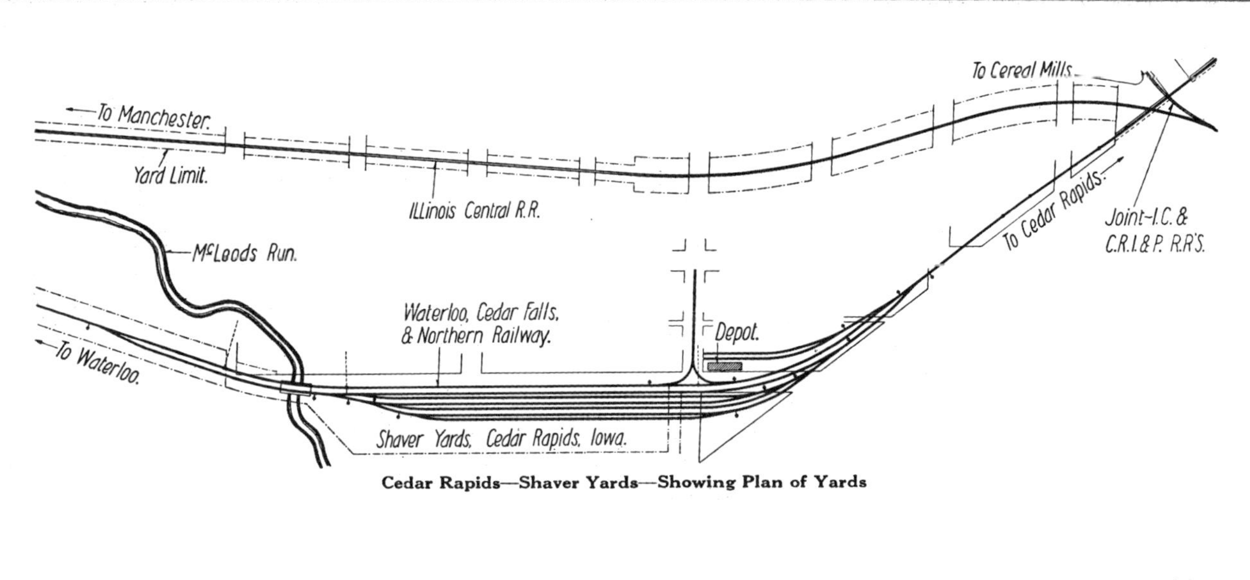

Cedar Rapids—Shaver Yards—Showing Plan of Yards

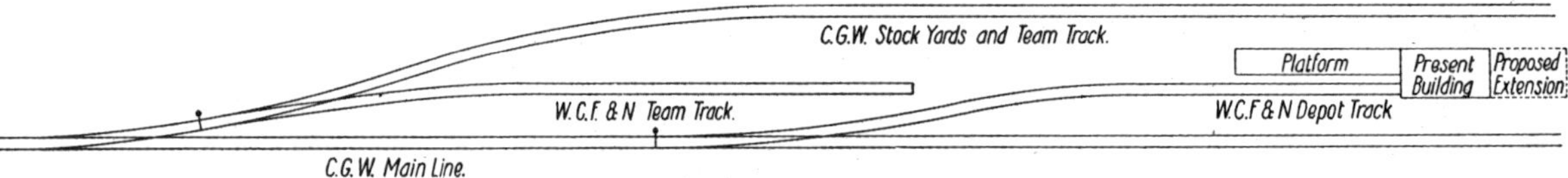

Cedar Falls—Chicago Great Western R. R. Freight Interchange

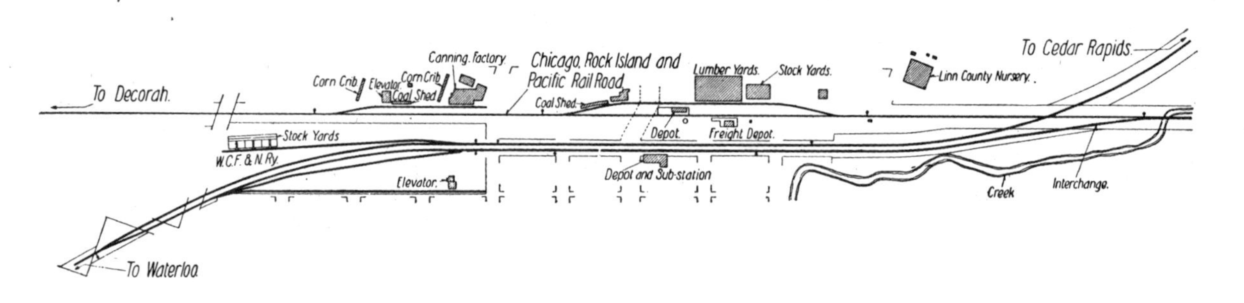

Center Point—Station Layout and Chicago, Rock Island & Pacific Railway Freight Interchange

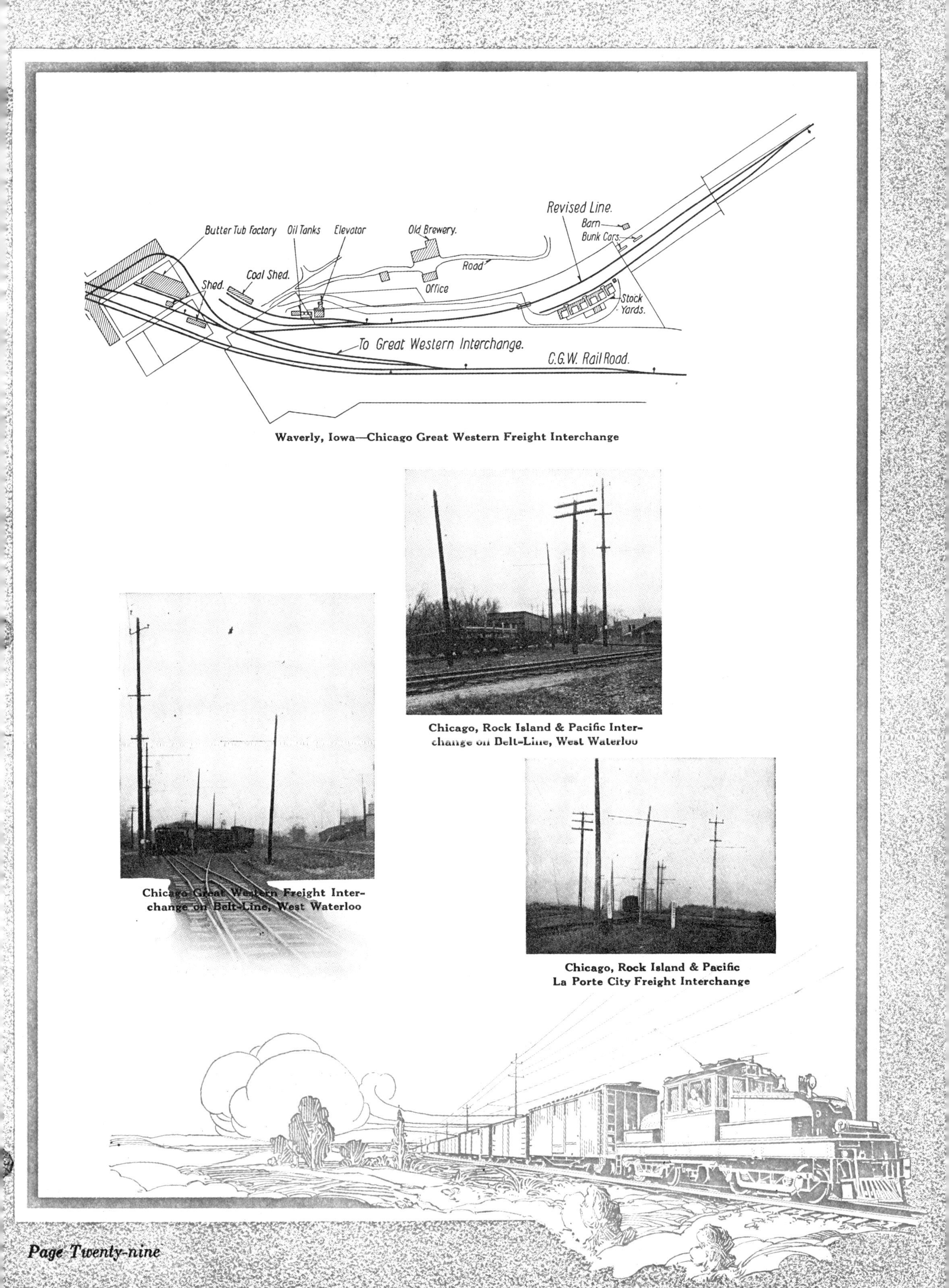

Waverly, Iowa—Chicago Great Western Freight Interchange

Chicago, Rock Island & Pacific Interchange on Belt-Line, West Waterloo

Chicago Great Western Freight Interchange on Belt-Line, West Waterloo

Chicago, Rock Island & Pacific La Porte City Freight Interchange

Interchange Inspection

IN operating the extensive interchange service of the "Cedar Valley Road" it is very important that the M. C. B. rules of interchange are properly carried out, and in addition to this, proper account be made of car repairs.

The object of the M.C.B. rules is to make car owners responsible for repairs made necessary by ordinary wear and tear in fair service; the company handling cars responsible for damage done by unfair usage, derailment or accident, and to provide proper loading and handling of cars interchanged.

To facilitate this inspection, there are two inspectors at Waterloo and one at Shaver Yard (Cedar Rapids). These inspectors look over all cars passing their respective interchanges, and make out a ***Daily Interchange Report*** showing the condition of all cars passed. This report is mailed to the Master Mechanic. In case repairs are required, the inspector makes note in a ***Repair Book*** from which a record is sent to the Master Mechanic each month. Cars that cannot be repaired at the interchange, or those requiring heavy repairs, are "carded" according to the M.C.B. rules and disposed of by either turning them over to the receiving road or by sending them to the ***"Cedar Valley"*** shops for the repair work required.

At the shop yards of the Waterloo, Cedar Falls & Northern is located a ***"Riptrack"*** and on this track all defective cars that must be repaired by the "Cedar Valley Road" are placed in first class condition. Repairs are made according to the ***"Defect Cards"*** that are placed on rolling stock inspected by the "Cedar Valley" Interchange Inspectors.

"Bad Order Cards" are placed on cars that should not be reloaded, but if possible, sent home.

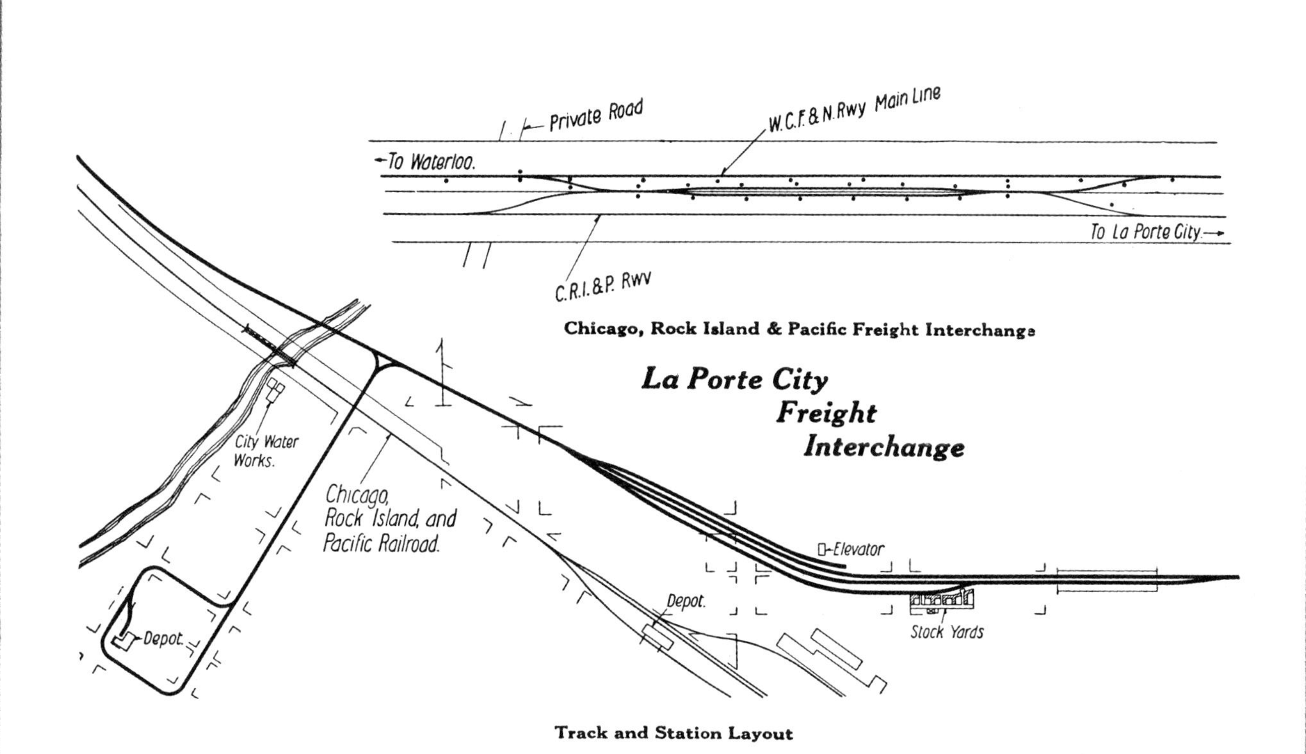

Track and Station Layout

AT other interchange points the trainmen inspect the cars, making out the ***Daily Interchange Report*** which is sent to the Master Mechanic. If the train crew make repairs (such as new air hose, knuckle, etc.) a ***Trainmen's Repair Card*** furnished in book form is made out stating the repairs made; this is also sent to the Master Mechanic.

In general all repairs are made by inspectors and trainmen as far as possible. If the repairs cannot be done by the inspector or trainmen and can be handled by sending one or two car repairers with tools, they are handled without bringing the car to the "riptrack," thus avoiding the unnecessary movement of the car, which is always undesirable.

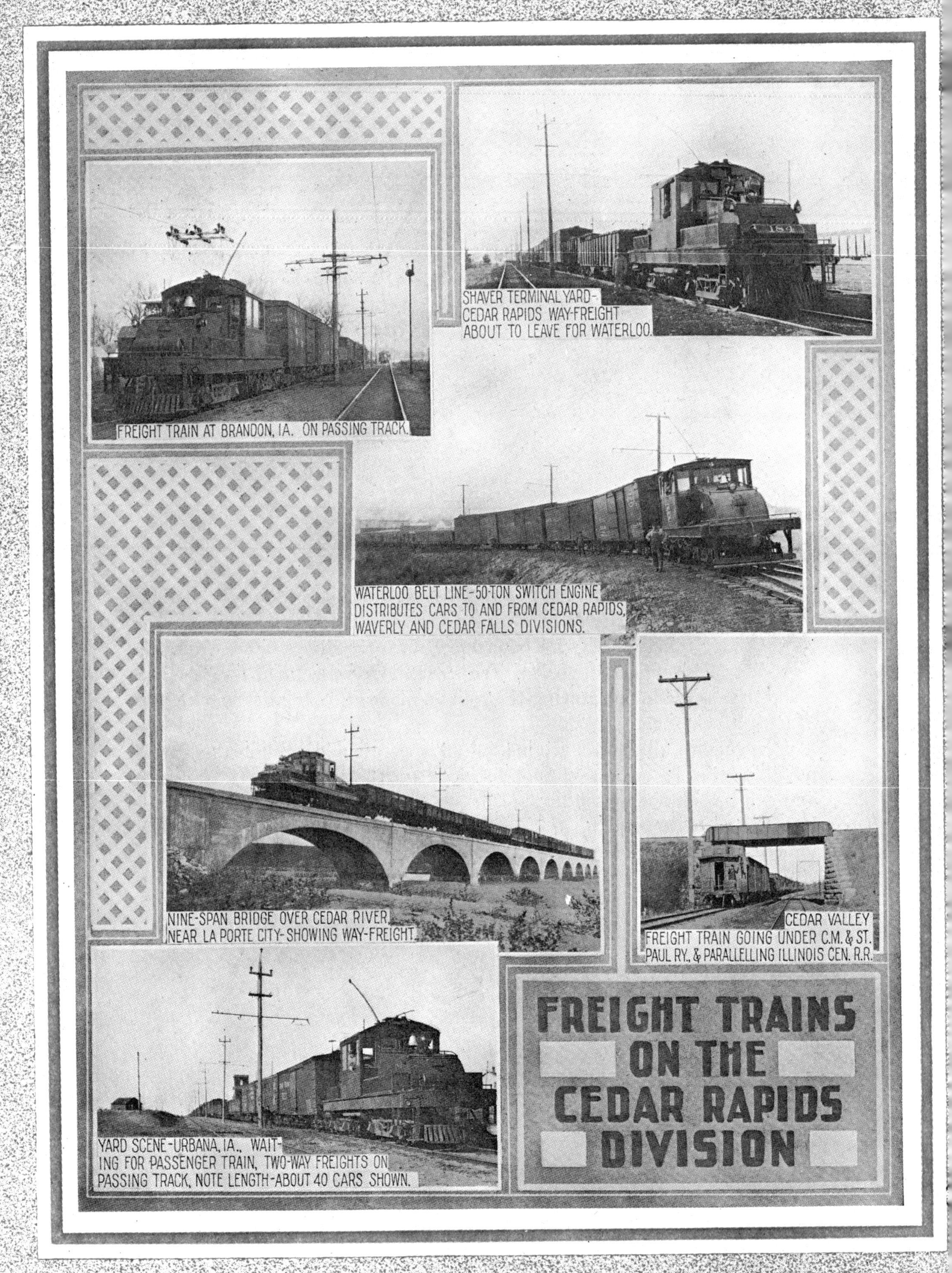

FREIGHT TRAIN AT BRANDON, IA. ON PASSING TRACK.

SHAVER TERMINAL YARD-CEDAR RAPIDS WAY-FREIGHT ABOUT TO LEAVE FOR WATERLOO.

WATERLOO BELT LINE-50-TON SWITCH ENGINE DISTRIBUTES CARS TO AND FROM CEDAR RAPIDS, WAVERLY AND CEDAR FALLS DIVISIONS.

NINE-SPAN BRIDGE OVER CEDAR RIVER NEAR LA PORTE CITY-SHOWING WAY-FREIGHT.

CEDAR VALLEY FREIGHT TRAIN GOING UNDER C.M. & ST. PAUL RY., & PARALLELLING ILLINOIS CEN. R.R.

YARD SCENE-URBANA, IA., WAITING FOR PASSENGER TRAIN, TWO-WAY FREIGHTS ON PASSING TRACK, NOTE LENGTH-ABOUT 40 CARS SHOWN.

One of Fifty Standard 80,000 Lbs. Capacity, 40-Ft. Automobile Car

Freight Service

FREIGHT is handled by the company to and from connecting steam trunk-lines. Freight from eastern points is delivered on the system early the fifth morning out, while second morning delivery is made from Chicago in conjunction with the Chicago, Milwaukee & St. Paul and the Chicago Great Western Railroads.

The freight business of the Waterloo, Cedar Falls & Northern Railway has been enjoying a very substantial growth, especially the ***inter-line freight,*** and about 75 per cent of the company's entire freight business is of this class.

Heated Car Service is run twice a week during the winter and more often if necessary to handle the business. Charcoal heaters are used in refrigerator cars to heat them in cold weather.

TERMINAL FREIGHT YARD.

AUTOMOBILE UNLOADING PLATFORM. OLD FREIGHT DEPOT AT RIGHT USED BY OVERHEAD LINE DEPT.

RECEIVING DOORS.

FREIGHT HOUSE INTERIOR— NOTE 5-TON SCALES AND ADJUSTABLE DOORS WHICH ELIMINATE "SPOTTING".

LOADING TRACKS. CARS ARE ELECTRICALLY LIGHTED DURING LOADING.

WATERLOO FREIGHT HOUSE – FREIGHT OFFICE & HEADQUARTERS FOR OVERHEAD LINES, ROADMASTERS, TRAINMASTER, AND DISPATCHERS.

FREIGHT HOUSE INTERIOR-LOADING CARS.

FREIGHT TERMINAL-WATERLOO

Waterloo Freight Terminal

IN view of the extensive freight operation of this company, and the high state of efficiency which its service has attained, it was necessary to build a large freight house, with yards. This is located on the east side of the city, at Utica and Lafayette Streets in Waterloo, adjacent to the wholesale district and only five or six blocks away from the principal business district of Waterloo.

The freight house and yard tracks are located on a 32-acre site, near the Cedar river, with a small creek running through the property. This land was very low so that it was necessary to fill in that part occupied by the freight house and tracks, about 100,000 cubic yards of "fill" being required. This "fill" material consisted in large part of rock, removed from the cuts on a new line which has been built from the freight house to the main line, to the north of the city and from the site for the proposed new shops.

The freight-house is a one-story brick structure with two-story office end with basement. The roof is tar and gravel composition, laid on steel trusses and purlines. The foundation is of concrete and part of it was carried down 20 feet to bed rock. On the team side of the building there are five sliding steel doors, sufficient in width to accommodate two teams, while the track side is composed entirely of rolling steel doors, eighteen in all. This arrangement makes it possible to run in a string of empty cars onto the siding without "spotting" them, so that the car and freight house doors will be exactly opposite, a procedure which is not always easy or convenient.

NDER the present arrangement it is possible to open up the freight house to a car wherever it may be located on the track. On the team side is a wide wooden canopy covered with Barrett specification roofing and fastened by heavy rods to the building columns. On either side of the building, above the doors, extends a row of windows so that the interior is excellently lighted. Two Fairbanks dial scales of four-ton capacity each have been installed with hard maple floor.

One of Forty 100,000 Lbs. Capacity Gondolas

The company handles a large amount of perishable freight and for taking care of it has built in one corner of the freight house a special storage room about 17 feet square. The walls of this compartment, which are 10 feet high, are made of nine inches of brick and four inches of cork, with a cement plaster lining on the inside. The room is used for refrigeration in the summer months, and for warming during the winter.

The two-story portion, which is trimmed in cut limestone, houses the freight office on the first floor, and on the second, offices for the line superintendent, roadmasters and the dispatcher. Floors in this part are of reinforced concrete. In the basement are located the boiler room, coal storage and fire-proof vault, 15 x 20 feet in size. The heating plant consists of a Kewanee tubular boiler of sufficient capacity to heat the offices, the warming room and five standard road cars in the yard.

A 60-Ton Westinghouse Equipped Freight Locomotive

Three tracks extend into the freight yard. Five cars can be "spotted" at the freight house per track, or by trucking through cars 15 cars can be loaded at one time.

EAST WATERLOO TERMINAL YARD-SWITCHING.

FREIGHT TRAIN, FARMER,-WAVERLY DIV.

PACKAGE CAR ON LEASED ELECTRIFIED SECTION OF CHICAGO, GREAT WESTERN R.R. CEDAR FALLS, IA.

YARD-CEDAR FALLS,-UNLOADING L.C.L. FREIGHT DIRECTLY TO DRAY FOR DISTRIBUTION.

FREIGHT TRAIN, DENVER, IA., WAVERLY DIV. HANDLING CAR-LOAD & L.C.L. FREIGHT.

FREIGHT TRAINS CEDAR FALLS AND WAVERLY DIVISIONS

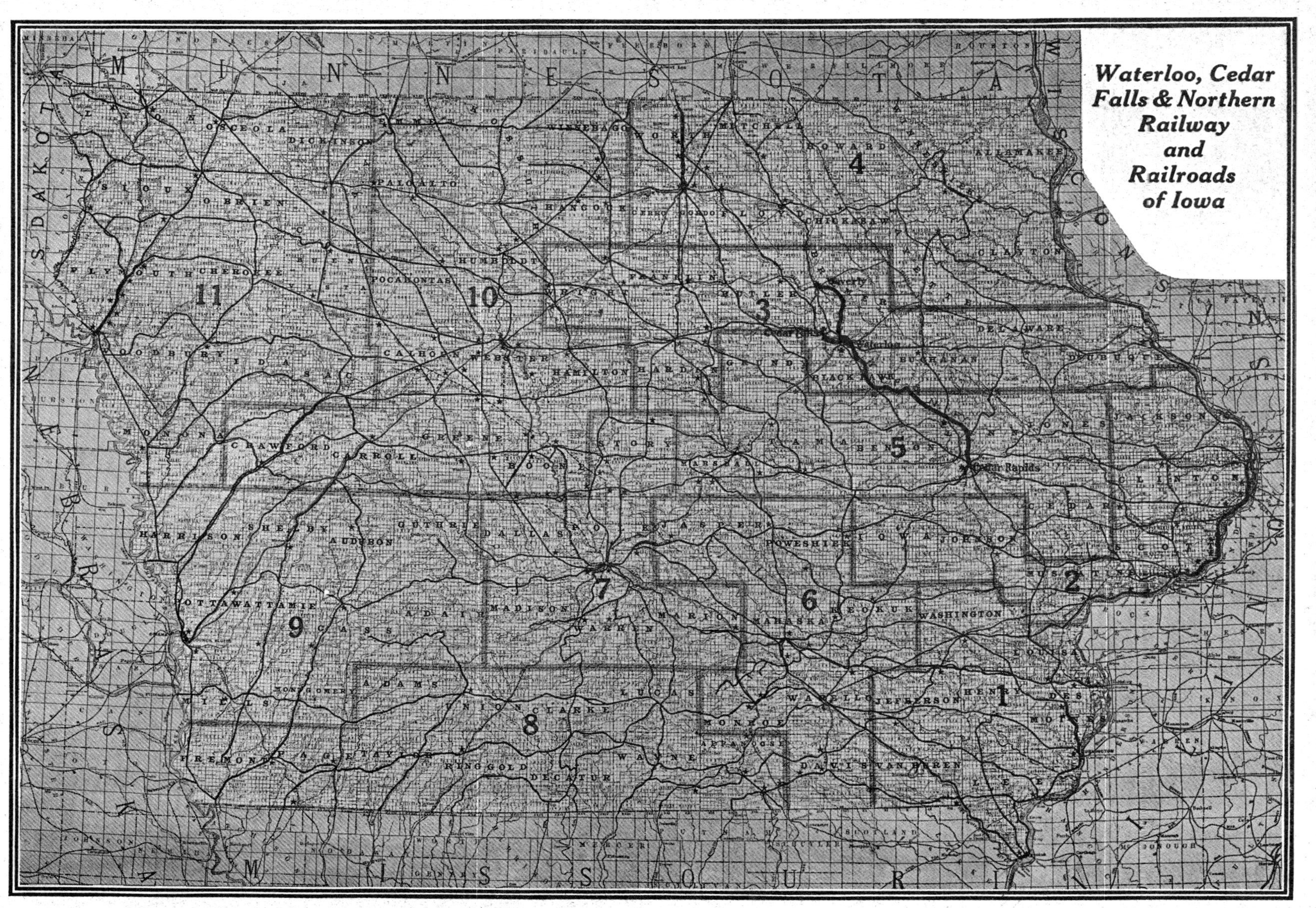

Map Showing Location of Electric Line Successfully Handling Heavy Freight and Passenger Traffic in a Highly Competitive Steam Road Territory

One of Two 40-Ton Switcher Locomotives Used for Belt-Line Yard Service

Waterloo Belt-Line

A ***Freight Belt-Line*** belonging to the Waterloo, Cedar Falls and Northern Railway, 7½ miles long, extends around the factory district of Waterloo, covering both East and West Waterloo, tying together all steam lines entering Waterloo. This belt-line starts at the eastern city limits of Waterloo where the Cedar Rapids Division enters the Waterloo limits, and extends around the northern section of Waterloo to the shop yards of the company. At these yards the Belt-Line connects with the Waverly Division, and runs thence to the East Waterloo freight house of the "Cedar Valley Road." From here it branches westward, running to Park Junction where connection is made with the Cedar Falls Division.

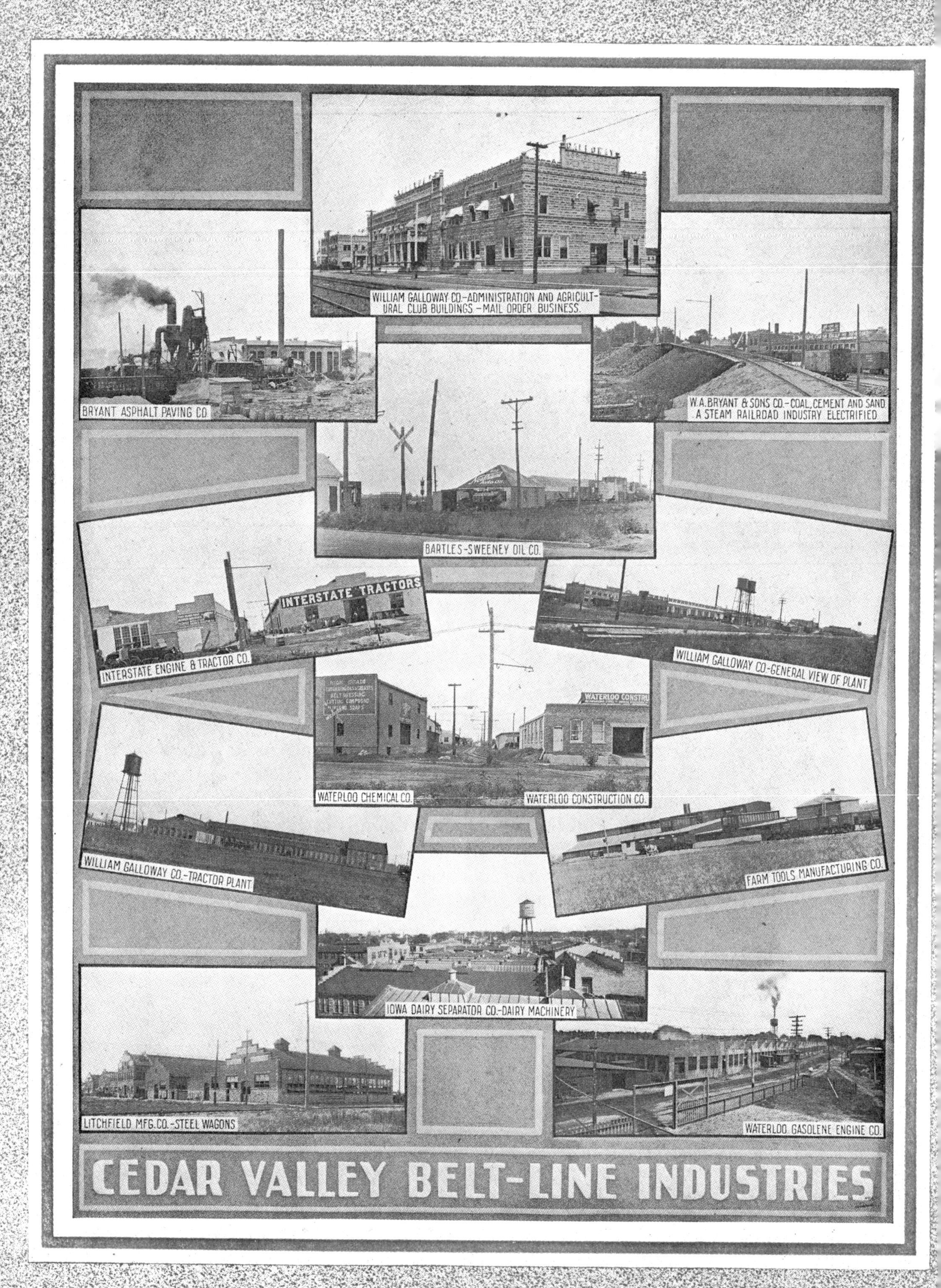
WILLIAM GALLOWAY CO.-ADMINISTRATION AND AGRICULTURAL CLUB BUILDINGS - MAIL ORDER BUSINESS.
BRYANT ASPHALT PAVING CO.
W.A. BRYANT & SONS CO.-COAL, CEMENT AND SAND. A STEAM RAILROAD INDUSTRY ELECTRIFIED
BARTLES-SWEENEY OIL CO.
INTERSTATE TRACTORS
INTERSTATE ENGINE & TRACTOR CO.
WILLIAM GALLOWAY CO-GENERAL VIEW OF PLANT
WATERLOO CONSTRU
WATERLOO CHEMICAL CO.
WATERLOO CONSTRUCTION CO.
WILLIAM GALLOWAY CO.-TRACTOR PLANT.
FARM TOOLS MANUFACTURING CO.
IOWA DAIRY SEPARATOR CO.-DAIRY MACHINERY
LITCHFIELD MFG. CO.-STEEL WAGONS
WATERLOO GASOLENE ENGINE CO.
CEDAR VALLEY BELT-LINE INDUSTRIES

One of Two 16-Ton "Trap" Cars, Collecting L. C. L. Freight on Belt-Line

FROM Park Junction this line continues around West Waterloo to Westfield and Aladdin Station, and thence to West Waterloo. In West Waterloo, the Belt-Line terminates at the C. R. I. & P. and C. G. W. transfers, and at the Waterloo Gas Engine Company and Iowa Dairy Separator factory. This freight belt-line serves all of the important manufacturing plants in Waterloo.

The serving of these industries requires ***two yard engines and two "trap" cars.*** These electric locomotives are used to deliver carload raw materials, and take away finished products in car-load lots from the various manufacturing plants on the Belt-Line, to and from the different divisions of the W. C. F. & N. Railway System, and Steam Railroad Interchanges.

Cedar River Sand and Material Company, Golinvaux, Ia. Capacity, 25 Car-loads per Day

TRAP Car Service is provided by flat cars of 30 tons capacity equipped with four 40-h.p. motors. These cars call at definite periods at the various manufacturing interests of Waterloo, securing the smaller freight shipments, which are conveyed to the East Waterloo freight house for mixed, or less-than-carload shipments.

The "trap" car also calls at the East Waterloo freight house of the Chicago Great Western R. R. and the office of the Wells Fargo Express Company for l. c. l. and express shipments, respectively. These last two points are reached by track existing on city streets and used by the "trap" car only.

From the foregoing it can be easily seen that the manufacturing interests of Waterloo are provided, by the "Cedar Valley Road," with one of the highest class and most efficient services that could be desired for both car-load and less-than-carload shipments.

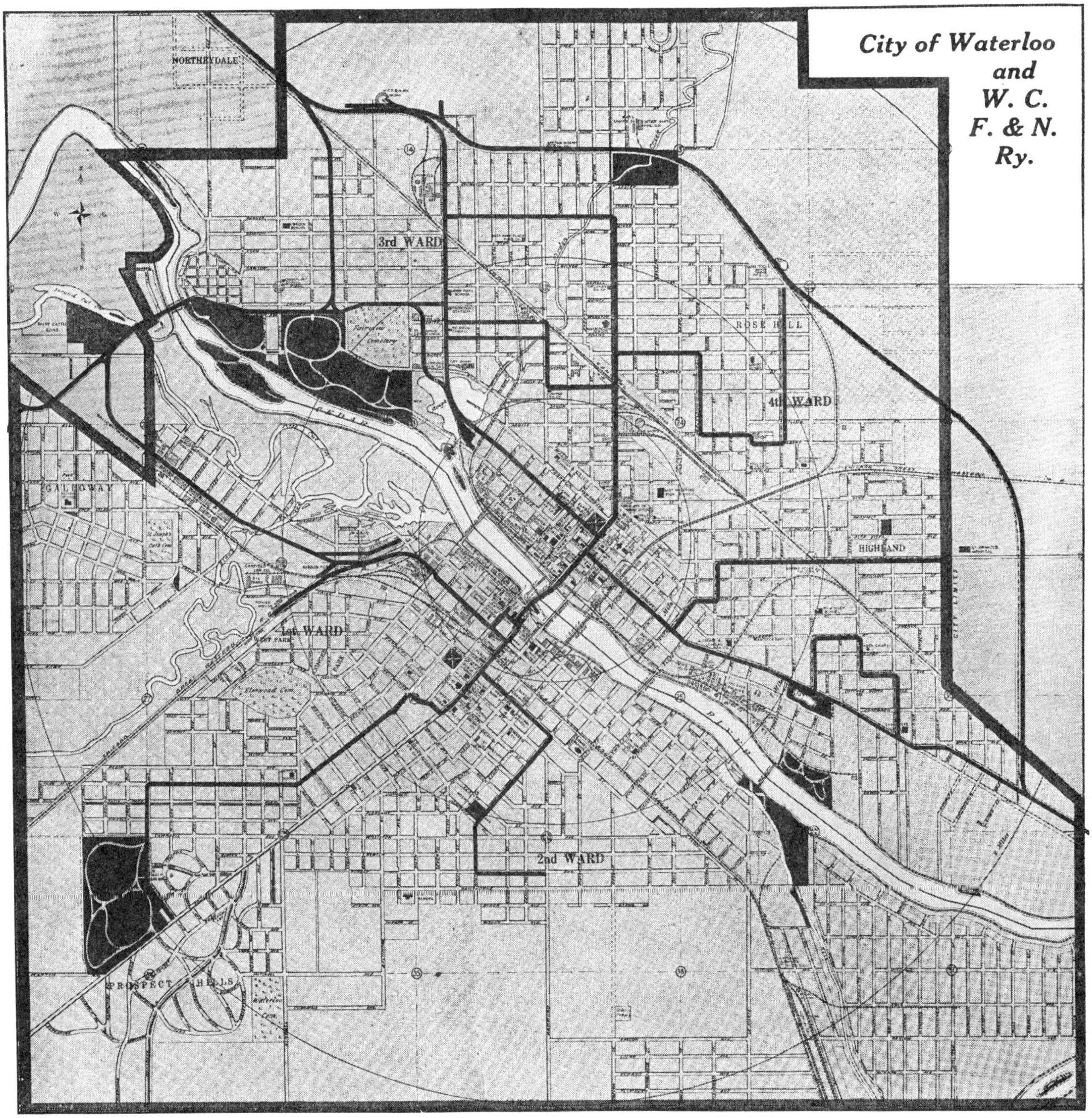

City of Waterloo and W. C. F. & N. Ry.

NORTHEYDALE
3rd WARD
ROSE HILL
4th WARD
GALLOWAY
HIGHLAND
1st WARD
2nd WARD
PROSPECT HILLS

Freight, Interurban and City Lines

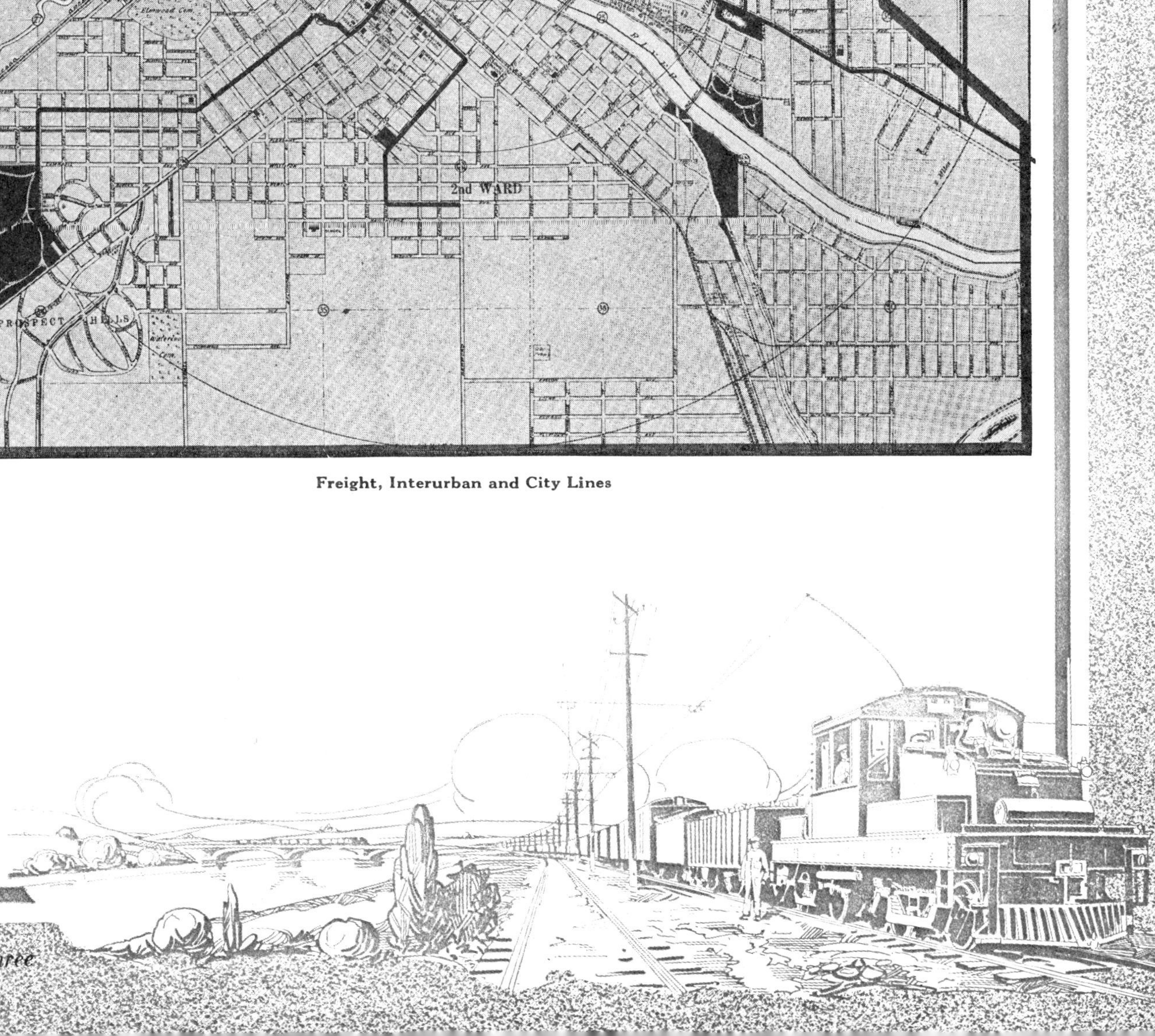

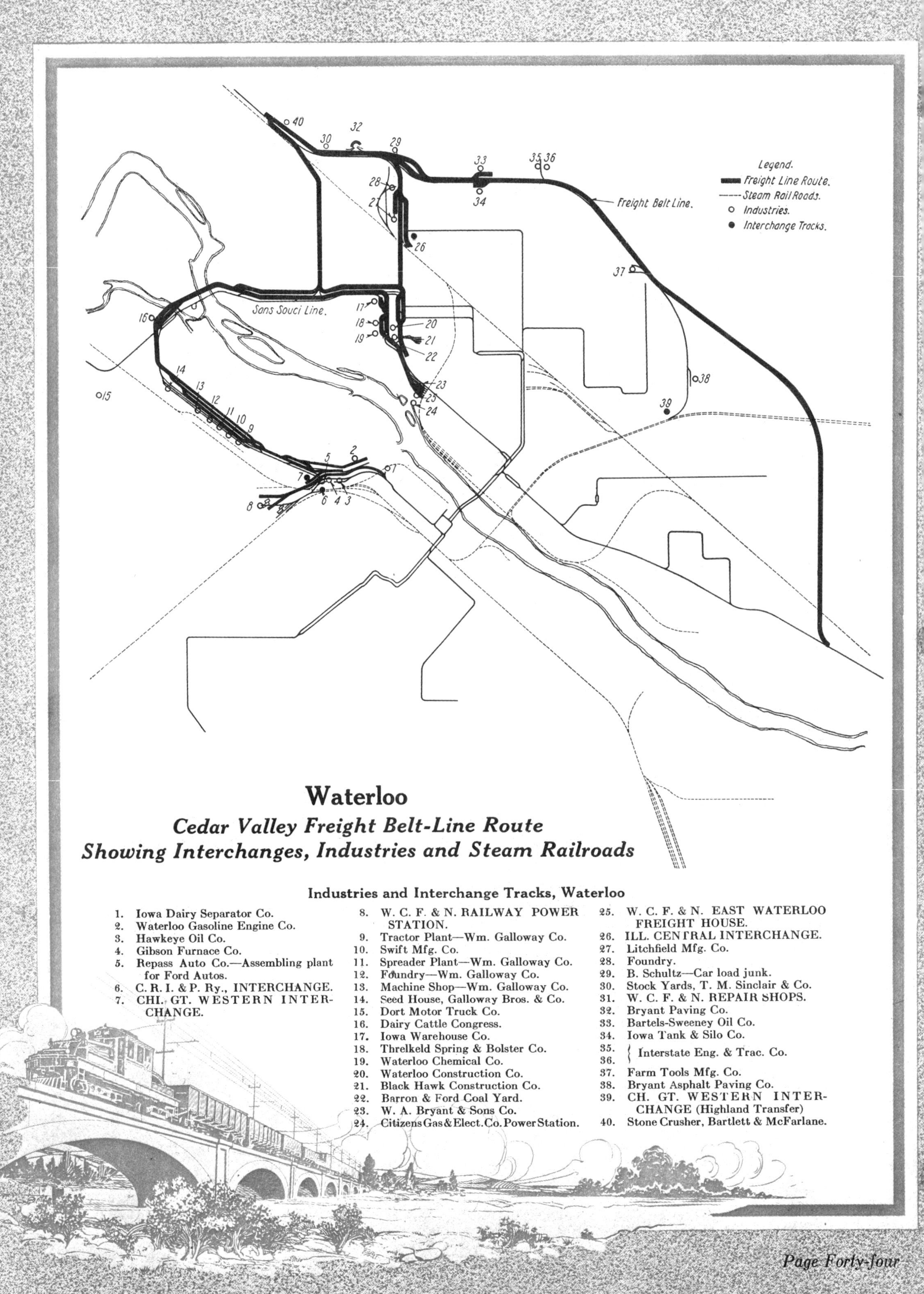

Waterloo

Cedar Valley Freight Belt-Line Route Showing Interchanges, Industries and Steam Railroads

Industries and Interchange Tracks, Waterloo

1. Iowa Dairy Separator Co.
2. Waterloo Gasoline Engine Co.
3. Hawkeye Oil Co.
4. Gibson Furnace Co.
5. Repass Auto Co.—Assembling plant for Ford Autos.
6. C. R. I. & P. Ry., INTERCHANGE.
7. CHI. GT. WESTERN INTERCHANGE.
8. W. C. F. & N. RAILWAY POWER STATION.
9. Tractor Plant—Wm. Galloway Co.
10. Swift Mfg. Co.
11. Spreader Plant—Wm. Galloway Co.
12. Foundry—Wm. Galloway Co.
13. Machine Shop—Wm. Galloway Co.
14. Seed House, Galloway Bros. & Co.
15. Dort Motor Truck Co.
16. Dairy Cattle Congress.
17. Iowa Warehouse Co.
18. Threlkeld Spring & Bolster Co.
19. Waterloo Chemical Co.
20. Waterloo Construction Co.
21. Black Hawk Construction Co.
22. Barron & Ford Coal Yard.
23. W. A. Bryant & Sons Co.
24. Citizens Gas & Elect. Co. Power Station.
25. W. C. F. & N. EAST WATERLOO FREIGHT HOUSE.
26. ILL. CENTRAL INTERCHANGE.
27. Litchfield Mfg. Co.
28. Foundry.
29. B. Schultz—Car load junk.
30. Stock Yards, T. M. Sinclair & Co.
31. W. C. F. & N. REPAIR SHOPS.
32. Bryant Paving Co.
33. Bartels-Sweeney Oil Co.
34. Iowa Tank & Silo Co.
35. } Interstate Eng. & Trac. Co.
36. }
37. Farm Tools Mfg. Co.
38. Bryant Asphalt Paving Co.
39. CH. GT. WESTERN INTERCHANGE (Highland Transfer)
40. Stone Crusher, Bartlett & McFarlane.

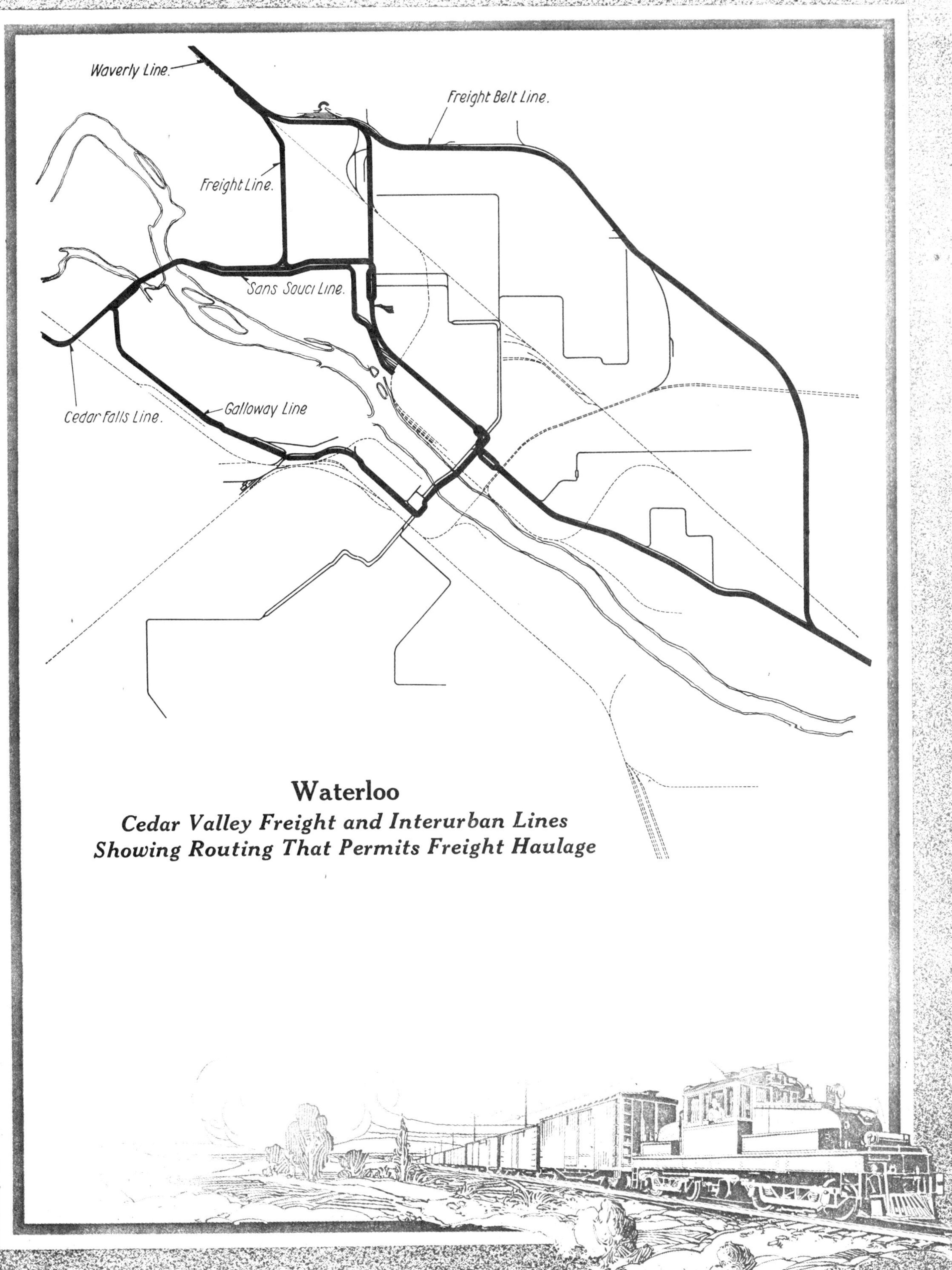

Waterloo

Cedar Valley Freight and Interurban Lines Showing Routing That Permits Freight Haulage

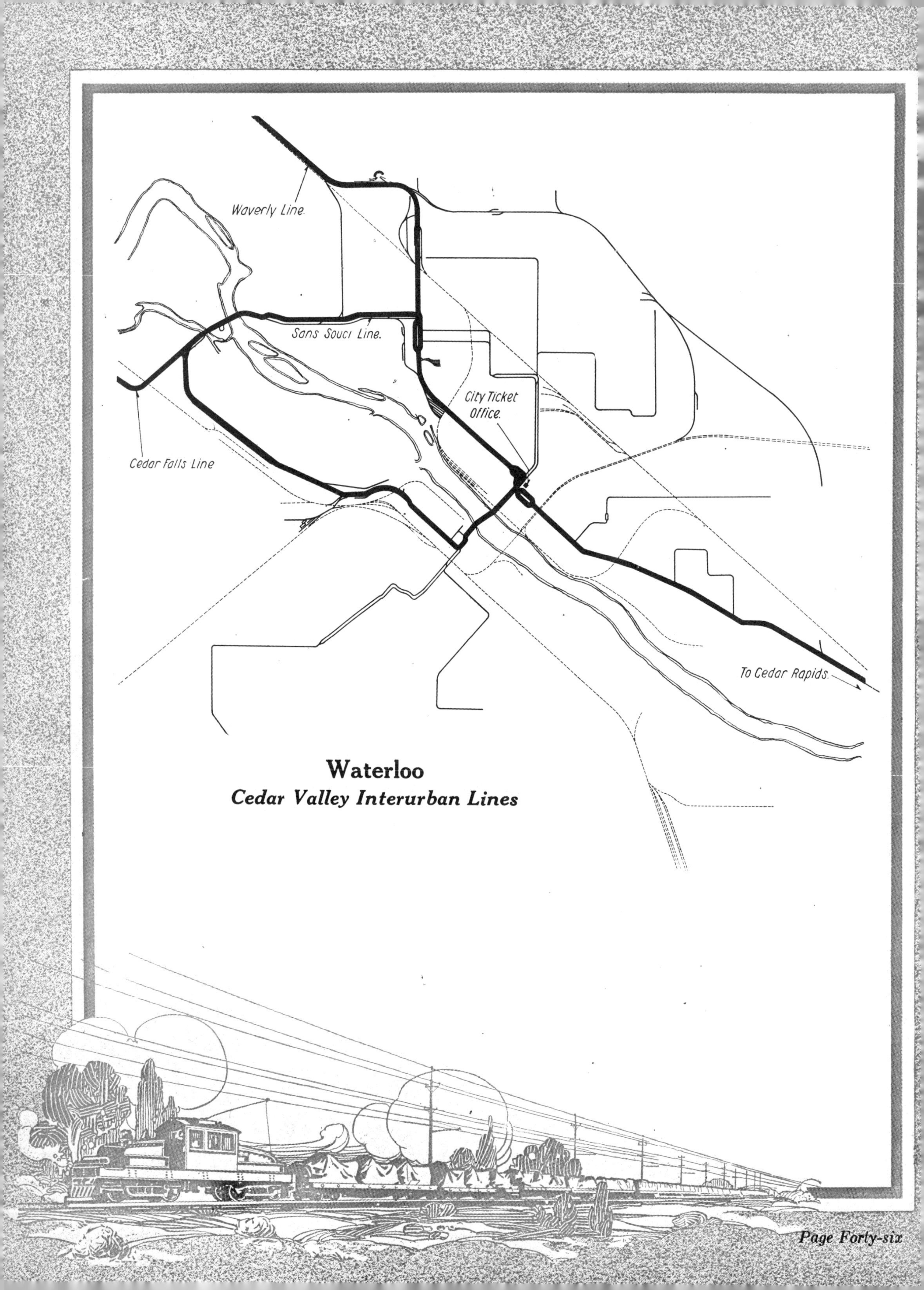

Waterloo

Cedar Valley Interurban Lines

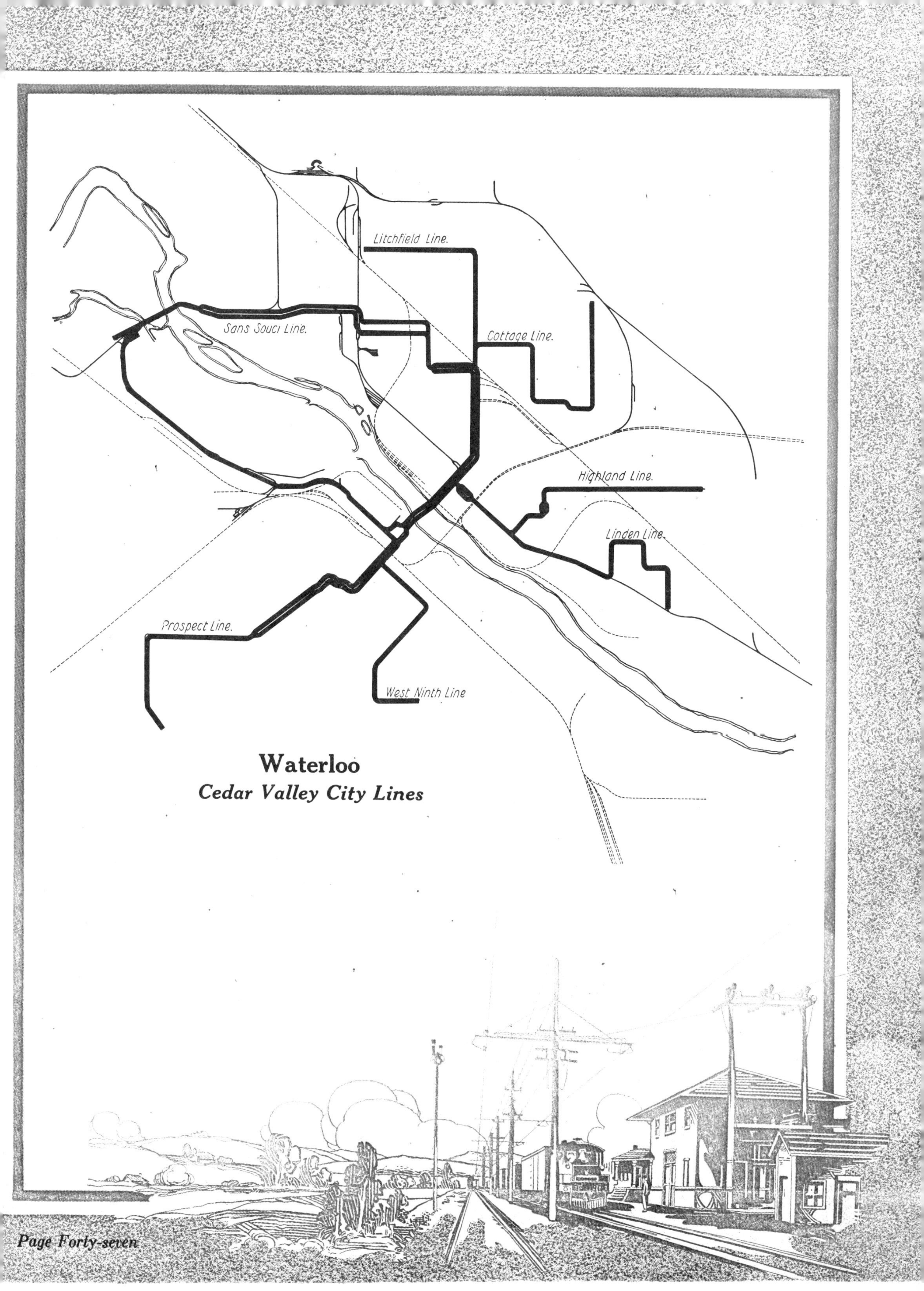

Waterloo
Cedar Valley City Lines

Industries and Stations

Glory, Iowa—Rock Crusher Capacity 800 Tons per Day

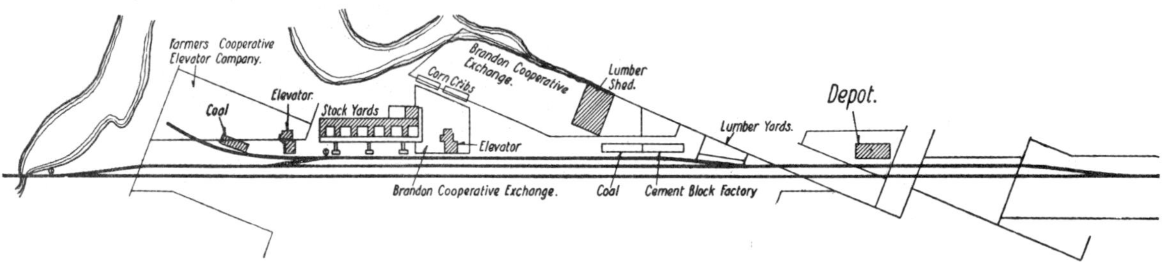

Brandon, Iowa—Showing Main Line, Passing and Industry Tracks, Also Industries

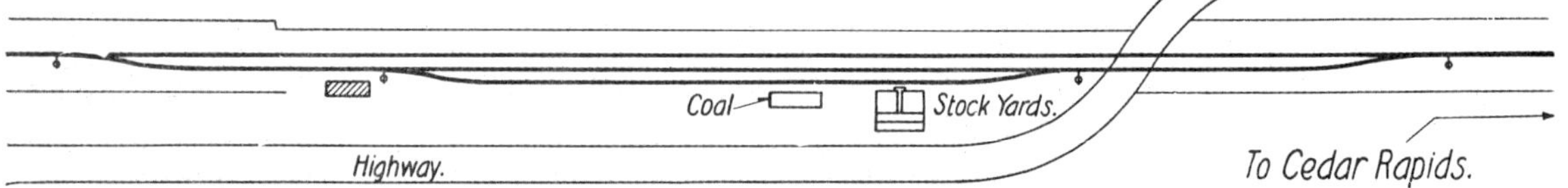

Robins, Iowa—Station and Industry Layout

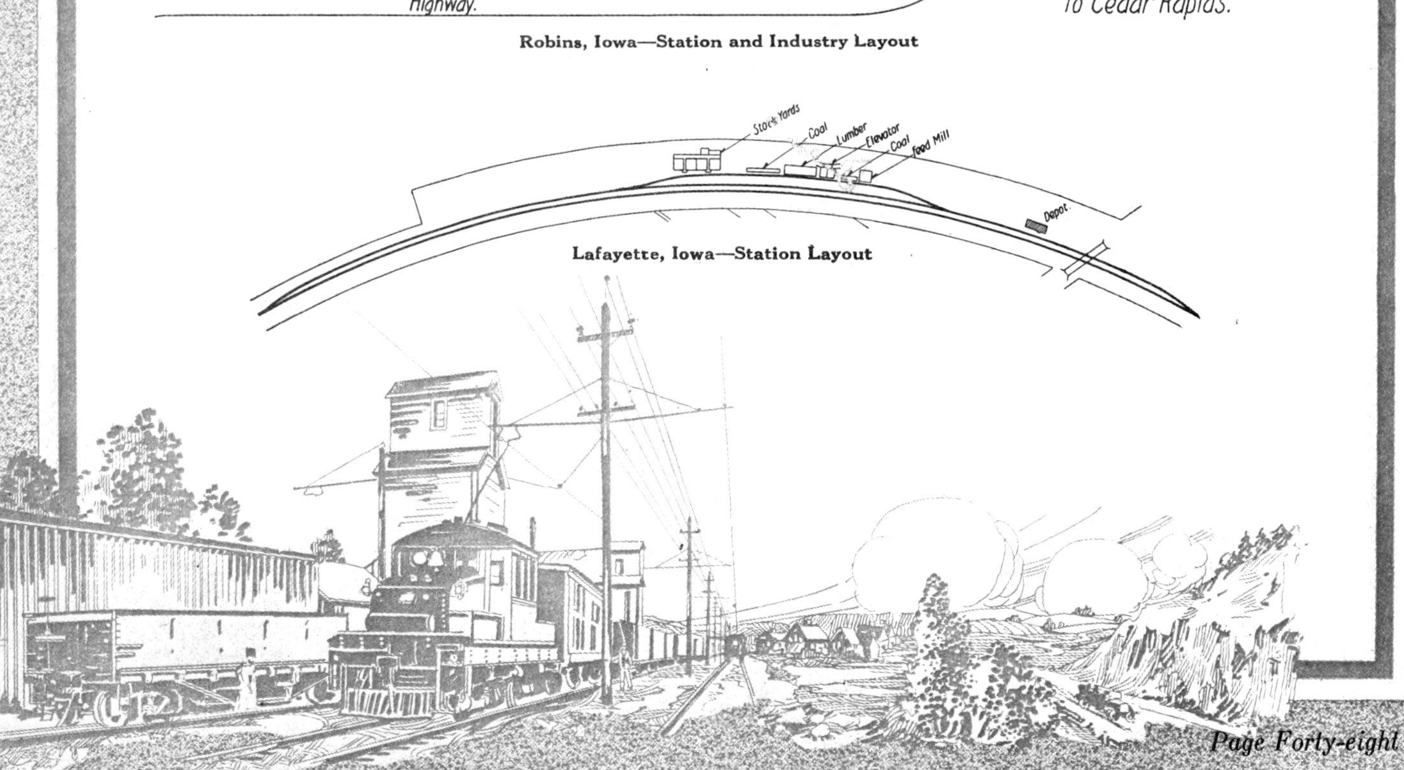

Lafayette, Iowa—Station Layout

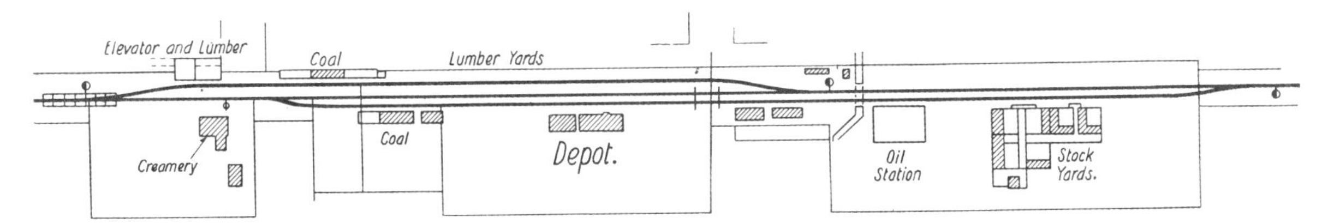

Denver, Iowa—Station and Industry Layout

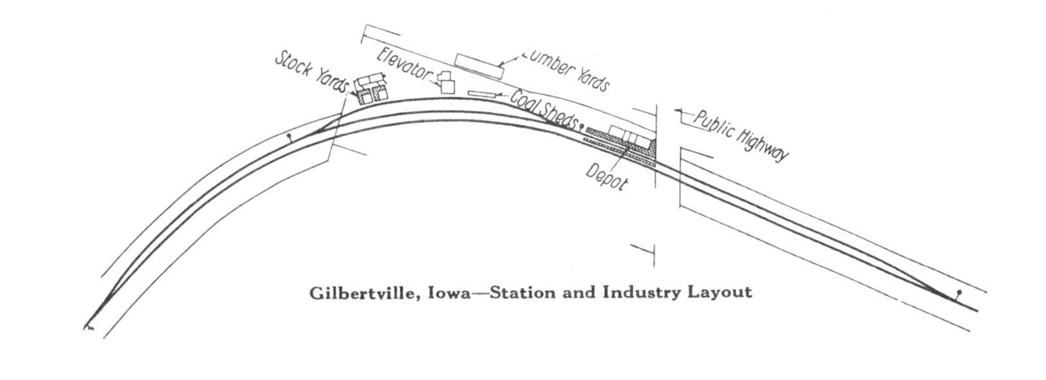

Gilbertville, Iowa—Station and Industry Layout

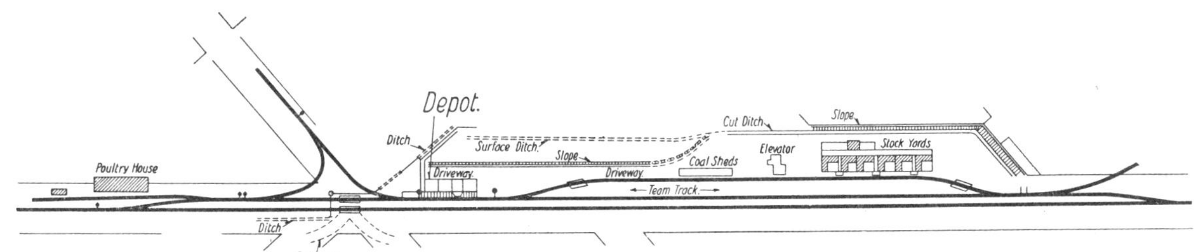

Station and Industry Layout

Team Track on the Left, Note Stock Cars, Two Way-Freights of 20 Cars Each on Passing Track

"Cedar Valley" Road on Left, Chicago, Milwaukee, and St. Paul Over Head and Illinois Central Paralleling

General Freight Service

Heavy Tonnage Train

THE Waterloo, Cedar Falls and Northern Railway follows standard steam railroad practice in handling its freight service, in addition to the ***standard road freight trains hauled by locomotives there is switching service, and also trap car service.*** The various methods of freight handling on this system are outlined by divisions in the following paragraphs.

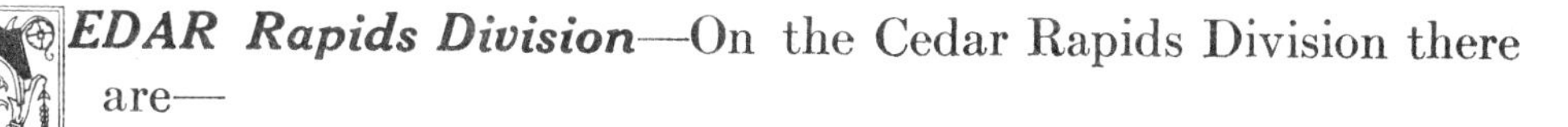

CEDAR *Rapids Division*—On the Cedar Rapids Division there are—

Two regularly scheduled local freight trains per day, one each way. In addition to this local service there are—

Two regularly scheduled time-freight trains per day, one each way. The schedule of these is as follows:

Local Freight Trains

****Train No. 401***—Local freight daily except Sunday—Cedar Rapids to Waterloo. On road from 7:40 a.m. to 2:05 p.m. This train is handled by one 60-ton electric locomotive equipped with four Westinghouse No. 308-D-3 motors and 1300-volt HL control. This train acts as a way freight picking up all local cars and l. c. l. freight from Cedar Rapids to Waterloo. This train generally has about 900 tons trailing load.

Train No. 402—Same as the above mentioned train, except Waterloo to Cedar Rapids. On road 5:35 a.m. to 12:25 p.m. This train picks up stock along the road for next morning delivery at Chicago. Also, on certain days (twice a week) a butter and egg loading car is picked up, for Eastern market. This train also handles all l. c. l. east-bound freight between towns or for foreign points.

Time Freight Trains

Train No. 498—This is a time freight run daily except Saturday night between Waterloo and Shaver (Cedar Rapids), leaving Waterloo at 7:10 p.m. and arriving at the latter point 10:45 p.m. The equipment of this train is the same as that on No. 401. Also, there is a general merchandise car movement for Chicago and Milwaukee.

**Train No. 401* runs as way freight from Cedar Rapids to Waterloo, west-bound, and peddles out empty stock cars for loading and l. c. l. business to local and foreign points.

Cab Interior of Locomotive Showing HL Master Controller

Way Freight Switching, Leaving Main Line Clear for Regular Traffic

Heavy Tonnage Freight Train

Train No. 498 (Ctd)—One of the main objects in running this train is to so handle the schedule that the Cedar Rapids transfer is reached with home-bound empty cars so that these can be "set out" before midnight, thus eliminating any chance for addition per diem charges. Thus by this train reaching the transfer, as above noted, considerable amount in per diem charges is saved each day. This is a heavy tonnage train and takes home-bound cars via Chicago, Milwaukee & St. Paul, and Chicago & Northwestern Railways, to be delivered at Cedar Rapids interchange. Empty coal cars are also handled on this train back to the mines in the Iowa and Illinois coal districts. This train also picks up carload shipments from intervening points outbound via Cedar Rapids, also eastbound l. c. l. merchandise cars are handled by this train. During the day this train does local switching at Waterloo.

To Cedar Rapids—(Also shipments over the Illinois Central and Chicago Great Western Railroads routed via Waterloo are handled on this train.)

Time Freight Train

Train No. 499—Time freight, daily except Sunday, leaves Shaver (Cedar Rapids) 1:30 a.m. arriving at Waterloo, 5:25 a.m. Also, a 60-ton engine which is standard equipment for road trains handles this schedule. The same crew that handles train No. 498 takes Train No. 499 to Waterloo. This is a heavy tonnage train and handles morning delivery merchandise car from Cedar Rapids to La Porte city and handles all through cars; such as automobile machinery and coal shipments and in fact everything of a competitive nature, to Waterloo. In addition to this service a merchandise car from Milwaukee and Chicago to Waterloo, is on this train and also cars for territory beyond Waverly and Cedar Falls. This train generally has about 900 tons trailing load, which is the daily amount hauled.

Type of Package Car Used for L. C. L. Freight and Also Hauling Car Loads. Equipped With Westinghouse Motors and HL Control

CEDAR Falls Division The freight service on this division is handled by the same type of equipment as used on the Waverly Division, namely; a 40-ton motor package car, which not only handles l. c. l. freight, but car-load shipments from the East delivered at Waterloo by the Cedar Rapids train. Also, outbound shipments from Cedar Falls to Waterloo for east-bound trains to Cedar Rapids. In Cedar Falls the W. C. F. & N. freight train operates over a leased track which is electrified, and belongs to the Chicago Great Western Railroad, but a separate freight house is used by the ***Electric Line.***

N the Cedar Falls Division there are—

Two local freight trains that make round trips twice a day between Waterloo and Cedar Falls, scheduled as follows:

Local Freight Trains

Trains No. 412 and 414—Local freights daily except Sunday, Cedar Falls to Waterloo, leaving Cedar Falls at 12:05 p.m. and 4:10 p.m. arriving at Waterloo, 12:26 p.m. and 4:26 p.m. The equipments consist of a 40-ton motor package freight car equipped with four Westinghouse No. 317 motors and HL control.

Trains No. 411 and 413—Local freight, daily except Sunday, Waterloo to Cedar Falls, leaving Waterloo at 9:55 a.m. and 2:40 p.m., arriving at Cedar Falls at 10:20 a.m. and 3:05 p.m. Equipment—Duplicate of that used on trains 412 and 414.

Waverly Division—On the Waverly Division all freight is handled by use of 40-ton package cars. In case there are any car loads these are hauled by the package cars to Waterloo and delivered to Cedar Rapids east-bound trains. Stock originating on the Waverly Division is turned over to steam railroads at Waterloo, either the Illinois Central or Chicago Great Western.

On the Waverly Division there are—

Two local freights each way, as follows:

Trains No. 408 and 410—Local freight daily except Sunday, Waverly to Waterloo, leaving Waverly at 9:30 a.m. and 2:55 p.m., arriving at Waterloo at 11 a.m. and 4:20 p.m. This service is handled by a 40-ton motor package car equipped with four Westinghouse No. 317 motors and HL control.

Trains No. 407 and 409—Local freight daily except Sunday, Waterloo to Waverly, leaving Waterloo 7:15 a.m. and 12:45 p.m., arriving at Waverly at 8:40 a.m. and 2:20 p.m. The same equipment is used as on trains Nos. 408 and 410.

Belt-Line District—Waterloo—Briefly summarizing, the operation of the Belt-Line, switching to and from industries and steam railroad interchanges on the Belt-Line, is handled by the following equipment:

1—40-ton Electric Locomotive equipped with four Westinghouse No. 317, 100-horsepower motors and HL control.

1—50-ton Electric Locomotive equipped with four Westinghouse No. 317, 100-horsepower motors and HL control.

Freight Motive Power Equipment—W. C. F. & N. Ry.

Number of Units	Designation No.	Type	Total Wgt. Lbs.	MOTORS			CONTROL (Single-End)		SERVICE		BODY		TRUCKS		Air Brake Equipment Cubic Feet Capacity
				Type	Gear Ratio	Hp.	Type	Volt.	Capacity	Class	Wdth.	Lgth.	Wheel Base	Wheel Dia.	
1	180	Double-Truck Locomotive	120000	Westinghouse 4 No. 308-D-3	16:57	250	Westinghouse HL	650 1300	†800 Tons	Fast Freight Ced.Rap.Div.	9'1"	35'0"	7'6"	36"	Westinghouse 2—38
1	181	Double-Truck Locomotive	120000	Westinghouse 4 No. 308-D-3	16:57	250	Westinghouse HL	650 1300	†800 Tons	Fast Freight Ced. Rap. Div.	9'1"	35'0"	7'6"	36"	Westinghouse 2—38
1	182	Double-Truck Locomotive	120000	Westinghouse 4 No. 308-D-3	16:57	250	Westinghouse HL	650 1300	†800 Tons	Fast Freight Ced. Rap. Div.	9'1"	35'0"	7'6"	36"	Westinghouse 2—38
1	183	Double-Truck Locomotive	120000	Westinghouse 4 No. 308-D-3	16:57	250	Westinghouse HL	650 1300	†800 Tons	Fast Freight Ced. Rap. Div.	9'1"	35'0"	7'6"	36"	Westinghouse 2—38
1	184	Double-Truck Locomotive	120000	Westinghouse 4 No. 308-D-3	16:57	250	Westinghouse HL	650 1300	†800 Tons	Fast Freight Ced. Rap. Div.	9'1"	35'0"	7'6"	36"	Westinghouse 2—38
1	5	Double-Truck Locomotive	95000	4 No. 73	17:73	100	Westinghouse HL	650	*500 Tons	Switching Waterloo Yard	9'8"	33'4'	6'6"	33"	50
1	4	Double-Truck Locomotive	83400	4 No. 75	16:82	75	MK	650	*500 Tons	Switching Waterloo Yard	9'0"	30'0"	6'0"	33"	50
1	25	Double-Truck Package Car	80000	Westinghouse 4 No. 317-A-3	16:73	90	Westinghouse HL	650 1300	*500 Tons	‡ Freight All Lines	9'2"	53'2"	6'11"	33"	Westinghouse 1—25 and 1—35
1	26	Double-Truck Package Car	80000	4 No. 73	17:58	100	M	650	*500 Tons	‡ Freight 650-V. Only	9'2"	53'2"	6'6"	33"	35
1	28	Double-Truck Package Car	56000	4 No. 80	17:69	40	K-28-B	650	——	Freight 650-V. Only	9'0"	52'2"	6'3"	33"	25
1	1	Double-Truck Trap Car	32200	4 No. 80	17:69	40	K-28-B	650	60000 Lbs.	L.C.L Freight	9'8"	41'0"	5'6"	33"	25
1	2	Double-Truck Trap Car	32200	4 No. 80	15:71	40	K-28-B	650	60000 Lbs.	L.C.L. Freight	9'8"	41'0"	5'6"	33"	25

†Continuous Rating on 1% Grade; Full Speed, 1300-Volts; Half-Speed, 650-Volts.

*One-Hour Rating.

‡Also Used for Special Train Movements Requiring Baggage Space with Coaches. Full Speed on Both Voltages.

Freight Car Equipment—W. C. F. & N. Ry.

Number of Cars	Car Nos.	Type	Cap. Lbs.	Service	Body Lgth.	Truck Wheels	Air Brakes
20	400—419	Stock	60000	General Interchange Freight	36'8"	33"	Westinghouse
40	500—539	Gondolas	100000	General Interchange Freight	40'0"	33"	Westinghouse
50	1000—1049	Automobile Box	80000	General Interchange Freight	40'8"	33"	Westinghouse
6	1994—1999	Refrigerator	40000	L. C. L. Merchandise Cedar Rapids Div.	35'0"	33"	Westinghouse
35	2150—2184	Flat	60000	Interchange Freight Stone, Gravel, Etc.	36'0"	33"	New York
2	12—13	4-Wheel Cabin Car	—	Main Line	—	33"	Westinghouse
1	19	4-Wheel Cabin Car	—	Main Line	—	33"	Westinghouse
1	10	8-Wheel Caboose	—	Main Line	42'0"	33"	Westinghouse
6	200—205	Box	40000	On Company Lines Only	34'	33"	Westinghouse
13	207—219	Box	40000	Bunk Cars	34'	33"	Westinghouse

Two-Car Multiple-Unit Main-Line Limited Train

Main-Line Service

Parlor Car on Limited Train

IN operating the 1300-volt line between Cedar Rapids and Waterloo, a high-speed, limited passenger service is rendered by four combination passenger and baggage cars and three standard parlor cars. The cars for this service weigh 94,000 pounds each and are equipped with four 120-horsepower Westinghouse motors. The parlor cars resemble those of a steam road in every detail with the exception that they are equipped with motors and will run in train with a combination passenger motor car. The time for the 60-mile run between Waterloo and Cedar Rapids is one hour and 45 minutes.

HE local passenger service between these cities is handled with 35-ton motor cars equipped with four 100-horsepower Westinghouse motors and when traffic demands a 20-ton trailer is hauled. These cars make the run in approximately two and one-half hours, making all stops.

The freight service is performed by five 60-ton locomotives equipped with four 250-horsepower Westinghouse motors, which make the run between Waterloo and Cedar Rapids in three hours. The capacity of these locomotives is such that an 800-ton train can be handled at 24 miles per hour. Westinghouse Unit Switch Control employed on all of this equipment, successfully meets the demands of the heavy traffic on this line.

Type of Car Used for Main-Line Local Service

Day Coach Compartment of Combination Steel Motor Cars Used on Limited Trains

Motor Car Cab Interior—Showing HL Master Controller

LIMITED TRAIN EN-ROUTE-60 M.P.H.

TWO-CAR ALL-STEEL LIMITED TRAIN
EACH CAR EQUIPPED WITH WESTINGHOUSE MOTORS & HL CONTROL.

LIMITED TRAIN-
FLAGGING STEAM ROAD CROSSING

PARLOR CAR INTERIOR-LADIES COMPARTMENT.

PARLOR CAR INTERIOR-LADIES COMPARTMENT.
NOTE WRITING DESK ALSO TABLES FOR BUFFET SERVICE.

ALL-STEEL LIMITED TRAIN, SHOWING WESTINGHOUSE HL CONTROL UNDER CARS.

PARLOR CAR INTERIOR-OBSERVATION END.

LIMITED TRAIN CROSSING ELK RUN BRIDGE.

LIMITED TRAINS
PARLOR CARS

Main-Line Limited Train with Parlor Car
Each Car Westinghouse Equipped

Parlor Car Service

TO furnish patrons a standard of service equal to that of the Chicago & Northwestern and Chicago, Milwaukee & St. Paul Railways with which it connects, the Waterloo, Cedar Falls and Northern Railway have in use three all-steel parlor-observation cars. These cars were placed in operation not only because of the demand for this service, but also because they cater to that class of people which desires the best accommodations, and is willing to pay for them.

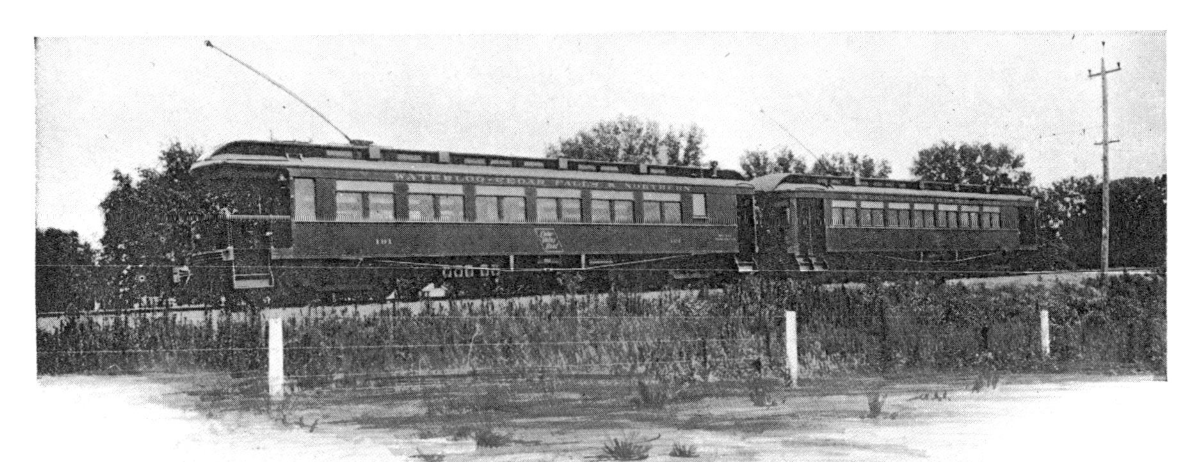

Two-Car Multiple-Unit Train—Combination Steel Motor Car and Motor Parlor Car

THE service adds considerable to the whole system in the way of publicity and putting snap into the whole operation. From these cars approximately $8,000.00 is derived each year. A buffet service is also provided, the menus being attractively printed cards, which were distributed through the mails at first to advertise the new service.

These parlor cars compare very well in length and width with the standard steam railroad equipment used for this purpose and were among the first all-steel cars of this type to be built for electric interurban service.

The principal dimensions are as follows:

Length overall	60 ft., ¾ in.
Truck centers	38 ft., 4⅜ in.
Wheelbase	6 ft., 10 in.
Width over belt rail	10 ft., 3 in.
Inside Width	8 ft., 10½ in.
Diameter of wheels	36 in.
Seating capacity	30

HE body is divided into a main passenger compartment and an observation compartment, and opening off the latter is a commodious observation platform. A kitchenette, 6 ft. x 3 ft. 2 in. in size, occupies a space in the partition between the two main compartments, portable tables being provided for the buffet service. At the front end of the main passenger compartment are two toilets, one on each side of an enclosed passage-way, and also an enclosed space provided with a hot water heater. Fourteen leather-upholstered wicker chairs are arranged along the sides and at the end of the observation compartment, and twelve chairs of the same design are located in the main passenger compartment. A built-in writing desk and a desk chair occupy one corner of the main passenger compartment, patrons being furnished with stationery bearing the trademark of the "Cedar Valley Road." A leather-upholstered davenport is located beside the partition between the two main compartments.

Couplings Used on Two-Car Limited Trains, Showing Automatic Draft and Air Couplers; Bus- and Train-Line Jumpers

Construction—Generally speaking the body framing of these cars is built as a truss side frame designed to carry the load transmitted to it by the cross-bearers in the underframe. The underframe is stiffened against buffing and pulling strains by two 8-inch, 18-pound, I-beam-sills which extend between buffers. The construction is of steel throughout, with the exception of the flooring and roof sheathing. Although this is a monitor-roof car, the side-post members are made continuous from side sill to side sill, being bent to the contour of the roof to form the principal carlines. The carbody flooring is formed of two layers of yellow pine with one thickness of building felt paper between them, and an insulating medium consisting of mineral wool is packed in the four-inch space between the bottom of the sub-floor and a plate which rests on the bottom flanges of the cross-bearers.

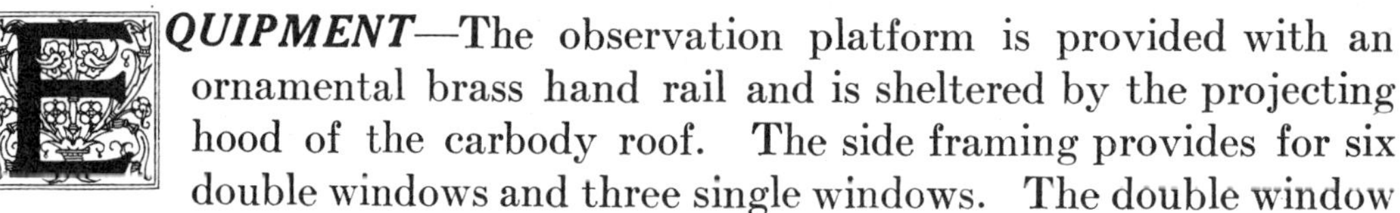

E*QUIPMENT* —The observation platform is provided with an ornamental brass hand rail and is sheltered by the projecting hood of the carbody roof. The side framing provides for six double windows and three single windows. The double window openings are fitted with single sashes above which are rectangular sashes continuous over each window opening. These upper fixed sashes are glazed with pressed prism plate glass. The floors in the main passenger compartments are covered with a green Wilton carpet. The observation platform floor, as well as that in the vestibule, at the opposite end of the car, is covered with interlocking rubber tile.

The bodies are mounted on No. 70 McGuire-Cummings high-speed interurban trucks, equipped with 36-in. M.C.B. rolled-steel wheels pressed on A.E.R.A. standard axles.

Electrical Equipment—Since these cars operate only as motor trailers in two-car trains, consisting of a standard motor car and trailer parlor car, no controllers are provided, but four Westinghouse No. 333-E-7 commutating-pole, 120-h.p. motors insulated for 1500-volts are installed on each parlor car and controlled by jumpers from the leading motor car.

The combination motor car of a two-car limited train has the same motor equipment and each of the cars has an HL control equipment which provides for full speed on 1300 volts and half speed on 650 volts.

Two circuits of ten 56-watt, 130-volt lamps (Alba shades) in series on 1300 volts and four circuits of five in series on 650 volts are provided for lighting each parlor car.

Each car has its individual change-over switch controlled by a master switch in the leading car. All circuits in each car requiring different connections for 650 and 1300-volt operation are changed by this switch. A change-over line is made continuous by means of standard receptacles and jumpers throughout a train made up of any number of types of cars. The light circuits are fused and switched in a cabinet located in the hot water heater space at the front end of the car, but the change-over function is supplied at the leading motor car in all cases.

This location of the master change-over switch was made so that by a single operation the motorman on the leading car could close all circuits in the two-car train when passing from 650 to 1300-volt power, or vice versa.

**EASIBILITY *of Equipment***—All cars will not successfully operate as multiple units, as their balancing speeds are different. However, the coach trailers can be worked in the middle or end of a train of any class of equipment, heating and lighting them from the equipment with which they are working. Motor cars can be hauled as "dead" trailers in any train using individual auxiliary circuits from the leading, or master, car of train.

On special days it is often necessary to make up trains consisting of almost any conceivable combination. The flexibility of Westinghouse apparatus as applied in this case under all possible conditions is well demonstrated. This scheme is carried out so that 1300-volt and 650-volt equipment can be operated in train on the 650-volt district without any care other than making the couplings.

Quick-Service Car Operation

Quick-Service car operation was instituted in December 1915 in Waterloo. During the year 1915 the city system had been gradually falling off in receipts, so that the management felt that some step should be made to curtail the cost of operation without impairing the service in any way. During the first part of December the company placed quick-service cars in operation on all but one line, and since that time all the city lines are served by these cars.

The car heretofore in use has been a single-truck double-end car, with short platforms, the forward left-hand side and the rear right-hand side having two-leaf doors, the opposite side being permanently closed. In remodeling the cars for quick-service operation it was of course necessary to equip the closed sides of the platform with folding doors and to install the necessary door operating mechanism. The doors which are not in use are locked by a latch on the outside. At the right of the motorman, at one side of the entrance, is located a fare box, mounted on a pipe standard. These boxes are fitted with a special casting which fits into the end of the pipe, so that the boxes are easily transferred from one end of the car to the other.

Interurban Passenger Equipment—W. C. F. & N. Ry.

Number of Cars	Car No.	Type	Wgt. Lbs.	MOTORS Type	MOTORS Gear Ratio	MOTORS Hp.	CONTROL (Single End) Type	CONTROL (Single End) Volt.	Service	Seats	BODY Width	BODY Lgth.	TRUCK Wheel Base	TRUCK Wheel	Air Brake Equipment Cubic Feet Capacity
3	100—102	Parlor Buffet Steel	94000	Westinghouse 4 No. 333E-7	25:52	125	Westinghouse HL	650 / 1300	All Lines Half Speed on 650-V. 2 Car Ltd. Train	26	10'3"	60'1"	6'8"	36"	Westinghouse
4	140—143	Combination Steel	94000	Westinghouse 4 No. 333E-7	25:52	125	Westinghouse HL	650 / 1300		52	10'3"	57'7"	6'8"	36"	Westinghouse 1-25 and 1-50
1	29	Combination	60000	Westinghouse 4 No. 317	19:70	90	Westinghouse HL	650	650-V. Lines	50	9'0"	59'4"	6'3"	37"	2-25
3	30—32	Combination	63000	Westinghouse 4 No. 317A-3	19:70	90	Westinghouse HL	650 / 1300	All Lines Full Speed on Both 650 and 1300 Volts	52	9'1"	57'7"	6'6"	37"	Westinghouse 1-25 and 1-35
2	33—34	Combination	73000	Westinghouse 4 No. 317A-3	19:70	90	Westinghouse HL	650 / 1300		52	9'1"	58'9"	6'11"	37"	Westinghouse 1-25 and 1-35
2	80—81	Combination	65000	Westinghouse 4 No. 317	19:70	90	Westinghouse HL	650	650-Volt Lines	44	8'11"	51'7"	6'10"	33"	2-25
1	20	Single Compartm't	46500	4 No. 80	17:69	40	K—28—B	650	650-Volt Lines	48	8'7"	45'4"	4'3"	33"	Westinghouse 16
1	22	Single Compartm't	41600	2 No. 301	15:69	40	*K—12	650	650-Volt Lines	46	8'10"	43'0"	5'6"	33"	Westinghouse 16
Interurban Trailers															
1	15	Coach Double Truck	50000	——	——	—	——	——	All Lines	52	10'0"	59'1"	7'1"	33"	Westinghouse
2	16—17	Coach Double Truck	50000	——	——	—	——	——	All Lines	58	9'8"	61'0"	7'0"	33"	Westinghouse
1	18	Coach Double Truck	50000	——	——	—	——	——	All Lines	62	9'8"	63'0"	7'9"	33"	Westinghouse

*Double-End Control.

City Passenger Equipment—W. C. F. & N. Ry.

Waterloo and Cedar Falls Local Lines

Nmbr. of Cars	Car Nos.	Type	Wgt. Lbs.	MOTORS Type	MOTORS Gear Ratio	MOTORS Hp.	CONTROL Type	CONTROL Volt.	Service	Seats	BODY Width	BODY Lgth.	TRUCKS Wheel Base	TRUCKS Wheel Dia.	Brake Equipment
6	306—311	Single-Truck Closed	28000	Westinghouse 2 No. 323	16:81	38	Westinghouse K-36-J	650	Local Lines	32	9'0"	32'2"	8'6"	33"	Hand
6	300—305	Single-Truck Closed	28000	2 No. 301	15:69	40	S	650	Local Lines	32	9'0"	32'2"	8'0"	33"	Hand
15	40—54	Single-Truck Closed	22000	2 No. 301	15:69	40	K-10	650	Quick-Service Operation	24	7'10"	30'8"	8'0"	33"	Hand
15	60—74	Single-Truck Open	22000	2 No. 301	15:69	40	K-10	650	Summer	50	10'0"	28'4"	8'0"	33"	Hand
City Passenger Trailers															
7	90—96	Single-Truck Open	—	——	—	—	——	—	Summer Local Lines	35	7'7"	24'7"	6'0"	33"	Hand
Work Cars															
1	21	Double-Truck Line Car	—	Westinghouse 4 No. 327-C	15:69	55	Westinghouse HL	650 / 1300	All Lines	—	8'7"	45'0"	6'6"	33"	Westinghouse Air Brake
3	X1 to X3	Snow-Sweepers Single-Truck	—	2 No. 80	—	40	K-10	650	650-V. Lines	—	9'0"	28'3"	8'6"	33"	Hand
1	1	Single-Truck Line Car	—	2 No. 301	15:69	40	K-10	650	650-V. Lines	—	8'0"	28'0"	13"	33"	Hand

Manually Operated Interlocking Signal Tower Crossing Illinois Central R. R. Conductor of Train "Clears" this Tower for His Train. No attendant located here.

Combination Type Interurban Car Used on 650-Volt Lines—Westinghouse Motors and HL Control. Mirrors on Cars Facilitate Loading Passengers and Baggage

First Interurban Car of "Cedar Valley Road"

Local En Route to Cedar Rapids Equipped to Operate Full Speed on 650 or 1300 Volts

DAIRY CATTLE CONGRESS GROUNDS — WATERLOO BELT LINE DISTRICT.
1916 ATTENDANCE, 50,000 PEOPLE IN ONE WEEK.

ELECTRIC PARK, WATERLOO—
OPERATED BY CEDAR VALLEY ROAD.

CEDAR RIVER—PUBLIC BATHING BEACH, WATERLOO.

THREE DEGREE CURVE AROUND ROCK BLUFF—
CEDAR RIVER ON THE LEFT, CEDAR RAPIDS DIV.

CEDAR VALLEY VIEWS

Sources of Revenue

Grain Elevator

THE fact that a number of the factories of Waterloo are located on the outlying districts of the city, has much to do with the rate of return from the passenger business. Waterloo has a number of small parks situated in different parts of the city, all of which are touched by the city lines.

A large Chautauqua Park is also touched by the "Cedar Valley" interurban line in the western part of East Waterloo. This park is situated on the bank of the Cedar River, which furnishes fine fishing and boating as forms of amusement. Although the Chautauqua proper lasts only three weeks a year, quite a number of permanent cottages have been built on the park grounds which serve as summer homes during four to six months of each year.

Just across the Cedar River from the Chautauqua Park is the railway company's electric park, and also Sans Souci Park, where a summer hotel is located, and this also serves to stimulate traffic on this section of the line.

Parks—To stimulate traffic on the city and interurban line, the company has equipped a large tract of land in the western portion of the city of Waterloo as an electric park. This park is situated on the Waterloo-Cedar Falls line on a cut-off channel of the Cedar River, and all attractions common to an electric amusement park have been installed. This park is owned and managed by a subsidiary organization of the railway company and is more than self supporting. In common with the experience of many other cities of the Middle West it has in fact become a center for the summer amusements of the townspeople.

Denver, Iowa—Substation on Left and Team Track on Right

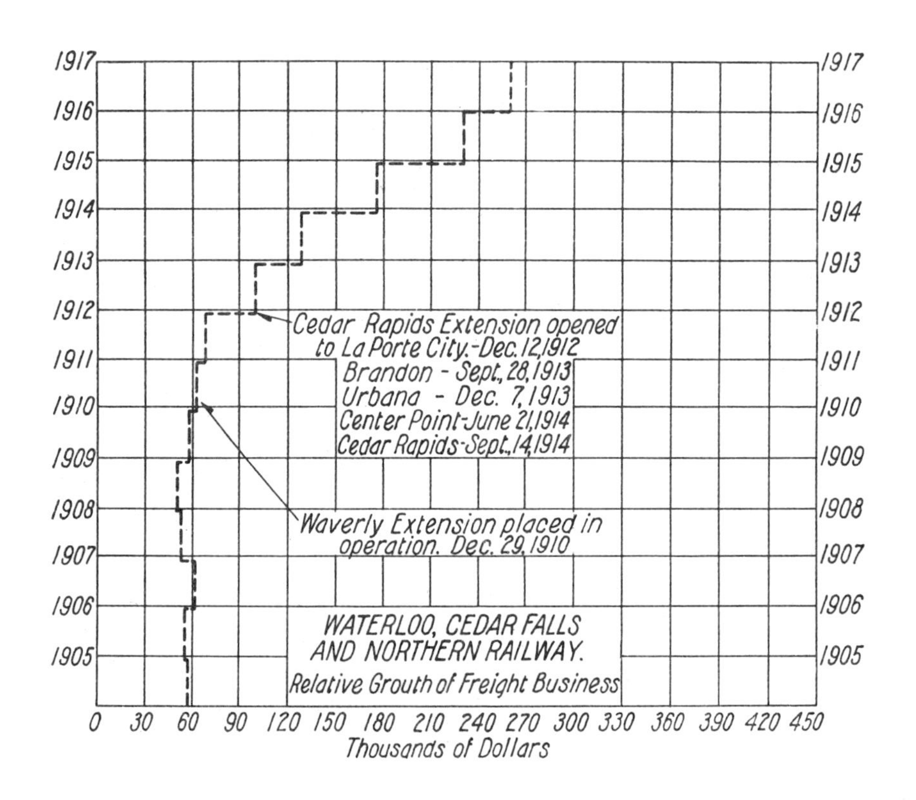

Operating Revenue

A statement of the gross earnings of the year ending June 30, 1917, and the cost of operation is given below:

Gross earnings	$868,268.39
Operating expenses and taxes	485,943.99
Net earnings	382,324.40

The passenger earnings cents per car-mile for city lines and interurban lines are described as follows:

Cedar Falls local lines, 5 miles	26.82 Cents, June 30, 1917.
Waterloo Local lines, 25 miles	21.55 Cents, June 30, 1917.
Cedar Falls Interurban, 8 miles	42.13 Cents, June 30, 1917.
Waverly, 23 miles	32.01 Cents, June 30, 1917.
Cedar Rapids Division, 60 miles	24.95 Cents, June 30, 1917.

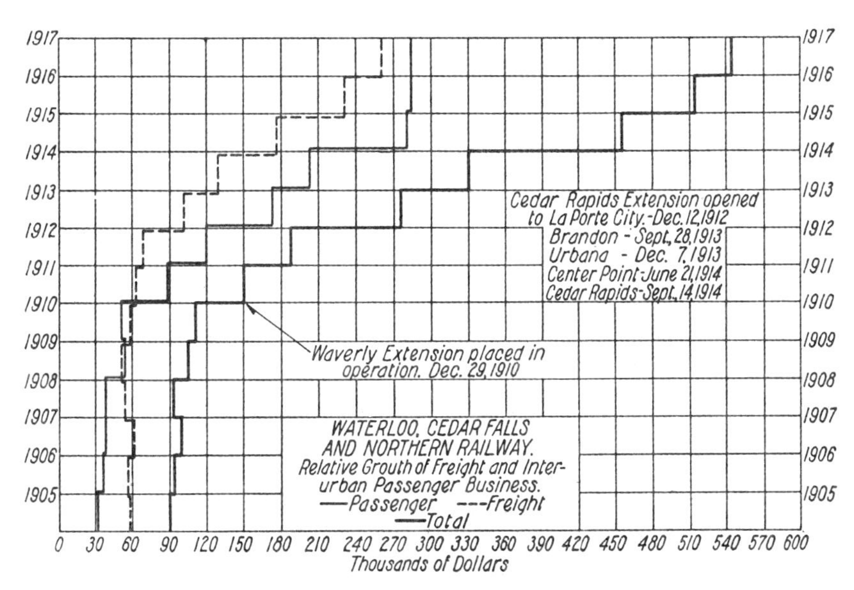

The freight earnings are about 40 per cent of the gross income

AS will be noted from the yearly earnings chart the gross earnings per month have shown a continual increase since 1905, but the results of the construction expenditures are just beginning to show their proportion of the net increase. The gross earnings for the fiscal year ending March 31, 1917, show an increase of 11.2 per cent over a corresponding period in 1910, and the present gross increase per month shows a net gain in excess of 6 per cent over the corresponding month of last year. As shown, July has invariably been the busiest month.

Division of Power

Sold power..	10%
Passenger cars including local service in Waterloo and Cedar Falls..........................	57%
Freight..	33%
	100%

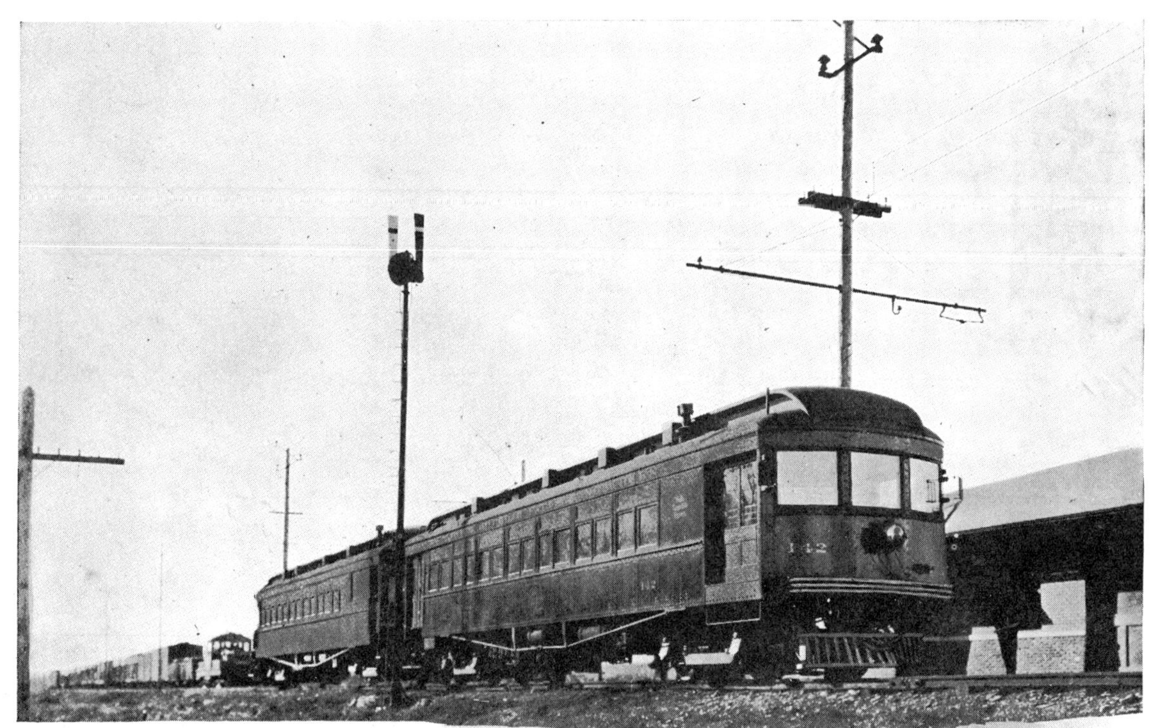

A Main-Line Limited Passenger Train and Heavy Freight Train Have No Trouble in Getting Under Way When Meeting at the Same Substation

Power Demands

Limited Train

PASSENGER train power demands are a certain definite fixed value per train, depending on the speed of the equipment, etc. All passenger cars having motors larger than 100 h.p. are equipped with an ammeter. The motormen are taught careful and economical acceleration by the use of the ammeter, thus reducing the big acceleration peaks on the substation.

Also, all locomotives are equipped with ammeters. When freight trains meet a large motored passenger train at a substation point, the freight train is required to remain idle until the passenger train has had time to reach a balancing speed and current before moving. The freight engineer is also taught to accelerate and operate his train intelligently, with the ammeter as his guide.

Waterloo Power House

HE is also taught to take advantage of the profile of the track in moving his train with as few peaks as possible and as economically as possible. Lastly, but not least important, the 19-point Westinghouse HL control, as applied, makes it possible to accelerate and run an 800-ton freight train until no larger peaks than are demanded by one two-car high-speed limited passenger train are obtained.

With all of the above equipment it is possible to handle heavy tonnage freight trains and limited and local passenger trains on time with 500-kw. substations without "pulling out" the circuit breakers at these substations. This is all the result of careful training of the trainmen and seeing that the proper equipment is placed in their hands.

Power Generation

A Deep "Cut."

THE power station site covers about 13 acres, and is situated near the junction of the Rock Island and Great Western Systems in the western part of West Waterloo. This location was selected because it was near the interchange connection of two steam roads that tapped the Iowa and Illinois coal fields, also because an excellent supply of water could be obtained from the Black Hawk Creek, which is within 600 feet of the power house.

The power station proper is centrally located with respect to the "Cedar Valley" System including the interurban and street railway lines. All power used by the Waterloo, Cedar Falls and Northern Railway is generated in this station.

In the boiler room six 500-horsepower boilers equipped with superheaters, economizers, and chain grate stokers operating with induced draft provide steam at 200 pounds pressure and 100 degrees superheat. All auxiliary apparatus in the boiler room is motor driven. Iowa coal, which is low in b.t.u. value and high in ash content is dumped into a track hopper from the cars and conveyed by means of industrial cars to an elevator where it is elevated to the bunkers in front of each boiler. The ash is conveyed in a similar manner from the ash pits to a hopper over the coal supply track. The condensers are of the jet type, taking water from Black Hawk Creek. The condenser pumps are turbine driven.

The generator room contains the following equipment:

Two 1500-kva., 2300-volt, 3-phase, 1500-r.p.m. steam turbines with direct-connected, commutating-pole exciters.

One 3000-kva., 2300-volt, 3-phase, 1500-r.p.m. steam turbine with direct-connected commutating-pole exciter.

Condensing equipments are of the jet type and in the basement.

One engine driven, 50-kw. auxiliary exciter.

Two banks of 500-kva. and one bank of three 1000-kva., 2300 to 44,000-volt water-cooled transformers. On the outgoing 44,000-volt circuit protection is provided by Westinghouse electrolytic lightning arresters.

***Powerhouse Substation Equipment*—Two 500-kw., 500-r.p.m., 650-volt rotary converters with 2300 to 405-volt transformers.**

A 27-panel switchboard controls all the power house and substation equipment located in the power house proper, using a double bus scheme on both the 2300 and 44,000-volt sides.

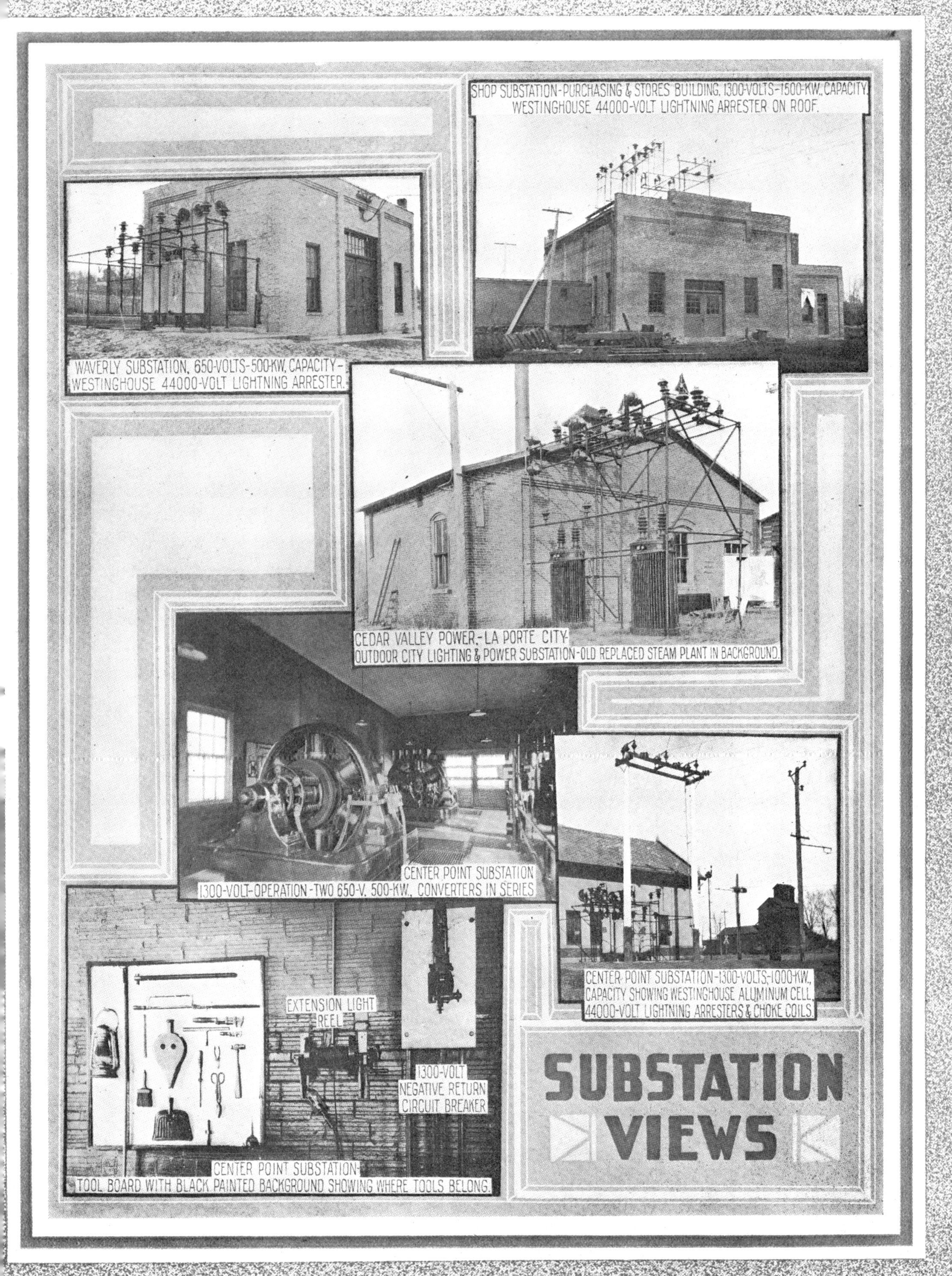
SHOP SUBSTATION-PURCHASING & STORES BUILDING, 1300-VOLTS-1500-KW., CAPACITY
WESTINGHOUSE 44000-VOLT LIGHTNING ARRESTER ON ROOF.
WAVERLY SUBSTATION, 650-VOLTS-500-KW., CAPACITY-
WESTINGHOUSE 44000-VOLT LIGHTNING ARRESTER.
CEDAR VALLEY POWER,-LA PORTE CITY-
OUTDOOR CITY LIGHTING & POWER SUBSTATION-OLD REPLACED STEAM PLANT IN BACKGROUND.
CENTER POINT SUBSTATION
1300-VOLT-OPERATION-TWO 650-V. 500-KW., CONVERTERS IN SERIES
CENTER POINT SUBSTATION-1300-VOLTS-1000-KW.,
CAPACITY SHOWING WESTINGHOUSE ALUMINUM CELL,
44000-VOLT LIGHTNING ARRESTERS & CHOKE COILS
EXTENSION LIGHT
REEL
1300-VOLT
NEGATIVE RETURN
CIRCUIT BREAKER
CENTER POINT SUBSTATION-
TOOL BOARD WITH BLACK PAINTED BACKGROUND SHOWING WHERE TOOLS BELONG.
SUBSTATION
VIEWS

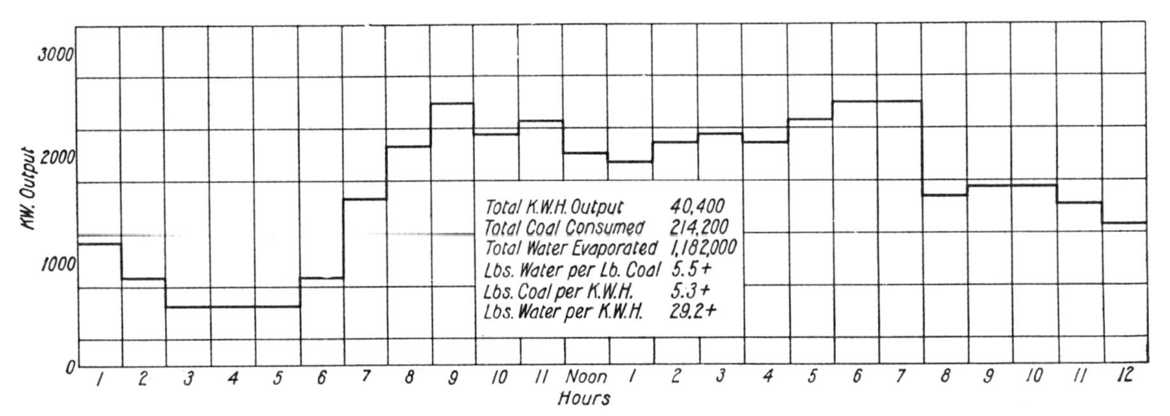

Power Station—Typical Twenty-Four Hour Load Curve

Substations

FORMERLY the entire system of the "Cedar Valley Road" was operated with a trolley voltage of 650 volts, the substations being placed from 6 to 10 miles apart. When the new line from Waterloo to Cedar Rapids had progressed part way, the freight and passenger business had grown to such an extent that it was found advisable to change the trolley voltage to 1300 volts, and to space the substations from 16 to 17 miles apart. This arrangement gives better average trolley voltage, higher substation load factors and lower operating labor and maintenance cost. The new 1300-volt substations are located at Gilbertville, Brandon, Center Point and Cedar Rapids yards, replacing the 650-volt substations at Gilbertville, La Porte City, Brandon, Urbana and Center Point, and also the proposed substation at Lafayette, Louisa and Cedar Rapids Yards. Substations operating at 650 volts are still located on the old lines at Cedar Falls, Farmer, Denver and Mills.

In addition to the above 1300-volt substations there is also a 650-1300-volt substation located at the car shops at Waterloo.

Waterloo, Waverly and Cedar Falls Substations, Each equipped with 650-volt, 500-kw. rotary converter and transformers.

Waterloo Shops, 650-1300-volt, three—500-kw., 650-volt, rotary converters with transformers and switchboards.

Gilbertville, Brandon and Shaver, each equipped with a 1300-volt single commutator, 500-kw. rotary converter with transformers and switchboards designed for 200 per cent overload.

Center Point, equipped with two 650-volt, 500-kw. rotary converters in series and otherwise similar to the above 1300-volt substation. All substations are protected with 44,000-volt Westinghouse electrolytic lightning arresters.

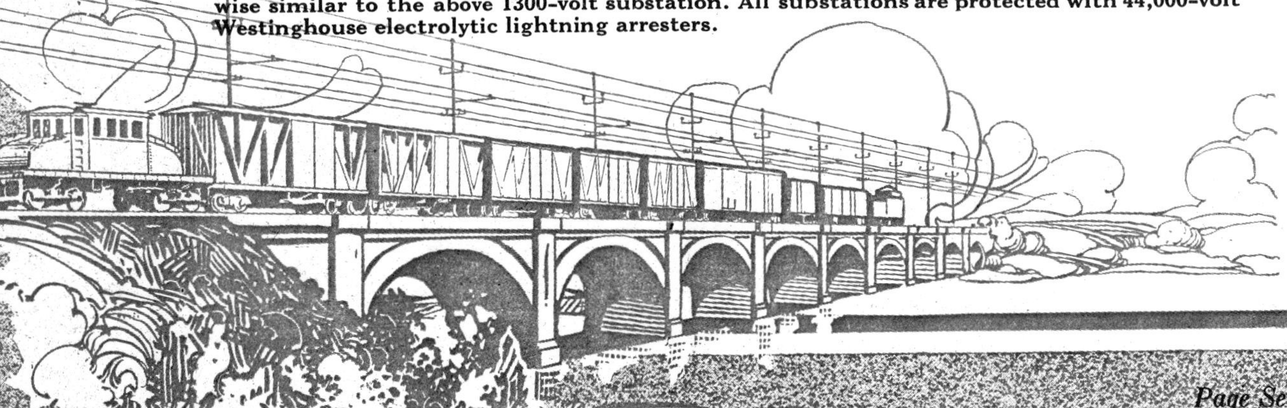

Cedar Rapids Division—Showing Cliff Cut for Right-of-Way

Right-of-Way

Line View

THE direction of this road is southeast from Waterloo and the entire line is constructed on the private right-of-way varying from 100 feet between stations to 200 feet or more, as required, for station facilities. To obtain a desirable outlet from Waterloo in constructing this line, it was necessary to extend a belt-line for three miles around the eastern limits of the city for freight traffic, one of the street railway lines serving as an entrance for passenger trains.

For a distance of between three and four miles in the line between La Porte City and Brandon practically all excavation was through limestone. There is so much of this limestone along this part of the line, that there is a stone-crusher plant located where the line runs along the Cedar River.

Cedar River Bridge near La Porte City—744 Feet Long with Abutments and Nine 78-Foot Spans

Physical Data (Cedar Rapids District)

Weight of Rail—Main and passing tracks—85 lb. A.S.C.E.

Rail Joints—Main and passing tracks—24" Continuous.

Tie Plates, Shoulder Tie Plates used on all curves.

Maximum Grade, 1% with allowance for curvature.
(Max. Grade in yards 0.5%).

Maximum Curvature, Outside of yard limits, 5° (3° except for one curve).
Inside of yard limits, 7° 30".

Road Bed Width, On Fills, 16-ft. + 0.2 per ft. of height over 10'.
In cuts, 24 ft.

Wooden Bridges—Except for two or three at Cedar Rapids, all wooden bridges are built from creosoted material, and rock ballasted.

Steel Bridges—Three such bridges in this district, all on concrete abutments. All steel bridges are designed for Cooper's E-50 loading.

Concrete Bridges—First Cedar River Bridge, near Elk Run, consists of six 70-ft. concrete arch spans on pile foundation. Cost $36,000.
Second Cedar River Bridge, near La Porte City, consists of nine 70-ft. concrete arch spans on pile foundation. Soil, sand and quicksand. Bridge cost $75,000. Two other concrete bridges in this district, each consisting of two 40-ft. arches.

All concrete bridges and culverts designed for Cooper's E-60 loading, viz. two 213-ton locomotives followed by a train weighing 6000 lbs. per lineal foot.

Galvanized iron pipe is used for small waterways, concrete box and beam top culverts for openings of medium size, and wood, steel or concrete bridges for openings where the span is over fifteen feet.

Station Layouts—Minimum length passing tracks, agency stations 2000 ft.
Minimum length Industry tracks, agency 1000 ft.

Warning Signs

THE accident hazard has been greatly reduced at crossings in the city of Waterloo and the right-of-way by use of distance warning signs of the type shown in the accompanying illustrations. These signs are placed near the curb line on the highway at a distance of 300 feet away from the track; and serve to call the attention of the travellers to the fact that they are approaching a crossing. These are an important safety feature in that the belt-line around the city of Waterloo serves the industrial districts with many branch tracks to manufacturing plants, and they are used at all highway crossings on the entire system. The Belt-Line tracks in addition to the high-speed passenger lines have a proportionately large number of highway crossings.

Crossing Gong and Sign

300-Foot Crossing Sign

Overhead Line Construction

Way Station and Industries

ALL overhead lines with the exception of the one between Waterloo and Cedar Rapids are standard direct-suspension construction; in general a No. 0000 trolley and a No. 0000 feeder are used. The line from Waterloo to Cedar Rapids is of the five-point catenary construction with 150-foot pole spacing on tangent track, the poles being of Idaho cedar, 40 feet long. The overhead construction is insulated for 1500 volts, with porcelain insulators. The 44000-volt high-tension line is spaced in delta with 52 inches between wires and is sectionalized at each sub-station with Burke horn-gap switches.

Cedar Rapids Division—A single line of 40-foot, 8-inch wood poles supports the catenary trolley suspension, telegraph and transmission lines. These poles are placed at 150-foot intervals on all track up to one-degree curve; at 120-foot intervals between one- and two-degree curve, and 100-foot intervals for curves of a shorter radius.

The catenary line construction is arranged for five-point suspension and supports No. 0000 grooved trolley from a $\frac{7}{16}$-inch galvanized S & M seven-stranded messenger.

The catenary trolley construction is carried on 10-foot mast arms provided with special end casting for "pull-overs" on curves. The use of this special end casting permits the purchase of standard mast arm for the entire system.

Tangent Construction—"Rymco" Trolley Switch on Right

A SPECIAL feature in connection with the overhead line construction is that of the method of supplying the trolley across two reinforced concrete bridges over the Cedar River. At these points tubular steel poles with eight-inch base are made in 10-inch pipe openings which were put in the concrete parapet walls when the bridge was constructed. After the trolley pole had been placed in position it was set plumb and surrounded with a sand cushion. It also was necessary to insulate the mast arms from the pole. A treatment of creosote oil has been applied to all pole bodies, and all cross arms have been treated by a similar process.

HE overhead line construction here is for 1300 volts direct current. No. 0000 grooved trolley wire, is used with $\frac{7}{16}$-inch S & M Messenger; 9—five-point catenary; flexible construction; and No. 0000 feeder on "J" brackets.

To facilitate repairs and construction of overhead lines a very novel line car has been built from an old four-motor, double-end passenger car, 40 feet 6 inches over all. The method of supporting and elevating the adjustable platform on the roof of this car is of particular interest.

The platform is approximately six feet square and is fitted with a folding rail on all sides. It is centered on a seven-inch pipe which extends down into the body of the car. The lower end of this supporting pipe is inserted in an eight-inch pipe mounted in a casting on the car floor, and the entire vertical structure is braced at the top with struts extending in four directions to the deck rail. The inner pipe has bar iron lugs which extend through slots in the surrounding pipe. Light cables attached to these lugs pass over pulleys supported on the braces near the roof of the car, thence to a winch bolted to the car floor beside the pipe platform supports. The slots in the interior pipe are long enough to permit a five-foot lift to the platform.

High Tension Telephone, Feeder and Trolley Suspension Layout. Also Right-of-Way Clearances

TO relieve the strain coming to the elevator cables when the platform is in use by repairman, two other slots near the top of the exterior pipe are fitted with small iron blocks which may be clamped against the interior pipe by means of an ordinary strap iron pipe clamp. In order to facilitate clamping and releasing of the adjustable platform in any position, this clamp is fitted with a hand wheel which is accessible from the car floor level.

Another feature of this line car equipment is the permanent reel rack mounted in the center of the car body near the elevated platform standard. The rack bearings are substantially supported on two wooden posts bolted to the floor and roof of the car body. The axle on which wire reels are mounted in this permanent rack is fitted with self-centering castings, and twelve-inch pulleys at each end permit the application of friction band brakes. Braking is obtained by way of turnbuckles attached to either end of the bands and to the floor.

A standard practice in overhead construction on this road is to string the trolley and messenger wires with current on the line, thus this car is a "hot-stringing" one. In order to accomplish this with the new line car, two twelve-inch sheaves must be in the front end of the car when doing this work, thence over the elevated platform where the linemen may tie the wire in at each pole.

Other equipment in this car includes an ordinary laundry stove for heating or cooking, a complete installation of shelves, bins, tool racks and a Johnson first-aid cabinet. A large cupboard for storing wire and a set of lockers for linemen's tools and clothing also form part of the interior fittings.

The remaining floor space in the car interior is sufficient to permit five miles of trolley or messenger wire on reels to be carried at one time. Two five-foot sliding doors, one at each end of car, just back of the bulkhead, facilitate the handling of line material from an elevated platform into the carbody.

SHOP and YARD VIEWS

GENERAL REPAIR SHOPS-WATERLOO.

SHOP YARD-SHOWING SHOPS, SUBSTATION AND STOREROOM.

MOTOR-DRIVEN SHOP DERRICK-USED FOR HANDLING MATERIAL & REMOVING MOTORS, WHEELS, ETS. FROM TRUCKS

RIP TRACK-M.C.B. INTERCHANGE REPAIRS.

RIP TRACK

In addition to the adjustable platform on the line-car roof, a 24-inch runway provided with a low pipe-guard rail extends the full length of the car body on both sides of the monitor deck. A set of two chains and hooks attached to the side of the door is so arranged that two poles may be fastened and transported over the road.

"Hot Stringing" Line Car

Repair Shops

TWENTY-THREE acres of plain farming land, at the north corporate limits of Waterloo, constitute the repair shop property.

The general equipment repairs are handled by these repair shops. The main repair shop building is in the form of a 12-stall roundhouse. Transferring to the different parts of the shop is done by a 55-foot motor driven turntable which is approached from the shop yard tracks by a single lead track. The shop building is 72 feet deep, and built on a 130-foot outside radius. This building has been arranged in three separate departments, one for blacksmith work, truck shop, and the machine and wood shop.

Snow "Fighters"

A general storeroom in another building adjoins the machine and truck shops and the armature repair shop is in a separate building adjoining the machine shop.

East Waterloo Car and Inspection Barn

Long Siding Track

The Great Northern Railway Electrification

Special Publication, 1857
September, 1929

Westinghouse Electric & Manufacturing Company
EAST PITTSBURGH, PA.

A 5000-Ton Freight Train on the Heavy Grades in the Cascade Mountains.

FOREWORD

JAMES J. HILL, founder of the Great Northern Railway, in his annaul report of 1890, said:

"When the Pacific extension has been completed, your company will have a continuous rail line from Lake Superior, St. Paul and Minneapolis to the Pacific Coast, shorter than any existing transcontinental railway and with lower grades and less curvature. . . . The policy of improving the property, which has given such good results in the past, is still being followed."

Through the efforts of his followers, the ideals and plans of James J. Hill, the Empire Builder, are being carried forward constantly and the Great Northern has continued to be the shortest and easiest line from the Great Lakes to Puget Sound. Leveling the towering Cascades was a part of such a program.

Safety, economy, and efficiency of operation over the mountain grades could best be promoted by being able to handle large trains at increased average speed. A long tunnel was found necessary in order to bring about the physical improvements in the line which were necessary and an essential feature of the operation of the long tunnel was electrification of the line.

On January 12, 1929, a Baldwin-Westinghouse locomotive pulled the first train through the new eight-mile Cascade Tunnel, the longest tunnel in the Western Hemisphere. This event also inaugurated all-electric operation between Skykomish and Wenatchee, a distance of 73 miles.

In connection with the tunnel, other improvements were made in the Cascade Mountain crossing which, together with the tunnel, shortened the route 8.9 miles, eliminated 7-1/2 miles of snow sheds and 3674 degrees of curvature, and reduced the height of the summit 502 feet.

As a result of electrification there has been developed the most powerful motor-generator type electric locomotives in the world, and an ideal unit of motive power for this operation. Passenger trains consisting of from nine to twelve cars are hauled over the mountain grades at twenty-five miles per hour; freight trains of over 5000 tons are hauled over these same grades at ten to fifteen miles per hour.

Improvements such as these have set up new standards of performance, resulting in benefits to all who are affected by the advancement of transcontinental transportation.

Ralph Budd

President
Great Northern Railway Company

The Great Northern Railway Electrification

IN the year 1890, John F. Stevens, one of America's foremost railroad engineers, located the pass over the Cascade barrier which now bears his name and laid out the survey for the right-of-way of the Great Northern Railway, which then was being pushed over the continent, through this gap. So abrupt was the rise that a series of switchbacks was required on each side of the saddle in order to surmount it with reasonable grades. At that time he appreciated the desirability for and anticipated the construction of a tunnel to eliminate these switchbacks.

Such a tunnel was constructed in 1900, being 2.63 miles long and extending from Tye on the Western side to Cascade Tunnel on the East. This tunnel did away with all of the switchbacks, shortened the distance 9 miles, reduced the maximum grade from 4 per cent to 2.2 per cent and eliminated 2332 degrees of curvature.

The First Step in Electrification

The first step taken by the Great Northern toward electrification in the Cascades was in 1909 when the tunnel from Tye to Cascade Tunnel station was electrified to eliminate the serious smoke and gas conditions which had developed with increased traffic, and to improve operating conditions for heavy freight trains through the tunnel.

The Only Example of Three-Phase Electrification in this Country

This was a three-phase, 6600-volt system with two trolley wires, the rails being used as the third conductor. The locomotives were equipped with three-phase induction motors and could be operated at two speeds, i.e., at approximately 15 miles an hour with passenger trains and light freight trains and at approximately 7½ miles an hour with heavy freight trains. The steam locomotives were not removed from the trains but were pulled through the tunnel along with the cars. The steam engines did not work in the tunnel and thus the greater part of the smoke was eliminated.

Electric power was supplied from a hydroelectric generating station operated by the railroad in Tumwater Canyon on the Wenatchee River, about 30 miles from the tunnel, on the now abandoned route of the system. Transmission was at 33,000 volts, three-phase, over a double circuit line on wood poles, with a step-down substation at Cascade Tunnel station, delivering 6600 volts, three-phase, to the trolleys. This is declared to be the only example of three-phase railroad electrification in this country. It was in operation nearly eighteen years and fulfilled the limited requirements for which it was intended.

From Three-Phase to Single-Phase Operation

Toward the end of 1925 work was commenced simultaneously on new locomotives, a substation at Skykomish and trolley construction from Skykomish to Tye. This was completed in time for the first train to run from Skykomish to Cascade Tunnel, hauled by an electric locomotive, on March 5, 1927. The change from operation with the old three-phase locomotives to operation with the new single-phase was accomplished without incident or delay. The single-phase trolley was installed between the two trolleys forming part of the old system and the pantagraphs were equipped with temporary short horns in order to clear the other wires.

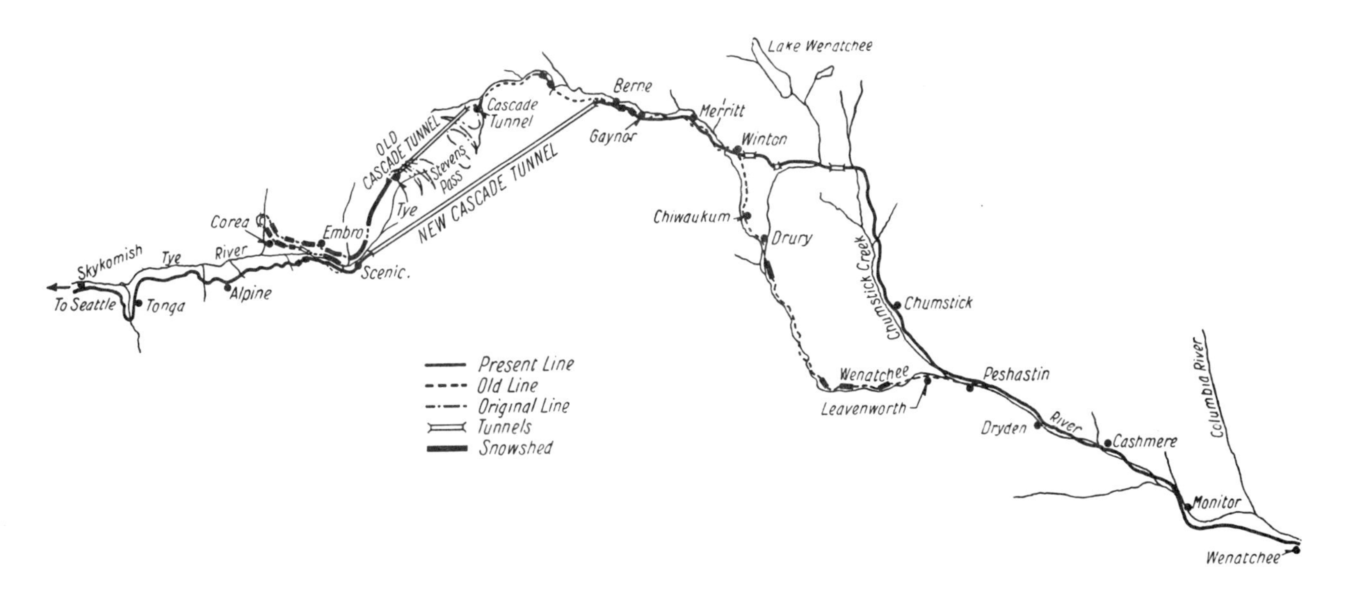

Fig. 1—Map of the Cascade Mountain Section of the Great Northern Railway Showing the Original and Now Existing Routes.

Electrification Completed between Berne and Wenatchee in 1928

All of the freight trains and the majority of passenger trains were hauled from Skykomish to Cascade Tunnel with electric locomotives until November 1928, when the trolley construction between Berne and Wenatchee was completed. During this period it was sometimes necessary to "smoke" a passenger train part way up the grade due to insufficient motive power, but it was always arranged to haul them through the old Cascade Tunnel by electric power.

After the line between Berne and Wenatchee had been electrified, a steam locomotive was used to haul the electric locomotive and train over the unelectrified gap between Berne and Cascade Tunnel and the electric locomotives hauled all passenger trains between Skykomish and Wenatchee. The steam locomotives were then cut off the train at each end of the electrified section and not hauled through as previously had been the case.

The Longest Tunnel in the Western Hemisphere

A long low-level tunnel which would eliminate practically all of the section subjected to the heavy snows and which would reduce the curvature materially, as well as the amount of heavy grade, seemed to be the answer to the problem presented by this mountain barrier.

Such a tunnel, extending from Scenic on the western slope to Berne on the eastern side, was commenced in November 1925. This tunnel shortened the distance 6.68 miles, six miles of which were covered with snow sheds, eliminated 1941 degrees of curvature and reduced the height of the summit 502 feet.

Line between Winton and Peshastin Relocated

Simultaneous with the building of the new tunnel, twenty miles of the old line between Winton and Peshastin on the east side of the mountains was relocated, shortening the distance one mile and eliminating a total curvature of 1,286 degrees. The maximum curvature was reduced from nine degrees to three degrees and the grade reduced from 2.2 per cent to 1.6 per cent. This line follows the valley of Chumstick Creek almost to its source, then goes through a tunnel back into the Wenatchee River valley.

On the completion of the new Cascade Tunnel, January 12, 1929, all-electric operation was initiated between Skykomish and Wenatchee, a distance of 73 miles.

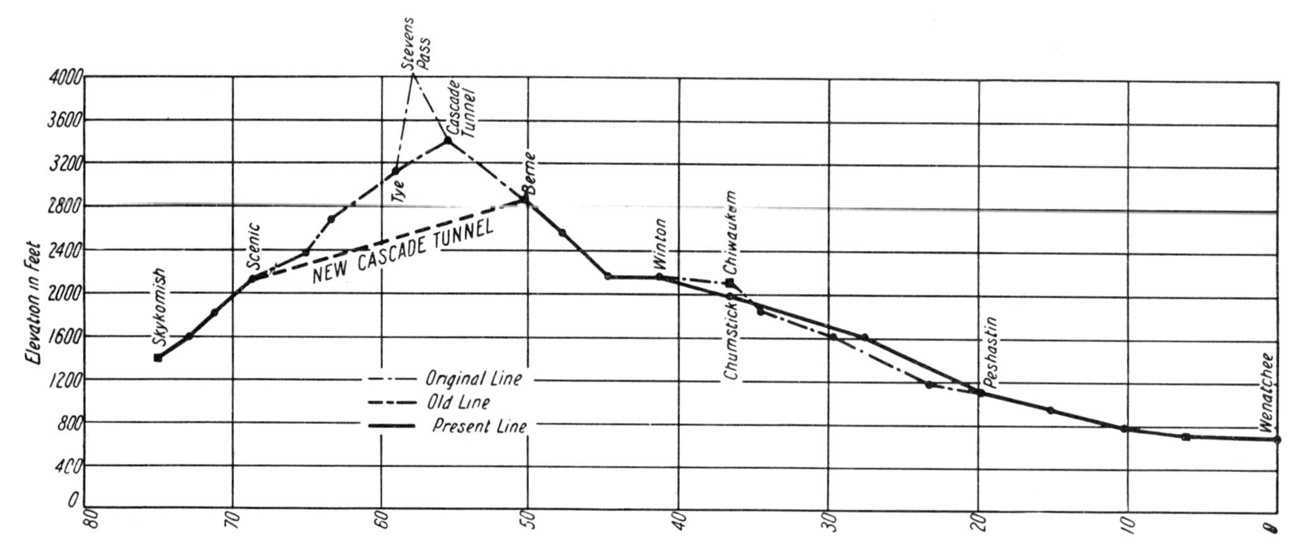

Fig. 2—Profile of the Electrified Section Showing the Reduction in Maximum Elevation Accomplished by the Successive Tunnels.

Distribution System

Power Distribution

Power Purchased from the Puget Sound Power & Light Company

POWER for the railway system is purchased from the Puget Sound Power & Light Company which in turn, leases and operates the railway company's hydro-plant at Tumwater. Power is delivered at three points: first, 110,000 volts, 60 cycles, at Skykomish (the west end of the electrification); second, 44,000 volts, 25 cycles, at Tumwater (about 30 miles west of Wenatchee); third, 110,000 volts, 60 cycles, at Wenatchee (the east end of the electrification). As the railway system has a frequency of 25 cycles, conversion of power delivered at Wenatchee and Skykomish is necessary. A map of the complete system with its related connection is shown in Figure 3.

Duplicate Frequency Converter Substations at Wenatchee and Skykomish

At Wenatchee and Skykomish, the two terminals of the electrification, frequency converter substations were erected and these two stations, which are duplicates with respect to equipment, are shown in Figure 4.

The layout of the wiring and arrangement of the equipment has been such as to provide for the installation, at a later date, of another frequency changer set in each station. The location of the frequency changer stations at the extreme ends of the present electrified section looks forward to the ultimate extension of the electrification westward toward Seattle and eastward toward Spokane.

In case of such an extension, the present stations would be in the most advantageous locations to supply the required power.

At these points 60-cycle energy is received at 110,000 volts from the power company's system. This energy is reduced to 13,200 volts and applied to the synchronous motor of the 7500 kv-a. frequency converter set. This motor is direct-connected to a single-phase generator which delivers 25-cycle single-phase energy at 13,200 volts through a three winding transformer to the 11,500-volt trolley system, and also to the 44,000-volt transmission system.

From the 44,000-volt bus at Skykomish substation, two single-phase transmission lines are run to the Tumwater hydro-station, and from there to the Wenatchee substation, both circuits being tapped at each trolley distribution station as shown in Figure 4. These two transmission circuits consist of 0000 stranded copper conductors mounted on pin-type insulators and are both carried on the same single-pole line, except from Scenic to Cascade Tunnel, over the Stevens Pass where snow fall conditions made it advisable to use heavy H-frame structures, suitably guyed; and suspension, instead of pin-type insulators.

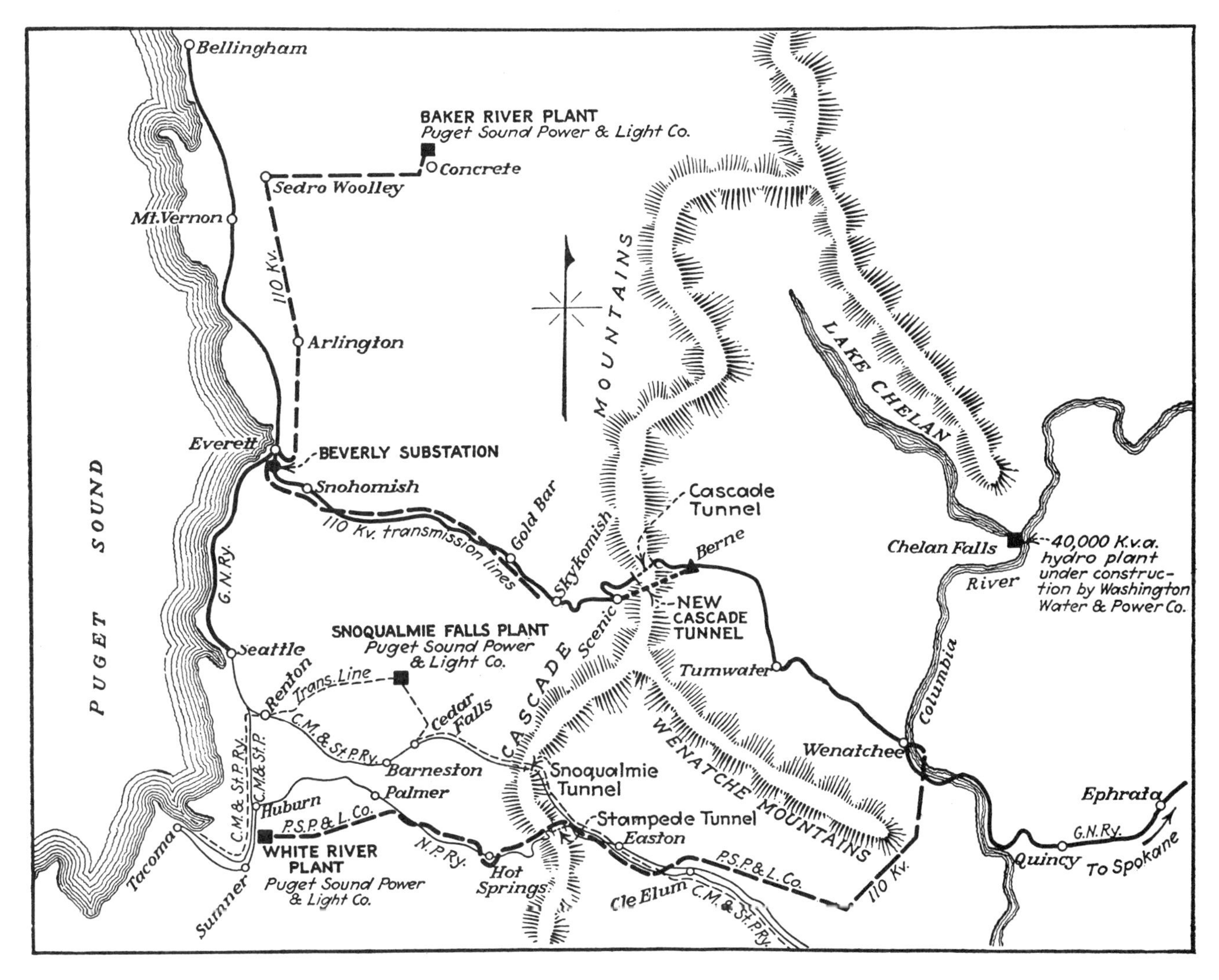

Fig. 3—Map of the Power Supply for the Electrified Section.

Trolley Distribution Stations Present Many Unique Features

The trolley distribution stations are simple in design and operation, and present many unique features. These stations are located directly opposite the depots at each location, with the exception of Cashmere, where the station is adjacent to the depot. This substation is shown in Figure 6. In the telegrapher's room of the depot is located a switchboard panel on which are mounted the oil circuit-breaker control switches, relays, and indicating instruments. A 60-cell sealed glass jar type storage battery is located in a small room in the depot and is kept charged by a bulb-type rectifier. A bell-alarm system and lamp indication is provided for the oil circuit-breaker control switches.

The relay system for the distribution stations was made as simple as possible without sacrificing reliability (Figure 5). The CZ type of impedance relay is used on the trolley feeders. This type relay, marked "A", operating with Westinghouse type O-221 oil circuit-breakers, which have a high rupturing capacity, clears the trolley rapidly in case of contact line trouble.

Balanced power relaying is provided on the 44,000 volt transmission lines, which normally operate in parallel.

From the schematic diagram (Figure 5), it is obvious that under normal operation the secondary currents from the current transformers located in the 44-kv. oil circuit-breaker leads, are circulating through both current transformer secondaries and no current flows through the current coils of relay B, which is of the duo-directional overcurrent type. In case a short cir-

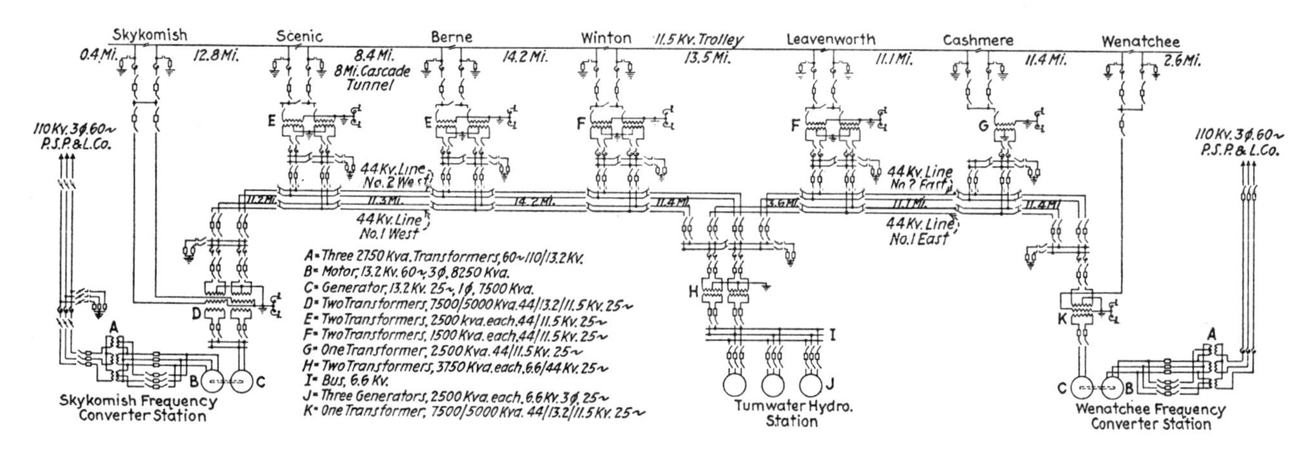

Fig. 4—Schematic Wiring Diagram of Electrification Circuits.

cuit occurs on one of the transmission lines between distribution stations, the secondary currents from the current transformers are in opposition and operate the overcurrent element of the relay, the directional element selecting the proper 44-kv. breaker to clear the faulty line from that station bus.

For line-to-ground faults, which constitute the majority of transmission line failures, as all insulator pins on the transmission system are grounded, the same operation is had with relay C (Figure 5), which is of the duo-directional over current type, the customary potential coil of the directional element being replaced by a current coil which is actuated from the secondaries of current transformers placed in the mid-point ground lead of the high-voltage windings of the power distribution transformers. It is apparent, then, that for a fault on the transmission lines, selective breaker operation is obtained for the faulty line, first, at the distribution stations adjacent to the fault, and then in sequence to the sources of power in either direction from the fault.

Each distribution station is also supplied with differential protection and temperature protection on the transformers, the latter consisting of contact-making thermometers which operate an auxiliary relay to open the oil circuit-breakers and take the transformers out of service when a predetermined temperature is reached.

Load Dispatcher Located at Skykomish Substation

The operation of the system is controlled by a load dispatcher located at Skykomish substation. He has direct communication with the telegraph operators at the distribution substations, and they do the switching operations under the load dispatcher's direction. Another function of the load dispatcher is to control the incoming and outgoing power between the railway system and power company system.

Tumwater Hydro Station

By reference to Figure 3 it will be seen that Tumwater hydro-station is between Skykomish substation

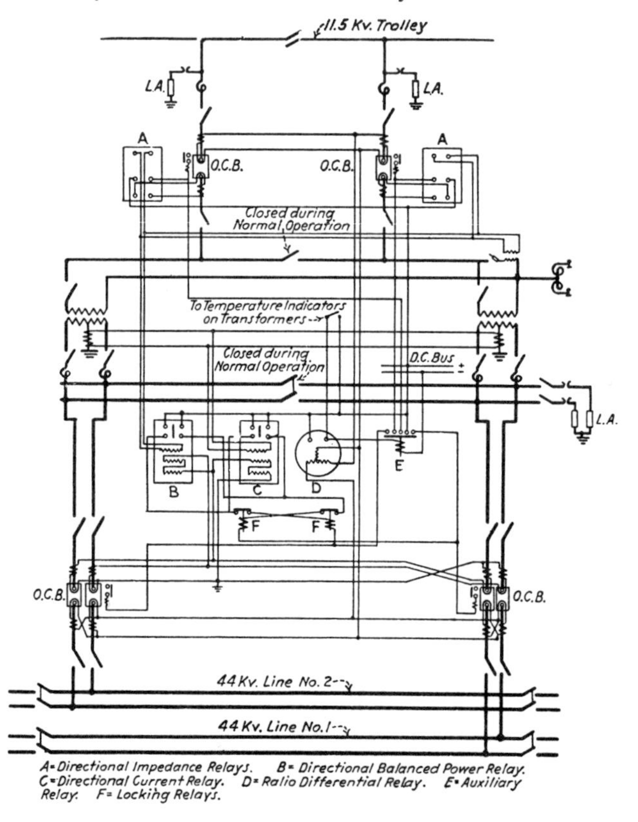

Fig. 5—Schematic Wiring Diagram of Distribution Substation

Fig. 6—Frequency Converter Substation Located at Skykomish.

and Wenatchee substation where, as before stated, power is received from two different 110,000-volt lines of the power company's system. Tumwater hydro-station generates full capacity (5500 kv-a.) at all times, water conditions permitting, and the total output goes into the railway system, where it is absorbed by train operation, or in event of light train movement the surplus is fed back to the power company's system through the converter sets at either Skykomish substation or Wenatchee substation. As the 60-cycle end of the converter sets are provided with stator shifting devices, the direction and magnitude of incoming or outgoing power between the two systems may be adjusted readily at both Skykomish and Wenatchee substations.

From the 25-cycle generating capacities of Skykomish and Wenatchee substations and Tumwater hydro-station, it is seen that the available power supply is approximately 20,000 kv-a. which is about 5,000 kv-a. in excess of the present traffic demands. Both Skykomish and Wenatchee substations were designed for the addition of second frequency converter units if additional capacity is required in the future. The capacity of the transmission and trolley system is sufficient to handle train movements at present, with the output of Tumwater hydro-station and either Skykomish or Wenatchee substations, providing that train movements are given careful dispatching.

Adequate Supply of Power Available at all Times

The 110,000-volt transmission line feeding the railway system at Skykomish substation comes from that part of the Puget Sound system which is supplied principally by the hydro-station at Baker River. The Wenatchee substation is supplied from the power company's system extending from the White River hydro-plant, which is also tied in with the Washington Water Power Company's new Lake Chelan hydro-station. From this it is seen that an adequate supply of power is available at all times, and a complete shut-down is very improbable.

From the foregoing descriptions and diagrams, it is evident that all the essentials of a power system for main line electrification have been met. The feed-in to the system from both ends and the center, together with two transmission circuits parallelling the trolley system, make for the utmost reliability. The type

Fig. 7—Track Substation at Cashmere.

of distribution stations, together with the method of operation and load dispatching, facilitates switching and the rapid restoration of service in event of line equipment failures. Simplicity has been maintained throughout without sacrificing in any way the reliability of the system.

Overhead Construction

Simple Inclined Catenary System Is Standard

THE contact system employed represents the application of a much-discussed type of overhead construction. The success of the initial construction between Cascade and Skykomish, where there is an unusual amount of curvature, much of which is ten degrees, has demonstrated the many economic advantages and satisfactory operating qualities of the simple inclined catenary system as furnished by the Westinghouse Electric and Manufacturing Company and employed as standard on this electrification.

The system used, which is illustrated in Figures 8, 9 and 10, comprises a 0000 cadmium-bronze contact wire, supported by a composite messenger consisting of a bronze or red brass high-strength core surrounded by pure copper strand. The total equivalent conductivity is approximately 450,000 c.m. Details of wires and cables used in the catenary system are given in Figure 11. Hanger fittings and clamps to the messenger and contact wires are non-ferrous. On tangent track the contact wire is supported by flexible loop hangers, the messenger and contact wire being solidly connected at 300-foot intervals. On curved track up to 3½ degrees, pure inclined catenary is employed which permits the elimination of all pull-off devices other than the hangers. On curves of greater degree than this, pull-off yokes are installed at the supporting poles, such curves employing catenary having arcs of 2½, 3, 3½ and 4 degrees curvature with a maximum offset of 6 inches at the center of the span and 9 inches at the support.

On tangent track, 150-foot pole spacing is standard where the 44-kv. circuits are carried. The pole spacing between Winton and Leavenworth, where the transmission lines are not carried, is 180 feet. On curved track, the pole spacing is reduced on curves sharper than six degrees to suit a 6-inch and 9-inch maximum offset of the contact wire. Where possible, the contact wire height from the rail is retained at 24 feet. Through tunnels, and the limited number of snow sheds now existing, this height is considerably reduced, the minimum height being 19 feet.

Bracket arms on wood poles are used to support the catenary system on single track construction, both tangent and curve. For more than one track, cross spans are used. The bracket arms are made up of two galvanized angle irons placed back to back and tied to the pole by means of a ⅝-inch steel tension rod. Standard bracket arm lengths are 7 feet 6 inches and 9 feet 6 inches on curves and 11 feet 6 inches on tangent track.

Fig. 8—Catenary Construction Using Suspension Type Insulators with Transmission Carried on Trolley Pole.

Fig. 9—Catenary Construction Using Pin Type Insulators.

On tangent track the catenary is supported on pin-type insulators, on curves and in cross span construction three standard ten-inch suspension-type insulators are used. The insulators are attached to the bracket arm through a triangular yoke on curves. This yoke transfers the curve pull of the catenary, as well as the weight, directly to the bracket arm and keeps the insulators at the desired distance below the arm to provide proper clearance.

Wood poles are used for supports in the open except on bridges and in some yard locations where guys could not be used. On bridges, expanded-steel poles are used, and in Appleyard and Cashmere yards, rolled H-section poles are used, to support cross spans without back guys.

Siemens-Martin galvanized steel strand is used for cross spans and back guys. Anchors are concrete discs or treated wood logs with galvanized anchor rod.

Messenger and contact wires are anchored and sectionalized at approximately two mile intervals. The air sections have jumpers except at distribution stations. In anchoring and dead-ending messenger and contact wires extra high-strength insulators are used. These insulators have an ultimate mechanical strength of 27,000 pounds and are recommended for working loads up to 9,000 pounds.

On sidings and in yards, inclined catenary is used, although some chord construction is used where special conditions made it advisable. At turnouts, spreaders are used to connect the contact wires and the contact

Fig. 10—Catenary Construction with Cross Span Covering Two Tracks.

wires are arranged so that they do not cross at any place. They are kept at the same level and a very simple and reliable turnout construction results.

The transmission line is a two-circuit, single-phase, 44-kv. line of 0000 copper conductors supported on pin-type insulators on wood cross arms. In addition to the power transmission line, there is a single-phase, 13-kv. line which supplies power for the signals and in many places there are several signal secondary wires on the poles.

In the construction of the new eight-mile tunnel, provision was made by means of inserts at 75-foot intervals to provide ample suspension insulator clearance so that, if necessary, a contact voltage of 22,000 volts eventually might be employed. By means of the inserts, maximum insulating distances are obtained without encroaching on the loading clearance of the tunnel. At suitable intervals, longitudinal troughs have been arranged in the tunnel for anchoring the messenger and contact wires so as to limit as much as possible any disarrangement of the catenary system.

The tensions in the contact and catenary wires are so related that, with inclined catenary, all hangers are parallel. This permits an easy mathematical analysis of the catenary system and is, no doubt, in a large measure, an explanation of the very perfect geometrical arrangement of the wires which resulted on this construction. It is interesting to note that on the 16 miles of line extending between Winton and Peshastin, where there is a considerable amount of curved track which does not, however, exceed three degrees—and where the structure spacing is practically uniform, there are no steady or pull-off devices attached to the contact wire on the main line. Figs. 8 and 9 explain the construction on this section of line, Fig. 10 representing cross-span construction at a siding.

With the excellent native wood poles used on this construction and with the well-aligned and simple system employed, the contact system on the Great Northern Railway represents quite a contribution to the art of contact system construction as applied to main line steam railways.

Size of Wire	Composition	Weight per Foot (lb.)	Breaking Strength (lb.)	Where Used	Normal Tension (lb.)	Equivalent Conductivity c.m.	Per cent Conductivity of Pure Copper
19—7	12 Copper 7 High-Strength Bronze	1.221	23000	Main Line Messenger	4700	261100	63
19—7	12 Copper 7 Red Brass	1.210	22000	Main Line Messenger	4700	287200	72
0000	Hitenso "C"	.641	12000	Main Line Contact Wire	2500	116500	55
0000	Hitenso "A"	.641	10300	Main Line Contact Wire	2500	170000	80
000	Hitenso "C"	.509	10000	Yard and Siding Contact Wire	2000	92000	55
0000	Copper	.641	7700	Tunnel Contact Wire	2500	211600	97
3/8-inch	High-Strength Bronze	.338	9000	Yard and Siding Contact Wire	1380	16830	15
0000	Cadmium Bronze Strand	.653	12460	Messenger	2500	162700	77
300000 c.m.	Copper Strand	.926	14160	Tunnel Messenger	3200	211600	97

Fig. 11—Table Showing Properties of Materials Used in Overhead Construction.

Locomotives

Fig. 12—Two Motive Power Units.

THE operating conditions on the Great Northern Railway required an electric locomotive which would be flexible enough to meet the limitations of the power supply system, and at the same time to give satisfactory train operation. It was necessary also to provide a locomotive which would be able to exert its rated tractive effort while remaining stationary without injury to the equipment, and which would brake by regeneration when descending the grade.

The Most Powerful Motor-Generator Type Locomotive in the World

The motor-generator type of locomotive (Figure 12) was chosen to meet these conditions. This locomotive combines low-voltage direct-current series traction motors with a high-voltage single-phase alternating-current power transmission system.

Each of the Baldwin-Westinghouse motive power units of the latest locomotive supplied has the following weights and dimensions:

*Total weight	368,600 lb.
Classification of wheels	$1\text{-}D_o\text{-}1$
*Weight on drivers	282,700 lb.
Number of driving axles	4
Number of idle axles	2
Total weight on idle trucks	85,900
Capacity at one-hour rating	2,165 hp.
Starting tractive effort	68,700 lb.
Maximum starting effort based on 50% adhesion (Limited by adhesion)	141,500 lb.
Tractive effort-continuous rating	44,250 lb.
Speed-continuous rating	14 mph.
*Maximum speed	45 mph.
Track gauge	4 ft., 8½ in.
Total wheel base	31 ft., 5 in.
Rigid wheel base	16 ft., 9 in.
Length overall between faces of buffers	47 ft., 2 in.
Width overall	11 ft., 0 in.
Height from rail to locked down position of pantagraph	15 ft., 10 in.
Diameter of driving wheels	56 in.
Voltage and type of conductor	11,000 volts—25-cycle, overhead
Number and type of motors	4 type 356
Method of drive	Double flexible gear
Gear ratio	18:91
Type of control	Electro-pneumatic HBFR
Years placed in service	1927, 1928, 1929
Number of motive power units	10

*These data apply to motive power units 5000, 5001, 5002, 5003, 5008-A and 5008-B. Motive power units 5004-A, 5004-B, 5006-A and 5006-B have a total weight of 357,500 lb., of which 274,800 lb. are on drivers and have a maximum speed of 38 mph.

High-Voltage Alternating-Current from the Trolley Wire Is Converted to Low-Voltage Direct-Current on the Locomotive

In this type of alternating-current locomotive, the current collected from the single wire trolley by pantagraphs is conducted through an oil circuit-breaker, shown in Figure 18, to the primary of an air blast transformer. The secondary of this transformer supplies the stator of the synchronous motor, rated at 2100 hp., 750 rpm., which drives the main generator supplying 1500 kw. at 600 volts direct-current to the traction motors. This will be clear from a consideration of the diagram of connections, Figure 14.

The control of the locomotive is accomplished by varying the voltage applied to the traction motors, which are permanently connected in parallel across the main generator armature. To reduce the number of steps required in the generator field resistance, a differential series field winding was introduced in the construction of the main generator. This series field, while having very little effect at normal voltage, is very effective at starting and makes the control extremely smooth.

The steps in the main generator shunt field resistor were designed to give equal increments of tractive effort. This increment is twenty per cent of the tractive effort at twenty-five per cent adhesive factor. At any lower value of adhesion the increments would be less than twenty per cent and at greater values slightly more, due to the shape of the curves. When the voltage of the main generator has been so increased as to approach full load on the synchronous motor, further increases in generator voltage are proportioned to give gradually decreasing increments of power in order to permit approaching closely the pull-out point of the synchronous motor.

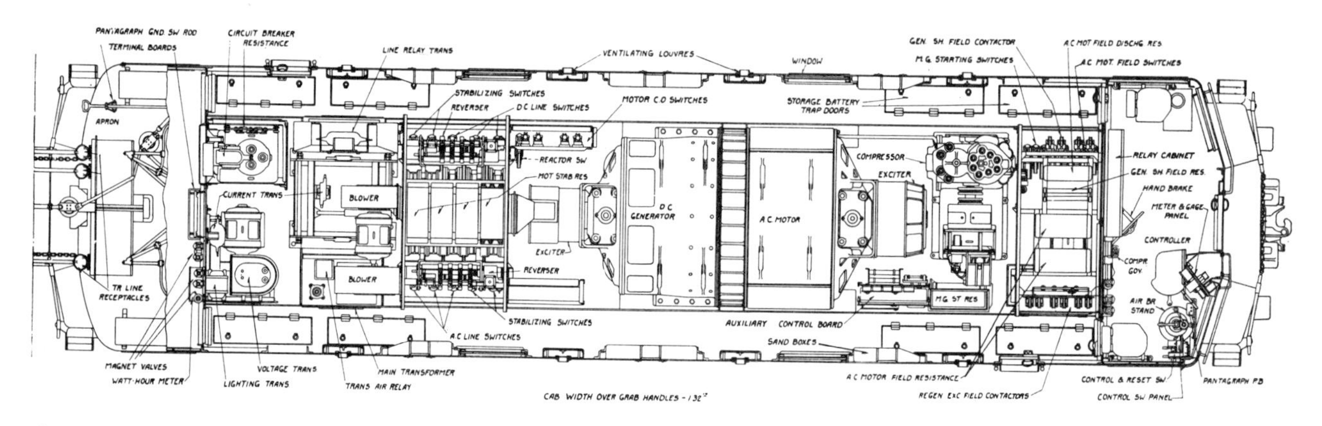

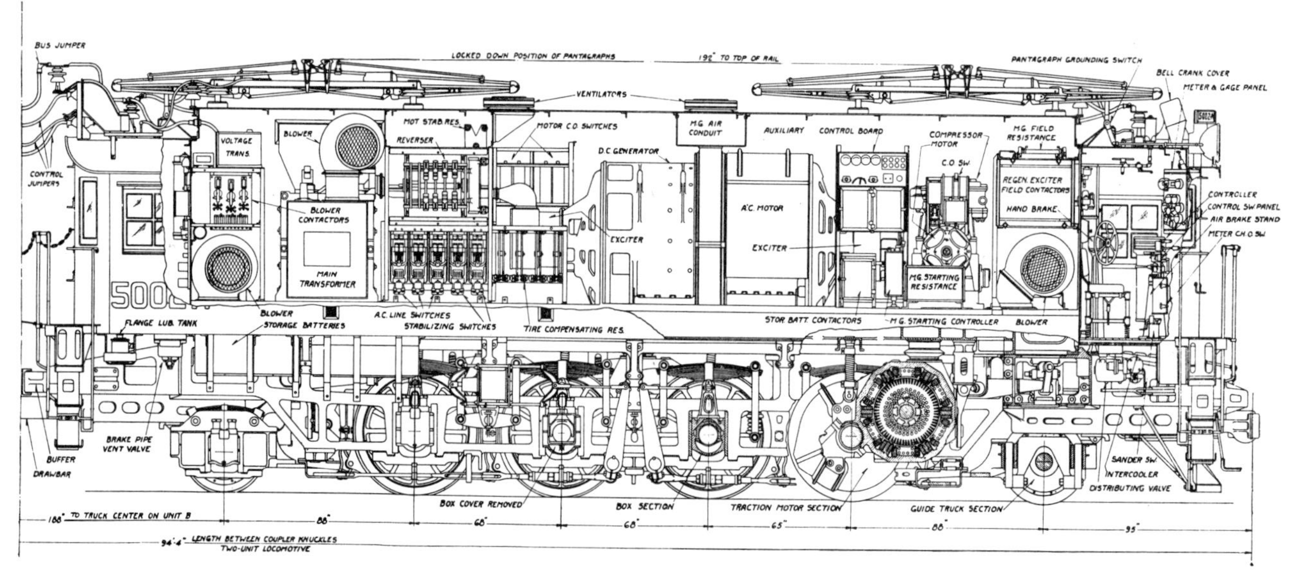

Fig. 13—Plan and Elevation of Motor-Generator Type Locomotive.

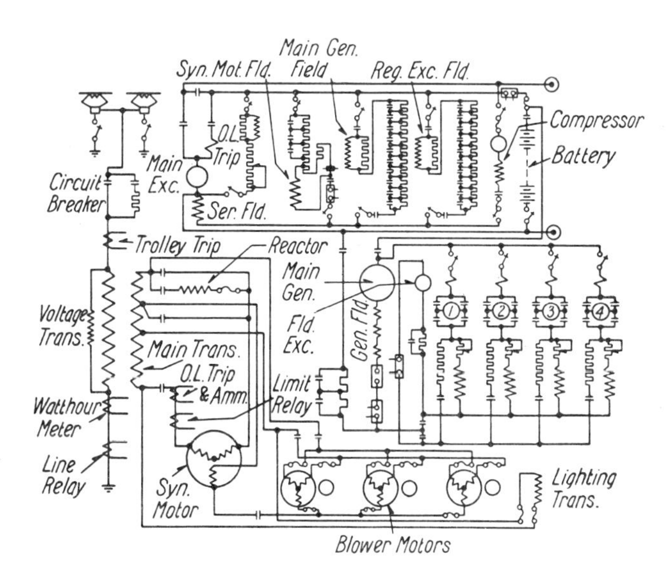

Fig. 14—Schematic Wiring Diagram of Main Circuits.

The external resistance in series with the main generator shunt field is arranged in ten steps, each shunted by a magnetic contactor, the value of these resistances varying in geometric progression making it possible to obtain the exact theoretical value desired for any number of steps by a proper combination of these ten contactors without resorting to any compromise values. This fact, combined with the time lag occasioned by the building up of the generator field, makes acceleration exceedingly uniform even when the locomotive is operating without a trailing load.

The four traction motors of each unit are permanently connected in parallel across the armature of the main generator and operate as simple series motors until full voltage has been applied to their terminals. Provided the trailing load does not require the locomotive to exert its rated tractive effort, the speed may be further increased by separately exciting the fields of the traction motors from a special field exciter, and weakening the field until the desired speed is obtained or the rated horsepower reached. This increase in speed is accompanied, of course, by a corresponding decrease in tractive effort.

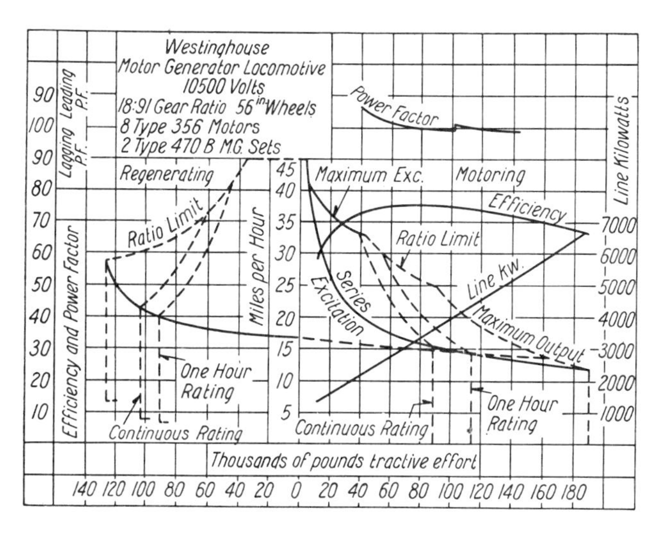

Fig. 15—Locomotive Characteristic Curve.

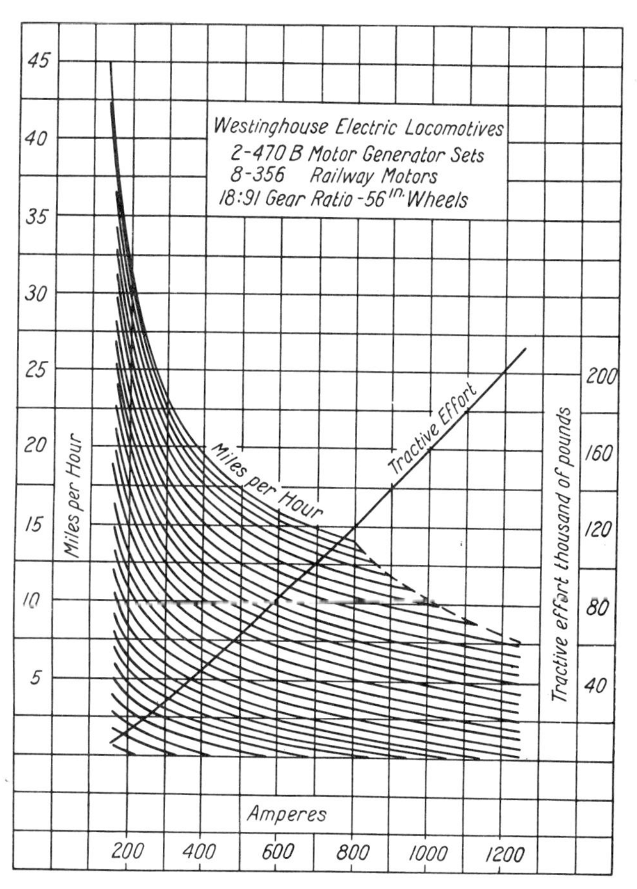

Fig. 16—Notching Curve with the Motors Series Excited.

The transition from a series excited to a separately excited traction motor can be made at any speed without any change in tractive effort, by first placing the armature of the field exciter in series with the traction motor fields and adjusting its voltage to be equal and in opposition to the voltage drop through the fields, and then closing the circuit around the fields. This circuit contains a resistor to stabilize the circuit.

The effect of this connection is to permit the locomotive to exert a constant tractive effort limited only by adhesion at any speed up to the point where it is exerting its full power, and to maintain this full horsepower constant almost to the maximum permissible speed. A consideration of the curves shown in Figures 15, 16 and 17 will make this point clear. The increments in traction motor field excitation are small, permitting the locomotive to be operated at its rated power output when using the separate excitation connections.

Should it be necessary to return to the series excited motor connections when operating separately excited, it may be done at any time without change in tractive effort by adjusting the field and armature current so as to be equal, and then opening the shunt circuit which under this condition is carrying no current. This transition is accomplished manually by the operator, meters being provided to indicate when a condition of balance is attained. Referring again to Figure 14, the changes in connections will be clear.

The field strength of the main generator and field exciter are controlled by the engine man from the master controller Figure 22. To allow compensation for differing motor characteristics or unequal tire diameters which would produce unequal load of the individual motors when operating separately excited, an adjustable resistor is installed in the field circuit of each motor.

In addition to the main generator and field exciter, the set includes the main exciter, a 75-kw., 125-volt generator, which supplies power for the excitation of all machines, charges the storage batteries, drives the compressor for the air brake system, and supplies power for the control and emergency lights.

The switches used to short circuit sections of the field resistor are magnetic contactors of 75-ampere capacity, while those in the field of the field exciter are magnetic contactors with a capacity of 25 amperes. All switches above 75-ampere capacity are electro-pneumatically operated, three sizes being used: 350, 1500 and 3000 amperes. The reversal of the traction motors is accomplished by two double motor cam type reversers. Except under emergency the main 3000-ampere switches are not called upon to open the main circuit which reduces arcing at the contacts to a minimum.

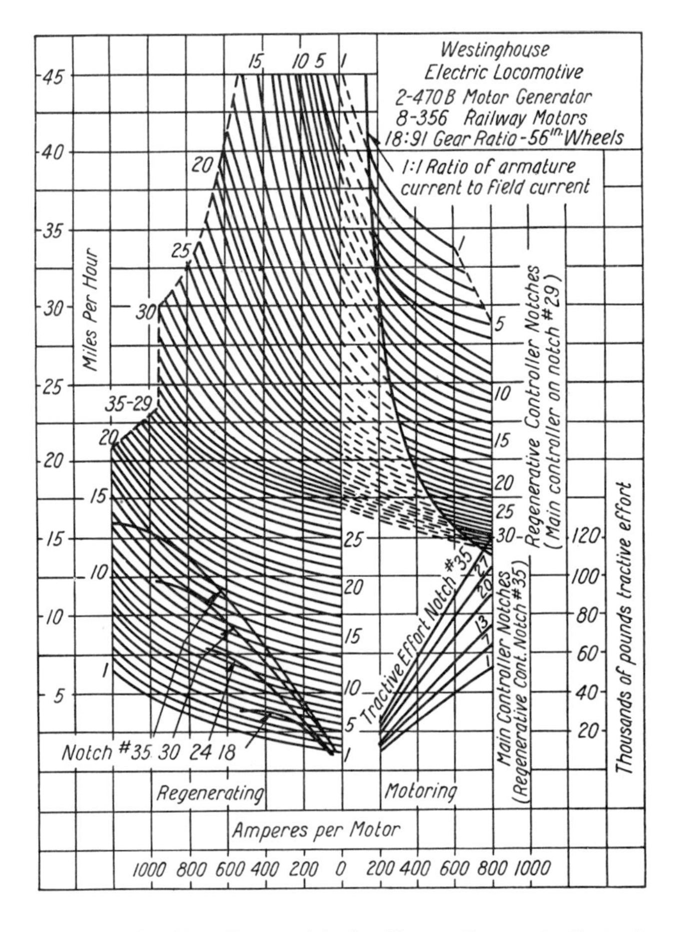

Fig. 17—Notching Curve with the Motors Separately Excited.

Auxiliaries Operated Directly from the Transformer

Insofar as is practicable, the auxiliaries are operated directly from the transformer. The principal of these auxiliaries, the blower motors for ventilating the traction motors and the transformer, are single-phase induction motors which are started by first connecting to a three-phase supply.

To obtain three-phase power for starting the blower motors the synchronous motor stator is equipped with a winding producing a voltage ninety electrical degrees out of phase with the voltage across the transformer secondary, and of such a value that when correctly inter-connected with the main transformer three-phase power is produced. These blower motors are started only when the synchronous motor is operating without load and as soon as load is applied to the synchronous motor the phase connected to that motor is opened leaving the blower motors operating single-phase directly from the transformer.

Fig. 18—Oil Circuit-Breaker.

A Storage Battery Is Used in Starting

The single-phase synchronous motor not being inherently self-starting, a storage battery is provided to supply power to the main generator which functions as a motor until the set is rotating at approximately one-third synchronous speed. At this speed the battery circuit is opened by the first synchronizing relay and the synchronous motor stator connected to a tap of the transformer giving roughly one-half normal voltage. The amortisseur winding on the synchronous motor field structure then causes the motor to act as a single-phase induction motor to accelerate the set to nearly synchronous speed.

The synchronizing relays, two in number, are connected across the armature of the main exciter, the field of which is connected across the battery. The voltage generated by the main exciter is, therefore, directly proportional to the speed of the rotor. The relays are single coil relays functioning on voltage only. The second synchronizing relay applies 30 per cent of full load excitation to the synchronous motor field when the set reaches 90 per cent synchronism.

The application of this low value of excitation insures that, as the set pulls into synchronism, due to the salient pole construction of the field, it will lock in step with the field poles of the correct polarity and will not have to slip a pole as the field excitation is increased to normal. Applying the field before the rotor is in synchronism causes the power taken by the stator to fluctuate approximately ten per cent.

Should the set come into step in such a manner that it had to slip a pole when excited, a rather heavy momentary current demand is occasioned. The closing of the second synchronizing relay starts the functioning of an air operated time element relay which, in eight seconds, by means of a reactor, transfers the connection of the stator from the low voltage tap on the transformer to full voltage without opening the circuit, and after this transition is completed applies normal no load excitation to the field of the synchronous motor. The eight-second interval allows sufficient time for the set to accelerate from 90 per cent of synchronism to synchronous speed. This completes the starting sequence and at this time the exciter is connected to the bus from which the battery is charged and all machines receive excitation from the main exciter.

The power factor at which this locomotive operates is kept at or near unity by increasing the excitation of the synchronous motor when the load increases to predetermined values. The first increase is made as the controller is advanced and the second by a relay as the set becomes overloaded.

Overload Protection Provided

Overload protection is provided by relays for the main exciter, the traction motors, and the synchronous motor. Appreciating the seriousness of the disturbance caused on the line if one of the synchronous motors was to be pulled out of synchronism with full load applied, several precautions were taken to prevent this occurrence.

First, when the load upon the synchronous motor reaches ten per cent above rated load a relay operated by stator current functions and reduces the main generator voltage ten per cent, thereby reducing the load ten per cent. The engineman can only regain full voltage by manually reducing the voltage by the same amount, when the relays are automatically reset.

Second, if the engineman permits the traction motor armature current to increase further until the synchronous motor is again overloaded ten per cent, the main generator voltage would not be further reduced, but instead the field excitation of the synchronous motor would be increased to the maximum available or 120 per cent of the normal full load value, which has the effect of increasing the pull-out torque from a value of 160 per cent to approximately 200 per cent of its full load value.

If, in spite of the above precautions, the load should reach the pull-out point or if, for any other reason, the set should be pulled out of synchronism with normal full load excitation, the design of the synchronous motor and the main transformer is such that the combined reactance will limit the instantaneous maximum current to approximately six times full load current.

One of the problems made more troublesome by the presence of synchronous machinery in the locomotive is that of properly caring for momentary loss of contact with the overhead trolley wire. While this loss of contact is made less serious by the limited power demand, it is guarded against as follows: first, by using two pantagraphs in contact with the trolley wire at all times, either one of which has sufficient collecting capacity for both units and which are connected together by means of a high tension bus line, and second, in addition to this, a relay is used functioning on the current taken by the high tension side of the main transformer.

When this current decreases to zero, during loss of contact or other interruption of power supply, the relay opens, disconnecting the load from the main generator and also disconnecting the synchronous motor from the transformer. When power is again supplied, this same relay closes connecting the stator of the synchronous motor to the starting tap of the main transformer at which time the regular starting sequence takes place as previously described.

Mechanical Features are Many

The wheel arrangement of each unit is 1-Do-1. The rigid wheelbase has been held to 16 feet 9 inches by a special arrangement of the traction motors. The driving wheels are 56 inches in diameter, all fitted with flanged tires four inches thick, shrunk on. The truck wheels are 36 inches in diameter with the tires shrunk and bolted in place.

The air brake system uses four brake cylinders, 12 inches in diameter with 10-inch stroke, and a hand brake is included for holding the idle locomotive only after it has been brought to rest.

The frame of the locomotive, of bar steel construction, supports a cast steel bedplate made in the form of two box like beams, with cross ties at intervals, on which are mounted all of the heavy pieces of equipment, such as the motor-generator set, the transformer, compressor, blower motors, etc. The box like beams form air ducts into which the traction motor blowers discharge and which in turn are connected to the traction motors. The cab is built integral with the frame and is arranged with the removable roof in two sections allowing any of the principal equipment to be placed and removed with an overhead crane. An operating compartment is provided at each end of the cab, although only one of these compartments is equipped with control equipment at present. The two motive power units are connected at the inside ends by means of a drawbar, supplemented by a safety bar.

Electrical Equipment Is Conveniently Arranged

The transformer, of the familiar railway air blast type, is cooled by a blower mounted on the top of the case discharging directly downward into the high tension end. The air after passing through the air ducts in the transformer is discharged outside the cab through the side of the clere-story. A picture of the transformer with its individual blower is shown in Figure 19.

Fig. 19—Main Transformer.

The motor-generator set is a self-contained unit, the rotor of which is carried on two self aligning bracket type bearings which have oil-ring lubrication. Located on the shaft between the motor field and generator armature is mounted a double inlet turbo-type exhaust fan which draws cooling air from both ends through the machines and discharges it through the clere-story roof and through the cab floor. The main exciter and field exciter are self cooling and are overhung on the bearing brackets at the motor and generator ends, respectively. This arrangement can be clearly seen in Figure 20. The outputs of these machines are as follows:

Main Generator	1500 kw., 600 volts, 2500 amps.
Main Exciter	75 kw., 125 volts, 600 amps.
Regenerative Exciter	25 kw., 10 volts, 2500 amps.

Fig. 21—Traction Motor.

Fig. 20—Motor-Generator Set.

Each of the four driving axles carries an axle hung, series, direct-current traction motor of 540 hp. as shown in Figure 21, receiving power at 600 volts from the main generator. These traction motors are connected through a 5¼-inch face spur pinion at each end to two flexible gears of the coil spring type, the gear ratio being 18/91 or 5.05 to 1. These motors are six-pole machines employing multiple-wound armatures and six brushholders. Instead of the conventional nose support, a link is used and the center of the motor is raised above the center of the axle in order to retain the short rigid wheelbase.

The compressor is driven by a 25-hp., direct-current series motor deriving power from the main exciter, when the motor-generator set is running, and from the storage battery, when the set is shut down. The storage battery furnishes power to the compressor motor to provide air pressure for raising the pantagraph and supplying control air when the locomotive has been standing for an extended period of time.

The traction motor blowers deliver approximately 5500 cubic feet of air per minute at a static pressure equal to six inches of water. The transformer blower is required to deliver 4600 cubic feet of air at three inches of water static pressure. In addition to the transformer blower, this motor drives a small blower which supplies air to ventilate the compartment housing the stabilizing grids, which requires about 1200 cubic feet of air per minute at one and one-half inches of water static pressure.

The storage battery is divided into eight trays each holding seven cells. Two trays are mounted under the aisle floor at each of the four corners of the cab

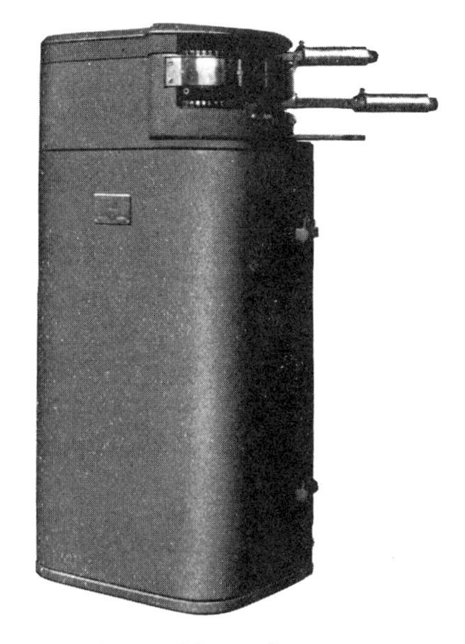

Fig. 22—Master Controller.

with trap doors in the floor for inspection and adding water. This battery has a fully charged capacity of 238 amperehours when discharged at the six-hour rate and is floated across the main exciter, no relay or amperehour meter being employed.

The total power used by the locomotive is recorded by a watthour meter, the voltage coils of which are connected to a dry type potential transformer with a ratio of 11,000/100 volts. The current coils are energized by a current transformer in the ground lead, resulting in accurate power measurements. The potential transformer is also connected to a voltmeter in front of the engineman, indicating the trolley voltage.

In locating this equipment within the locomotive cab, ease of maintenance has received foremost consideration while at the same time the apparatus is placed to facilitate installation and removal, to insure minimum length of connections and to obtain perfect weight distribution. The arrangement of apparatus is illustrated in Figure 13.

Under conditions of limited power, this type of locomotive has proven ideal permitting a freight train under limited speed to be on the line at the same time as a passenger train with no limit in its power demand. The operation of electric and steam locomotives in the same train has proved perfectly feasible with these engines, since it is possible for the electric engineman to adjust his speed to that at which the steam engine is designed to operate most efficiently.

Operation

OPERATION on the original line over Steven's Pass was slow and laborious under the most favorable conditions, but the heavy snowfall on this section of the line which was not protected by snowsheds made the task a formidable one. For the trip over the switchbacks, a steam locomotive was coupled to the rear end of the train and the engineman of this locomotive controlled the train when it was moving backward in its progress over the different levels. The trains were necessarily light and the frequent stops required made progress extremely slow.

Steam Locomotives First Operated Through Old Cascade Tunnel in 1900

Service through the first tunnel, now known as the Old Cascade Tunnel, was begun in 1900 using steam locomotives. Some time was required to clear the tunnel of these gasses before a following train could be allowed to enter. This limiting of the capacity of the tunnel, together with the ever-existing possibilities of the enginemen being overcome led to the electrification of the tunnel between Tye and Cascade Tunnel.

Traffic had increased up to the time the decision was made to extend the electrification to Skykomish until train movement on the Cascade division consisted normally of three passenger trains east and west each day, and from two to four full tonnage freight trains eastbound. Westbound freight movement had never been a problem. During the tourist season or under exceptional conditions of freight movement, additional trains were operated. There were frequently special express trains carrying fruit and silk.

Economy in railway operation requires large freight trains. Under steam operation eastbound freight trains from Skykomish to the Cascade Tunnel station (at east portal of the old tunnel) were generally made up to about 2500 tons trailing load and were drawn by a Mikado leading engine with two simple Mallett helpers, one about 30 cars from the front and one near the rear of the train. The total weight of the train, including engines, was 3256 tons. At Tye the steam helpers were replaced by the three-phase electric locomotives for the run through the tunnel. This operation from Skykomish to Cascade Tunnel station at the east end of the tunnel required a normal running time of about 4½ hours for the 25-mile run, including the time of taking water at Scenic and the delay for cutting in and out of the train.

Electric Helpers Used between Skykomish and Tye

In 1925, it was decided to take advantage of additional power available from the Tumwater power plant to improve operating conditions by electrifying the heavy grade section on the west slope between Skykomish and Tye, and utilizing electric helpers on this section. This is the ruling grade against east-bound traffic on the entire system. On account of the complication of a double trolley and the operating limitation imposed by constant speed operation inherent with the three-phase system, this system was not adapted for extension, and a single-phase 11,000-volt trolley system with locomotives of the a-c.—d-c. motor-generator type was selected by the railroad officials as suitable to utilize the available facilities and to give the required service which included operation of heavy freight and passenger trains.

For this service four electric locomotives were provided. The Mikado steam engines on this 25-mile run were not taken off the trains but were run through, assisting in pulling up the 2.2 per cent grade to Tye, operating on a drifting throttle through the tunnel, a 1.7 per cent grade. By this method the normal running time was cut about in half, and the normal freight train tonnage was increased from 2500 to 3500 tons. The electric locomotives were also used to take trains down the grade from Cascade Tunnel station to Skykomish utilizing regenerative braking.

All Electric Operation between Skykomish and Wenatchee in 1929

With the completion of the new tunnel, electric operation was initiated over the entire section between Skykomish and Wenatchee. Passenger trains consisting of nine to twelve cars are hauled over the 2.2 per cent grade at 20 to 25 mph. and make the run over this section in 2 hours and 50 minutes, consuming 3500 kw.-hr. not including regenerated power which averages 1000 kw.-hr. Freight trains vary considerably in size and power used; trains made up of 5000 tons, about 125 cars, are about the maximum handled These are made up with four motive power units on the head end of the train and three units placed about 45 cars from the head end.

While it is possible for this motive power to haul such a load at a speed of 15 mph. on the 2.2 per cent grade, the power demand would be approximately 15,000 kw. at the substation. The power situation at the present time is such that the speed of trains is limited to hold the demand to 13,000 kw. by reducing the speed of the train to 9 or 10 mph. The time usually consumed in making the 73 mile run is 5 hours; approximately 18,500 kw.-hr. of energy is used not including 5000 kw.-hr. regenerated.

The actual energy used is subject to wide variations depending on whether it is necessary to cut out all but one of the head end locomotives at the summit for use on other trains, hauling the train down the grade under combined air and regeneration, or whether sufficient motive power is available to enable the train to be handled without the use of the air brakes.

Experience has shown that it is possible to place a helper locomotive at the head end of a 3000-ton train, making a total of four units at this point, without subjecting the first cars to too great a strain. However, the possibility of trouble is always greater under these conditions and if trouble is experienced the damage to the equipment is usually greater. Placing the motive power in this manner saves considerable time in making up the train and cutting out the helper locomotive.

The power charts show that between 25 and 30 per cent of the power required to haul a train to the summit of a grade is recovered when the same train is hauled down the same grade at a reasonable speed by regeneration without the use of air brakes. Trains of 2500 tons have been held easily on the 2.2 per cent grade and of 3500 tons on the 1.6 per cent grade with one two-unit electric locomotive. The speeds in both cases are between 20 and 25 mph.

Wheel Slippage Negligible in Starting

No trouble has been experienced due to wheel slippage. This is accounted for by the parallel connection of the traction motors and the good mechanical layout. When slippage does occur, it is in the nature of a creeping, not rapid spinning, and the locomotive continues to exert a tractive effort equivalent to approximately 10 per cent of the weight on drivers. No difficulty is experienced in starting a full tonnage train anywhere on the grade.

In case an over-tonnage train is to be handled in regeneration either combined brakes and regeneration are employed or two electric locomotives are placed in the train, the leading engineman controlling the brakes and governing the speed of the train by holding more or less as the grade demands, the second engineman holding a constant load irrespective of speed or grade. This method of operation has been very successful and usually results in the second locomotive carrying the greater portion of the load, leaving the leading engineman considerable power available for actually decelerating the train.

Maintenance

THE inspection and maintenance of these locomotives is cared for in the main shops at Appleyard, Wash., about 2.5 miles east of Wenatchee, the eastern end of the electrified section.

These shops, located in a new building (Figure 23), are provided with four tracks, three of which are equipped with inspection pits. Three tracks are provided with cradles which can be dropped by a motor-operated screw type table running on tracks in a cross pit. These cradles and drop table allow any wheel and axle with motor complete to be dropped for replacement. While one cradle is in use, the table may be released for use on other tracks.

Two cranes are provided, one of 40-ton capacity over the two center tracks, for use in handling the heavy pieces of equipment and a smaller one of 7½-ton capacity for use in serving the machine tools which can be seen on the extreme right of Figure 24. A steam heated baking oven, large enough to accommodate the largest piece of equipment, is provided next to the force ventilated battery room, above which is located the office space. The usual lathes, shapers, drill presses, grinders, etc., are provided with individual motor drive.

Locomotives are moved in and out of the shops using low-voltage direct-current from a three-unit motor-generator set which can be plugged into special receptacles on the locomotive and applied to two of the traction motors in series. The electrical and mechanical equipment is inspected and reconditioned on a regular schedule by a force of twenty-five men. Pits are located at Skykomish for light inspections at that end of the section with a force of seven men.

Fig. 24—Inspection Pits and Machine Tools.

In spite of the very small number of locomotives purchased, the availability has been so high that at no time since the completion of the electrification has it been necessary to use steam power for any appreciable part of the traffic.

Fig. 23—Locomotive Maintenance Shop at Wenatchee.

New York, New Haven & Hartford Railroad Electrification

Special Publication 1698
June, 1924

PUBLISHED BY
Westinghouse Electric & Manufacturing Company
EAST PITTSBURGH, PENNA.

WITH THE COOPERATION OF OFFICERS OF
The New York, New Haven & Hartford Railroad

HELL GATE BRIDGE

THE LONGEST AND HEAVIEST STEEL ARCH EVER CONSTRUCTED

Colonial Express—Boston to Washington—Crossing the East River with Electric Power

A Bit of History

THE thought of railroad electrification today suggests a modern railroad with perfectly aligned tracks on a well dressed roadbed, surmounted by graceful caternary structures, and brilliant daylight signals, with a highly varnished train, propelled by a powerful electric locomotive operating on a high-speed schedule such as the "Merchants Limited." This present day achievement is the result of steady and rapid development in methods of transportation during the past two hundred and fifty years.

The First Post Rider in 1673

It was only as far back as 1673 that the Governor of New York dispatched the first post-rider bearing a letter to the Governor of Connecticut, requesting that the latter assist the rider on his journey through to Boston. This post-rider, leaving the fort at the foot of Broadway in New York, travelled north, crossing the ferry at Spuyten-Duyvil, and followed the old Indian trails to the settlements along the shore of the Sound. From here he followed the wagon roads through New Haven and Hartford to Springfield, where he crossed the Connecticut River and followed the trail to Massachusetts Bay over which the first white man had ventured in 1633. The time required was two weeks.

First Stage Coach Was in 1750

Although New England grew rapidly, it was not until 1750 that the first stage coach was run between New York and Philadelphia, and even then the post-rider between New York and Boston was not replaced by the stage coach until 1772. Following the Revolution and during Washington's administration, which began in 1789, two coaches and twelve horses were enough to carry all of the passengers and goods between New York and Boston, then the two great commercial centers of the country. This stage coach, from which our modern railroad coach was developed, accommodated eleven passengers and the driver, and was scheduled to leave New York on Monday morning and to arrive in Boston Saturday night, returning the following week.

The "Granite Rail" in 1825 and Steam Engine in 1830

Although the steamboat, perfected in 1807, came in to compete with the sailing vessel for water transportation when the "Fulton" began her regular trips between New York and New Haven in the year 1815, the demand for improvement in inland transportation became insistent. The turning point in the transportation industry seems to have been reached when, in 1825, the idea was conceived of moving granite for the construction of the Bunker Hill monument by placing it upon a platform suspended from wheels running upon rails. Although horses and oxen were used as a motive power, the "Granite Rail" occupies a most important position in the development of railroad transportation. The first steam locomotive for actual service in America was "The Best Friend" built at West Point (New York) Iron Foundry in 1830. It weighed four and one-half tons, and reached a speed of 35 miles an hour with load, and 20 miles an hour with four cars.

New Haven System Began in 1834

The Boston and Worcester, now the Boston and Albany, the first steam railroad in Massachusetts, opened in April, 1834. Two months later the Boston and Providence Railroad, the first in the New Haven System, was opened from Boston to Readville and to Providence within the year. Thus Boston was connected by rail and by steamer from Providence with New York. The field now was opened; and in 1837, the New York, Providence and Boston pushed along from Providence to Stonington, there connecting with steamers for New York. What is now the Boston and Albany Railroad reached Springfield in 1838, and the Hartford and New Haven was constructed from tidewater at New Haven to Hartford in 1839. Worcester was connected with Norwich and tidewater in 1840, and the line serving the Housatonic Valley reached tidewater at Bridgeport in the same year.

The next ten years up to 1850 showed a tremendous increase in railroad mileage. In southern New England, over one thousand miles of railroad had been completed, of which 835 miles now are included in the New Haven System. This period saw the extension of the line from Stonington to Groton and the opening of the line from New London to New Haven. In 1847, the New York and New Haven Railroad inaugurated service, thus completing the all-rail route from Boston to New York, broken only by the ferries at Providence, Thames River, Connecticut River, and the Housatonic.

New York, New Haven & Hartford Railroad Formed in 1872

The New York, New Haven & Hartford Railroad Company was formed August 6, 1872, by the union of the Hartford and New Haven Railroad with the New York and New Haven Railroad. The latter controlled by lease the Shore Line Railway from New Haven to New London. There was thus united into one system the operation between New York, New Haven, Hartford, Springfield and New London. In 1887, the New Haven added to its system three north-and-south lines known as the Northampton, the Naugatuck and the Valley. In 1892 and 1893, it added the remaining north-and-south line in Connecticut, the Housatonic, and reached into the eastern territory by its leases of the New York, Providence and Boston, Providence and Worcester, Old Colony, and Boston and Providence. In 1898, by its lease of the New England System extending from Boston to the Hudson River, with branches, the consolidation of the present New York, New Haven and Hartford Railroad System was substantially completed —a system of something over 2,000 miles, embracing a unique network of railroads serving every town and hamlet in southern New England, and representing also the combined efforts up to that time of 175 different corporations.

Fig. 1—An Early Electric Locomotive Installed on the Manufacturers Railroad. This Locomotive was Exhibited at Chicago World's Fair in 1893

CHAPTER I

Early Electrification Experience

THE New York, New Haven and Hartford Railroad very early became interested in electric operation. It was in 1891 that President Clark called attention to the large part electrification probably would take in steam railroad operation. The competition from trolley cars in the early Nineties showed signs of presenting a serious situation, and the management felt that the problem would best be met by electrifying and operating a service as closely as possible competitive with the trolley companies which paralleled the right-of-way. It also was in their minds that the experience gained in branch line service would be of great value in subsequent main line electrification.

Electrification of First Branch Line in 1895

On June 30, 1895, electric operation on the Nantasket Beach Branch was inaugurated. This is about nine miles long, extending from Nantasket Junction to Pemberton, Massachusetts, and the summer traffic is extremely heavy. The initial operation was the first example of electric power applied to steam railroad operation in this country, antedating by a few months the electric operation of the Baltimore Tunnel of the Baltimore and Ohio Railroad.

The power distribution for this electrification was by means of an overhead trolley and much of this wire still is in service. The wire was directly suspended from double brackets, the poles being between the tracks. The power was furnished by two tandem compound Greene engines, operating at 110 r.p.m. This steam plant was operated until 1920, when it was discontinued and power purchased, and the building now is a substation.

The original rolling stock equipment consisted of ten motor cars, six open and four closed, together with a number of trailers. The motor cars weighed 60,000 pounds and hauled three or four trailers, each weighing 43,000 pounds. Much of this equipment still is in service.

The Annual Report to the Stockholders in 1895, in referring to the electric operation, contains this statement:

"The experiment has demonstrated that power generated in a stationary plant and transmitted by electrical energy can be successfully used in the operation of a standard railroad. The current expenses for fuel indicate that this result is economically obtained.

"The use of the existing power station will be extended presently, and it is probable that electricity will be promptly adopted by the Company at other points on its lines. With a road free from grade crossings, it is not too much to expect its ultimate application wherever the business justifies frequent train service, and it is to be hoped without the use of an overhead trolley."

First Use of Third Rail in 1896

The idea of a third-rail distribution had long been in mind, and the second step in the electrification of the New Haven was undertaken on July 26, 1896, when the electric operation of the Nantasket Beach Branch was extended, three and one-half miles to East Weymouth. The power distribution on the extension was by means of a third rail instead of by overhead. This was the first time in the history of railroads that a surface railroad was operated electrically from a conductor laid "on the ground," just as, in the year before, the operation on the Nantasket Beach Branch was the first application of electricity for the sole motive power on a section of standard steam railroad.

The third rail was in the form of a flattened "A" and was installed midway between the track rails. This rail, which weighed 93 pounds per yard, was designed to shed water and to protect the wooden blocks which supported it. The top of the third rail was one inch above the top of the running rails, and its lowest point 1⅝ inches above the ties. The contact was thus high enough that the shoe cleared the rails at turnouts, crossovers, etc., but not high enough to foul steam equipment which, necessarily, operated over the electrified tracks.

The splice plates for the third rail were wrought iron, and between the splice plates and the third rail were copper plates acting as bonds. Wood blocks beveled on the ends assured that the shoes would ride up over the third rail.

At gaps in the rail at crossovers and turnouts, street crossings, etc., jumpers were installed between sections of rail and were placed in creosoted wood ducts, which were filled with asphalt compound, the whole laid in creosoted wood troughs and filled with the same compound. The connections in the third rail were sweated and bolted.

From East Weymouth to Pemberton, a distance of ten and one-half miles, seven miles of which was overhead trolley construction and three and one-half miles third rail, a sixty-four ton train, making thirteen stops, reached a maximum speed of about 37 miles an hour between stations and maintained an average speed of 17 miles an hour. A maximum speed of 57 miles an hour was attained by the express trains. In the summer of 1896, six thousand train miles were made and 700,000 passengers were carried electrically on this electrified section.

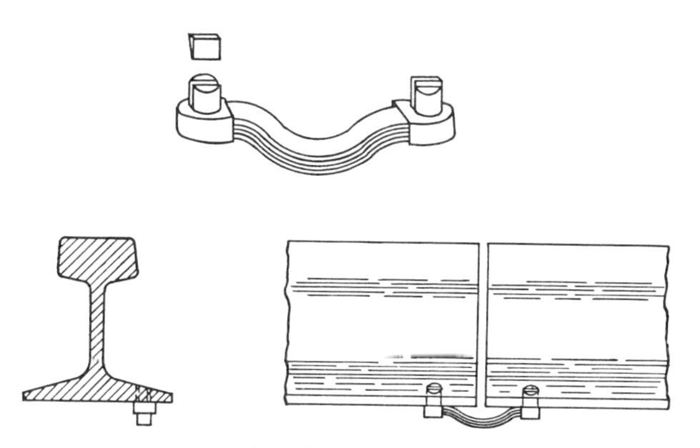

Fig. 2—Rail Bond Used on First Electrification—1895

Use of the Third Rail Extended

In December, 1896, work was started on the electrification of another section between Hartford and New Britain (9.3 miles) and between New Britain and Berlin (3 miles), a total distance of 12.3 miles. The power plant for this section was located at Berlin, Connecticut, and the same type of third rail was installed as between Nantasket Junction and East Weymouth. Storage battery substations to help out on peaks were installed at New Britain, Forestville, and under the platform of the Hartford passenger station. Later these were replaced by rotary substations.

The trains were normally operated with two cars 51 feet long, each seating 96 passengers. The forward car aways was the motor car, but both cars were equipped with shoes, not only to bridge the gaps in the third rail at switches or highway crossings, but to provide for the possible loss of a shoe.

The next electrification development occurred in 1898, when the New Britain electrification was extended to Bristol, Connecticut, and to Cooks, Plainville, and Forestville, a distance of 8.8 miles. This was all on the basis of a third-rail distribution. While the steam railroad was double-tracked in this section, only one track was electrified, the other one being retained for steam operation.

At about this time also the Nantasket Beach electrification was extended by third rail from East Weymouth to Braintree, 4.4 miles of double track; and, in 1899, the electrification was extended from Nantasket Junction to Cohasset, a distance of about 3.4 miles. Operation was continued thus until 1904, when the traffic between Braintree and Cohasset had fallen off so greatly that operation was not profitable and the third rail in this section was abandoned. The Nantasket Beach Branch, however, still is operated as formerly, except that with purchased power it is practicable to operate the year around, and freight is also handled electrically on this branch.

The steam road from Stamford to New Canaan, Connecticut, a distance of 8 miles, was electrified at about this same time. This "New Canaan Branch" was equipped with an overhead trolley in order that interchange of operation could be made with the local street railway, the tracks of which for about 2 miles ran parallel to the steam railroad main line at Stamford. The trolley wire was 3/0 and 4/0, suspended direct from brackets, but there were no trolley frogs, the pole being shifted by hand at turnouts and crossovers. The bonds were installed under the rail flange as with the other installations.

The power plant for this electrification was located at Stamford, and now is used as a substation by the Connecticut Company for its street railway operation at Stamford, the power being supplied from the New York, New Haven & Hartford Railroad traction distribution system at 11,000 volts, three-phase, 25 cycles.

First Locomotive in Service in 1896

The Manufacturers Railroad, organized by the Bigelow Boiler Company and other manufacturers in New Haven and now owned and operated by the New York, New Haven and Hartford Railroad, had been operated initially by means of "horse power," the horses taking cars from the steam railroad interchange tracks and hauling them along the right-of-way of the Manufacturers Railroad and through the streets of New Haven to the individual factories. This railway was two miles long and contained grades as high as 2½ per cent.

In 1896, a 30-ton locomotive was purchased, and was placed in operation on December 11th. This locomotive, which had a drawbar pull of 7,000 pounds and a length over all of 16 feet 6 inches, with a wheel base of 5 feet 6 inches, was equipped with four drivers 44 inches in diameter, and two gearless motors mounted on quills around the main axles. The locomotive hauled two loaded cars up a two per cent grade at a speed of seven miles an hour.

In 1901, the electrification of the Providence, Warren and Bristol Branch, which was projected in 1898, was completed. That road is double-tracked from Providence, R. I., to Warren, and divides at that point—one single track extending to Bristol, R. I., and another to Fall River, Mass. The power plant was located at Warren, R. I., and two storage battery stations of 800 ampere-hour capacity

250 cells each, were installed at East Providence and Brayton, R. I., respectively, to assist on peaks. Car repair facilities with transfer table and other necessary tools were installed at Warren near the power plant.

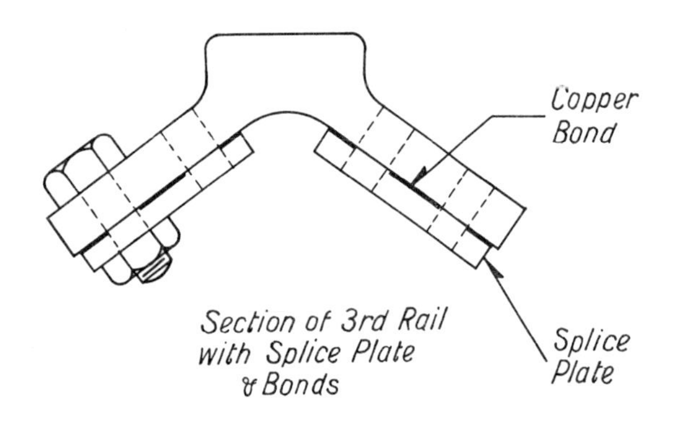

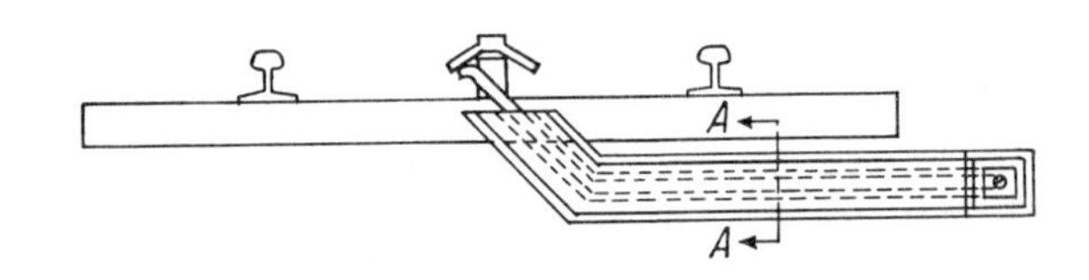

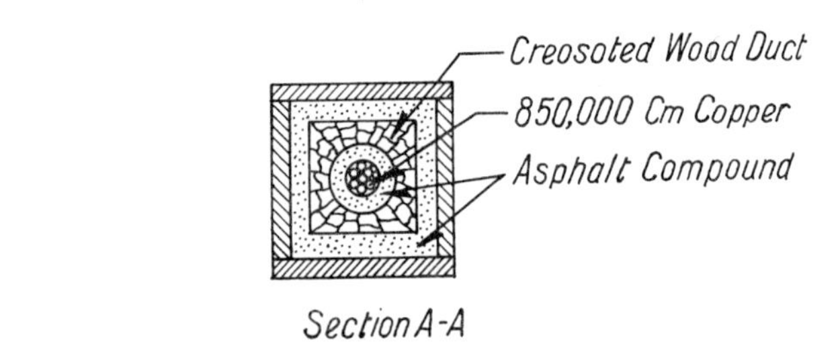

Fig. 3—First Surface Third-Rail Installation—1896

The rolling stock for this branch consisted of 46 cars, of which 24 were trailers; and of the motors, nine were combination baggage and passenger coaches and two were combination baggage, express and mail cars. These cars were 40 feet long over all and weighed 20,000 pounds. There also was one standard 70-foot coach equipped with motors. The motor cars were equipped with four 80-horsepower motors each.

The power distribution system was overhead, the trolley wire being 4/0 round copper, suspended direct from span wire or bracket construction. The rail bonds consisted of leaf bonds installed as were those of the other electrifications.

The initial operation of this electrification included 112 trains. Under steam operation between Fall River and Providence, the schedule was 48 minutes with 14 stops; but with electric operation, the local trains, making 26 stops, accomplished the distance in 45 minutes; and the express trains, making seven stops, in 33 minutes. On the Bristol Branch, the scheduled time under electric operation was six minutes with four stops, while under steam operation, seven minutes with no stops had been required. The fare between Providence and Fall River was reduced from 50 cents to 20 cents.

Third Rail Removed by Court Order

In 1905, the Legislature of Connecticut ordered either the installation of some form of protection or else the removal of the third rail on the Hartford-Bristol electrification. The Railroad Company contended that it was the duty of the public not to trespass on the right-of-way and that it could not undertake to protect trespassers from injury on account of the third rail. The State insisted, however, and in July, 1906, the third-rail installation was discontinued by Court order.

In 1906 and 1907, there was considerable activity in branch line electrification, largely in connection with street or interurban operation. In 1906, track was electrified between Berlin and Middletown, Meriden and Westfield, Middletown and Cromwell, and Tafts and Central Village, all in Connecticut, and all with overhead trolley direct suspension, and generally wood pole bracket construction. In 1907, the tracks from East Hartford (Burnside), Connecticut to Vernon, Rockville and Melrose, were electrified, also with overhead trolley; and on the section between Burnside and Vernon, a catenary system was installed for high-speed operation. Due to a falling off in traffic, however, the Rockville-Melrose operation was discontinued in 1917 and the construction removed.

It was also in 1906 that the electrification of the suburban territory at New York was installed. The early branch line electrification experience of the New Haven was of inestimable value in determining the characteristics of the main line installation, although, both in character and extent of traffic, the New York suburban electrification, of course, far exceeded anything which had gone before. As is well known, the main line electrification, which was installed initially as far as Stamford, Connecticut, and included only passenger service, later was extended to New Haven and embraced not only local and through passenger service but freight and yard switching as well, and now comprises what is, perhaps, the most important and comprehensive electrification system in the world.

Fig. 4—Double Catenary Construction on Curve and Fabricated Steel Bents, Showing Auxiliary Control Transformers

CHAPTER II

The Track and Overhead System

The Track Layout

THE initial electrification of the New York, New Haven and Hartford Railroad extended a distance of 33 miles to Stamford, Connecticut, the terminal of the commuting service into New York City. Of this a distance of 12 miles from Grand Central Terminal to Woodlawn is over the tracks of the New York Central Railroad, and this section is third-rail construction at 600 volts direct current. The balance of the distance, 21.4 miles, is on the New Haven right-of-way, using overhead catenary at 11000 volts.

In 1912, the electrification was extended to the Harlem River Branch between New Rochelle and Harlem River, N. Y., and the New York, Westchester & Boston Railway between New York and New Rochelle and White Plains; and, in 1914, to New Haven. This gives a total of 120 electrified road miles, including 590 track miles. The entire distance over the New Haven right-of-way from Woodlawn to New Haven is four-tracked, using 107-pound rail, creosoted ties, and rock ballast.

Stamford is the junction with the New Canaan Branch which runs for eight miles through a high-class suburban section. Between Stamford and New Rochelle, a distance of sixteen miles, there are eleven passenger stations generally located in centers of considerable population.

The main line from New Rochelle Junction, five miles west of Woodlawn, is used mainly for passenger business The six tracks of the Harlem River Branch extend south westerly from New Rochelle Junction to Harlem River, a distance of about 12 miles. This is the main artery of freight traffic and through passenger traffic between New England, New York City, and the South, and West. The main line is either four, five or six track; and some of the yards are of large size, the Oak Point yard, with 35 miles of track, being the largest electrified yard in the world.

Freight is interchanged with other railroads by car floats at Oak Point and Harlem River Yards and is also floated to other piers in Greater New York. The float bridges are of modern type and, with the supporting yard tracks, permit the rapid loading and unloading of floats.

The yards at Westchester, N.Y., formerly were used for classification of freight received from western connections by float at Oak Point and include a large, l.c.l. transfer platform layout. Since the construction of the large modern hump classification yards at Cedar Hill, New Haven, the Westchester yards have not regularly been used for classification of freight.

Overhead Contact System

The nearly six hundred track miles of electrification are provided with 11000-volt single-phase overhead construction. This trackage, as already explained, comprises main line, siding, yard, and industrial tracks.

Pantagraph trolleys with single steel shoes are used on the locomotives and multiple unit cars. The high-speed passenger locomotives operate at a maximum speed of 60 miles an hour with a pantagraph pressure on the wire of from 15 to 18 pounds. Freight locomotives operate up to 35 miles an hour, while multiple unit cars operate up to 50 miles an hour. The quantity of current collected is small, seldom exceeding 100 amperes per shoe.

The normal height of the contact wire is 22 feet above the top of rail but there are many places where, on account of low overhead bridges, it is necessary to bring the contact wire as low as 16 feet. With the high-speed operation, it is necessary to carefully grade the wire approaching and leaving a low section so that the shoe shall not break contact.

Where it is necessary to sectionalize and also to operate at high speed, it is found advisable to install so-called air section breaks, which permit a safe and easy collector transition from one section to the next. This is shown in Fig. 36.

Triangular Construction, Woodlawn to Stamford

The high-voltage overhead construction begins at Woodlawn; and this section from Woodlawn to Stamford was installed in 1906, and commonly is known as the double-catenary type. The supporting structures are substantial fabricated steel bents on concrete foundations with a normal spacing along track of 300 feet. The most marked feature of this section of the overhead is the triangular-shaped hangers of gradually varying sizes strung along the two $\frac{9}{16}$-inch suspending messenger strands and the trolley wire, and serving to tie them into an integral structure.

The triangular type of hanger was selected to give lateral stability to the contact wire and to hold it in a proper position over the tracks. Practically the only modification that has been made in the original construction has been the addition of a 4/0 grooved contact wire below the original copper contact wire. The first material used was steel; but later, phono-electric contact wire was substituted. One comparatively short section, however, has retained its original form. On this section a 6/0 phono-electric wire originally was installed and it was not found necessary to add a second contact wire to it.

The triangular hangers are spaced 10 feet apart; and the clips supporting the bronze contact wire are similarly spaced but are located midway between the main hangers in order to provide maximum flexibility. The $\frac{9}{16}$-inch messengers are made up of 7 strands of extra-high strength steel. The vertical sag in a 300-foot span is 66 inches.

Chord construction is used on curves, the contact wire being held over the center line of track in a series of chords by means of pull-offs as shown in Fig. 4. The maximum displacement from center of pantagraph is 10 inches.

At each side of the track, extensions to the poles of the supporting bents carry the necessary feeder and signal wires. At intervals of approximately two miles, anchor bridges are installed. These, as the name would indicate, are for the purpose of anchoring all wires and are of much heavier construction than the ordinary intermediate bridges. Some of the anchor bridges, located at crossovers,

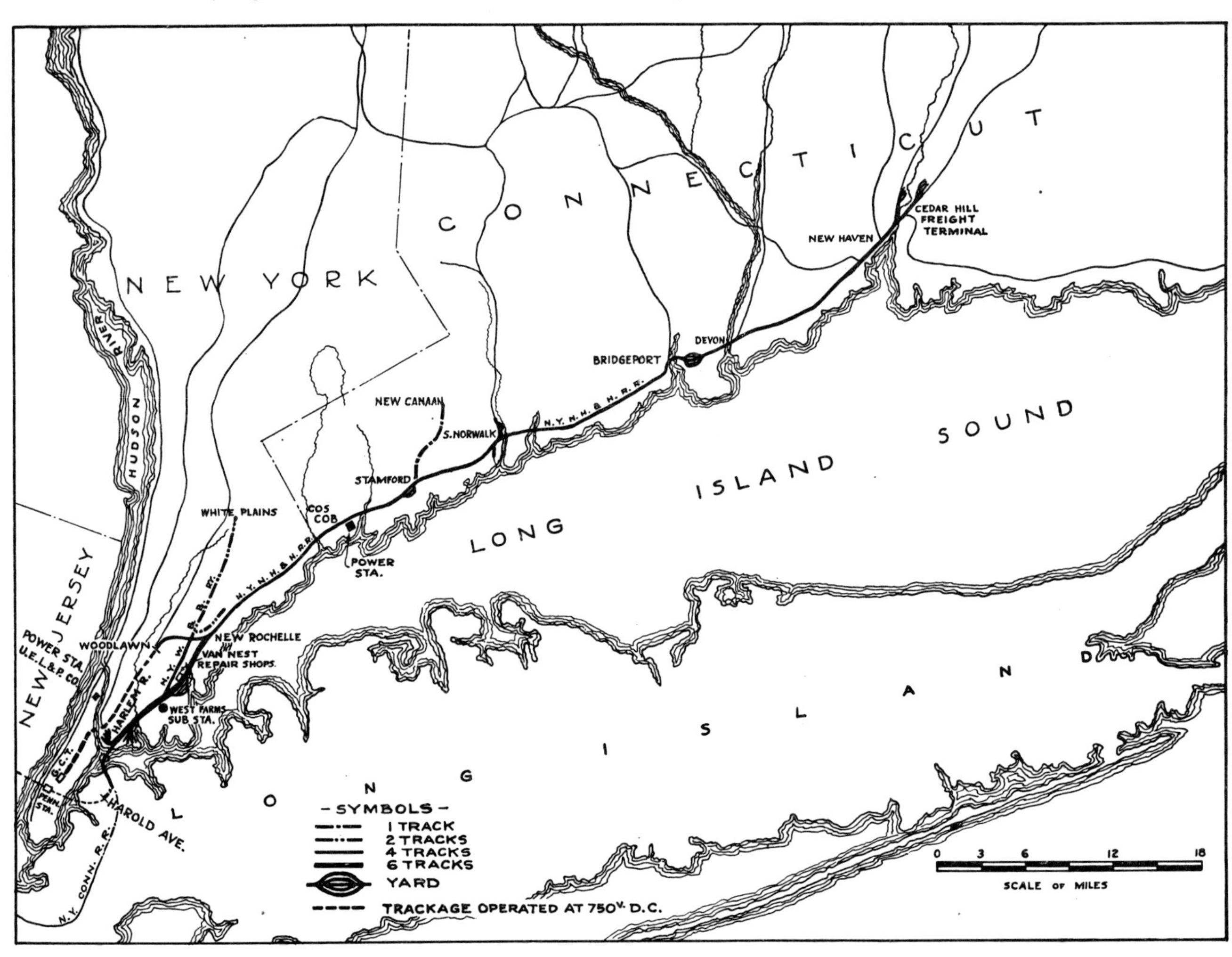

Fig. 5—Map of New Haven Electrification Showing Number of Tracks and Location of Yards, Substations and Power Plant

Fig. 6—Experimental Arch-Shaped Bents Near Glenbrook Supporting Compound Catenary as Originally Installed

Fig. 9—Construction on Harlem River Branch at Woodside Station Platform

Fig. 7—Curve Catenary Construction Near Bridgeport, Showing Telephone Cable and Signal Transformers

Fig. 8—Compound Catenary Construction West of Bridgeport

Fig. 10—Six-Track Section at West Farms Junction with New York, Westchester & Boston Railway on the Harlem River Branch

are also used for electric sectionalizing so that in case of trouble on the line, only a short section of track will be involved, and trains may cross from one track to another around the defect. This is illustrated by Fig. 26.

Experimental Arch-Shaped Bents and Compound Catenary

As a development, a short section of a radically different type of overhead was installed experimentally near Glenbrook. The supporting bents are arch-shaped as shown in Fig. 6. The catenary construction is of the compound type, the main messenger strands of which are 1 ¼ inches in diameter. The track catenary span is only 100 feet, there being three such spans per main span. The hangers on this section are simple rods and there is only one track messenger. A fundamental difference between this and the original construction is in the character of the construction on curves; in the later installation the contact wire lies in a position parallel to the rails instead of in a series of chords.

Fig. 11—Typical Catenary Construction in the Harlem River Yards Where as Many as 15 Tracks are Spanned Between Upright Supports

This construction was installed about 1909 and led to the development of the compound catenary construction which was installed some few years later, in 1912, on the New York, Westchester & Boston Railway and the 6-track Harlem River Branch and in 1913-14 on the main line between Glenbrook and New Haven.

Compound Catenary with Cross I-Beam Supports

This latter type has a main span of 300 feet and sub-spans of 150 feet. The main messenger is a ⅞-inch extra-high strength steel strand. It is grounded and carries no high voltage current. It supports the track messengers by means of 3-inch cross I-beams, and suspension insulators located every 150 feet, at the quarter points in the span. The messenger in the sub-span is a ⅝-inch extra-high strength 19-wire steel strand. As in the Glenbrook section, the 4/0 phono-electric contact wire and 4/0 copper trolley are supported by simple rod hangers. Similarly on curves, the contact wire and secondary are supported by a simple rod hanger; and the hanger is bent and inclined, and so supported by the messenger in a resultant vertical and horizontal direction, that the effect is to maintain the contact

Fig. 12—Typical Construction on the Hell Gate Bridge Approach

wire in a curve whose shape corresponds perfectly to the graceful curve of the track below. This is shown in Fig. 7.

The steel supporting structures also differ from the early type in that the posts taper downward, the horizontal loads across tracks being taken at the connections between posts and truss. This results in an effect of grace, lightness, and strength. The change in the type of design also reduces the size of foundations, as compared with the original type.

Extensive Yard Electrification

All large yards and important sidings along the electrified main line also have been electrified and completely equipped with an 11,000-volt overhead contact system. The type of construction used follows the same general design as on the main line. Simple catenary spans, 300 feet long, are standard with a ⅜-inch extra-high strength steel messenger and a 2/0 steel or phono-electric contact wire. The supporting structures usually are latticed steel poles with cross-spans, and in large yards may carry as many as 15 track constructions in a single span. Supporting posts are backguyed wherever this is feasible, but self-supporting posts also are employed.

Insulation an Important Factor

Insulation is, of course, an important factor in any overhead construction. To provide adequate and economical insulation always has been a problem on the New Haven because of the presence of steam locomotives. The standard practice now is to use three 10-inch porcelain suspension insulators in series at supporting points; and wood, or porcelain, or both, at strain points. Wood section insulators in the contact line are standard at slow-speed points and where air breaks are not practicable.

In the overhead design, ample factors of safety have been used both mechanically and electrically. Under the assumed maximum conditions of low temperature, ice load, and wind, a factor of safety of 5 determined the size of steel supporting cables; the factors for steel supports and concrete foundations being those customarily used in similar classes of railway structures.

A study of the New Haven overhead construction work reveals that many of the important developments in the art were initiated on this electrification. The duplex contact wires, the inclined hanger, the long cross-catenary yard support and the high-strength contact wire are some of the ideas which were conceived and given ample demonstration under large scale operating conditions.

This electrification is without question the pioneer installation in the heavy single-phase traction field.

Fig. 13—Two-Track Single Catenary Branch Construction, Bracket Supported, Showing How Contact Wire Conforms to Reverse Curve of Track. South Norwalk

Fig. 14—Cos Cob Power Plant with Telephone Cable in Foreground, Adjacent to the Main Line, Showing Coal Bunkers, Conveyor and Water Tanks

CHAPTER III

Power Generation and Distribution

THE generating station for the electrification is located at Cos Cob, Conn., five miles from Stamford, the suburban terminal. The first part of the power house was completed during the summer of 1907 and represents part of the first installation of single-phase equipment for the operation of trains on a trunk line railway. It is located adjacent to the main line of the railroad on the bank of the Mianus River, about a mile from Long Island Sound. The location is such that coal can be delivered by either water or rail, and an adequate supply of water is available for condensing purposes.

Initial Installation Three Turbine Generators

The initial installation of generating equipment included three multiple-expansion parallel-flow Westinghouse-Parsons steam turbines direct-connected to 3750-kv-a. (3000 kw., 80 per cent P. F.) single-phase, 11,000-volt, 25-cycle Westinghouse generators. The turbines operate at 1500 rpm. by steam at 180 pounds pressure and 100 degrees superheat. The generators are wound three-phase and are arranged for delivery of both three-phase and single-phase alternating current. This permits furnishing three-phase current to the local street railways and power companies at various points on the system, as illustrated by Fig. 19.

Shortly after the first three generators were installed, they were supplemented by one 4160 kv-a. (3330 kw., 80 per cent P. F.) unit, which is equipped with a turbine sufficient to deliver a three-phase output of 6000 kv-a.

When the electrification of the main line was extended to New Haven, a distance of 45 miles, the Harlem River Branch with its freight yards electrified, and the New York, Westchester & Boston Railway completed, it was necessary to add to the original power plant the turbines and boilers required to carry the increased load. Therefore, in 1912, four additional units, each of 5000 kv-a. capacity, were installed; this made a total of eight horizontal steam turbines with a combined rated capacity of 35,400 kv-a., 11,000 volts, single-phase, 25 cycles. The new turbines operate at 1500 rpm. with steam at 180 pounds pressure and 100 degrees superheat. Jet condensers were installed instead of the surface type used with the older generating units.

Fourteen new water-tube boilers, each having 6250 square feet of heating surface, were added to the original installation which consisted of fourteen water-tube boilers, each having 5200 square feet of heating surface. The transmission voltage was raised to 22,000 volts by six 7200 kv-a. oil-insulated, water-cooled 11,000-22,000-volt, single-phase, 25-cycle auto-transformers installed at the plant, shown in Fig. 16. The annual output of the Cos Cob plant now is approximately 100,000,000 kw.-hr. per year.

Power Also Purchased From New York Edison Company

On account of the heavy suburban traffic terminating at New Rochelle and Port Chester, together with the New York, Westchester & Boston Railway traffic, and the large freight yard operations on the Harlem River Branch, it was considered desirable to develop a power supply near the New York end of the electrification. Under a contract, which became operative September 1, 1915, this source of supply was obtained by purchasing a minimum of 40,000,000 kw.-hr. per year from the New York Edison Company. This energy is generated in the Sherman Creek power sta-

Fig. 15—Turbine-Generator Installation at Cos Cob

Fig. 16—7200 Kv-a. Auto-Transformers at Cos Cob

Fig. 18—Switchboard and Interior of West Farms Substation, Showing Current Limiting Grid Resistance

Fig. 17—West Farms Substation, Showing Lightning Arrester, Main Transformers and Teasers

tion of the United Electric Light and Power Company located at 201st Street and the Harlem River, and is transmitted in underground conduit to a substation operated by the power company and located at West Farms, New York, at the junction of the New York, Westchester and Boston Railway and the Harlem River Branch of the New Haven Railroad.

The transformers and electrolytic lightning arresters are located outside the substation building. There are three transformer units, each consisting of a main 5000 kv-a. Westinghouse oil insulated, self-cooled transformer and a 1000 kv-a. teaser transformer, the pair being connected to receive unbalanced three-phase supply and to deliver 11,000-volt, three-phase and 22,000-11,000 three-wire single-phase power. Power is transmitted to the substation by a three-wire, two-phase system with 24,600 volts between outer wires and 17,394 volts between each outer wire and the third wire through 350,000 cm. triplex underground cables. Single-phase power at 22,000 volts is taken off from the secondary of the main transformer and three-phase power from the teaser and main transformer secondaries.

Within the building is a switchboard for controlling the incoming power and distribution of three-phase power. A second switchboard controls the single-phase sectionalizing bridge-type switches. A motor-generator set and storage battery supply switch control current for the substation switches. That for the bridge type switches is A.C., 25-cycle power. Compartments are provided for the bus bars and all oil switches in the substation building, and grid resistances are installed for use in limiting the flow of current under short circuit as indicated in Fig. 18.

Two 20,000 kv-a., 6600-volt, three-phase, 25-cycle, turbine-generators, which have a single-phase rating of 14,000 kv-a. at 70 per cent power factor, were installed for single-phase power supply. At the United Electric Light & Power Company's Sherman Creek Plant, a 3750 kv-a. frequency changer set was installed to permit the 60-cycle equipment to reinforce the 25-cycle supply if necessary. The power is supplied to T—connected air blast transformers which are rated at 5000 kv-a. each.

Three-Wire System of Transmission Adopted

In the original installation, all the traction power was delivered by the generating station at Cos Cob over the

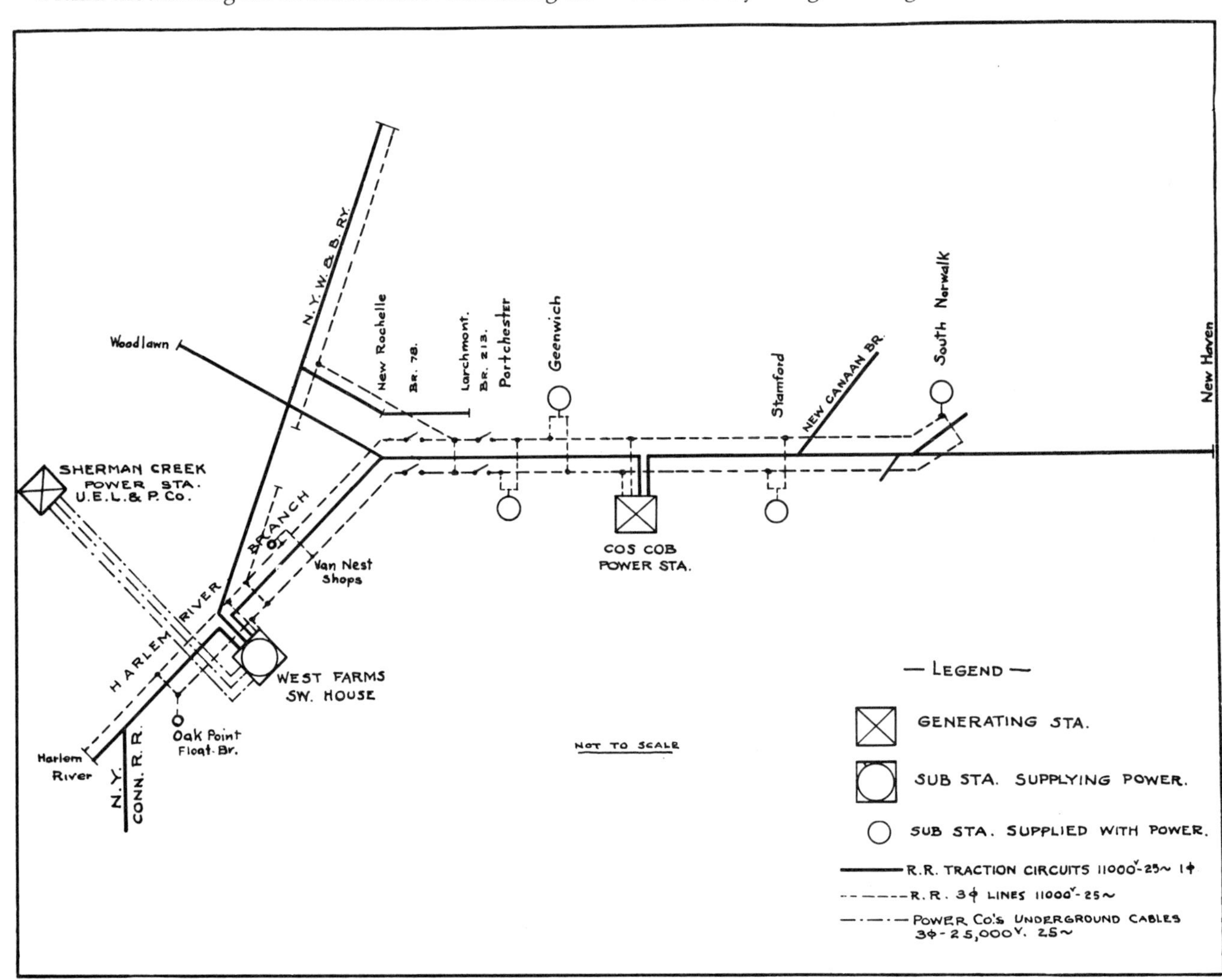

Fig. 19—Schematic Diagram of Line, Showing Three-Phase Substations and Power Supply

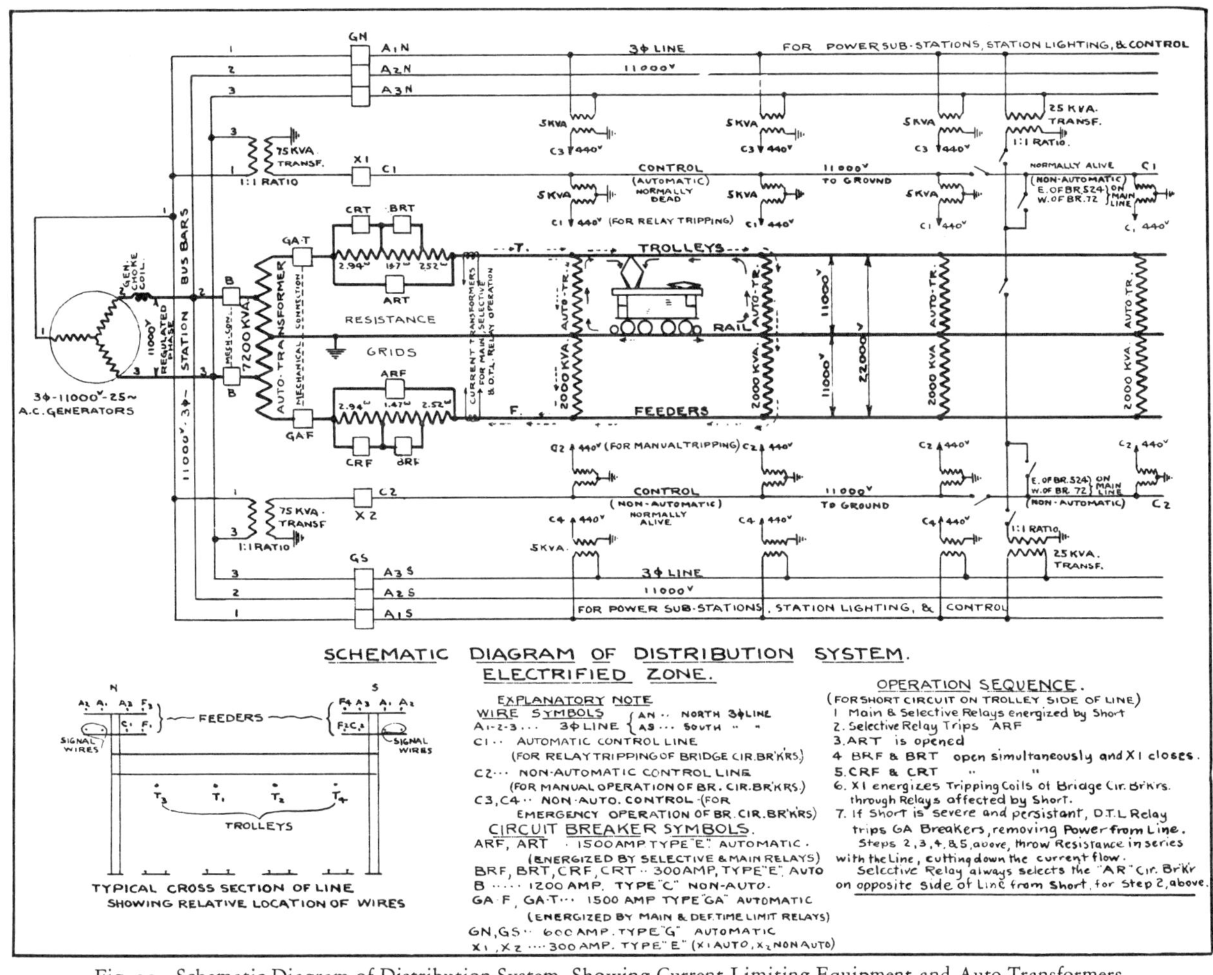

Fig. 20—Schematic Diagram of Distribution System, Showing Current Limiting Equipment and Auto-Transformers

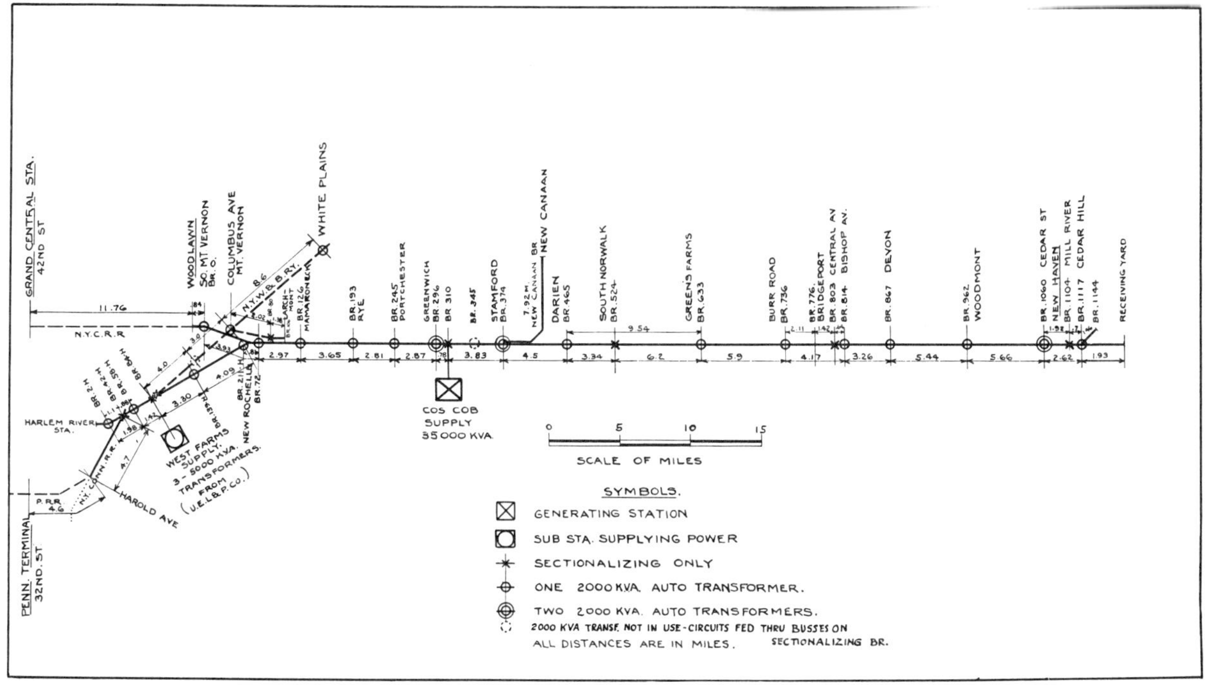

Fig 21—Schematic Diagram of Distribution System with Locations of Auto-Transformers

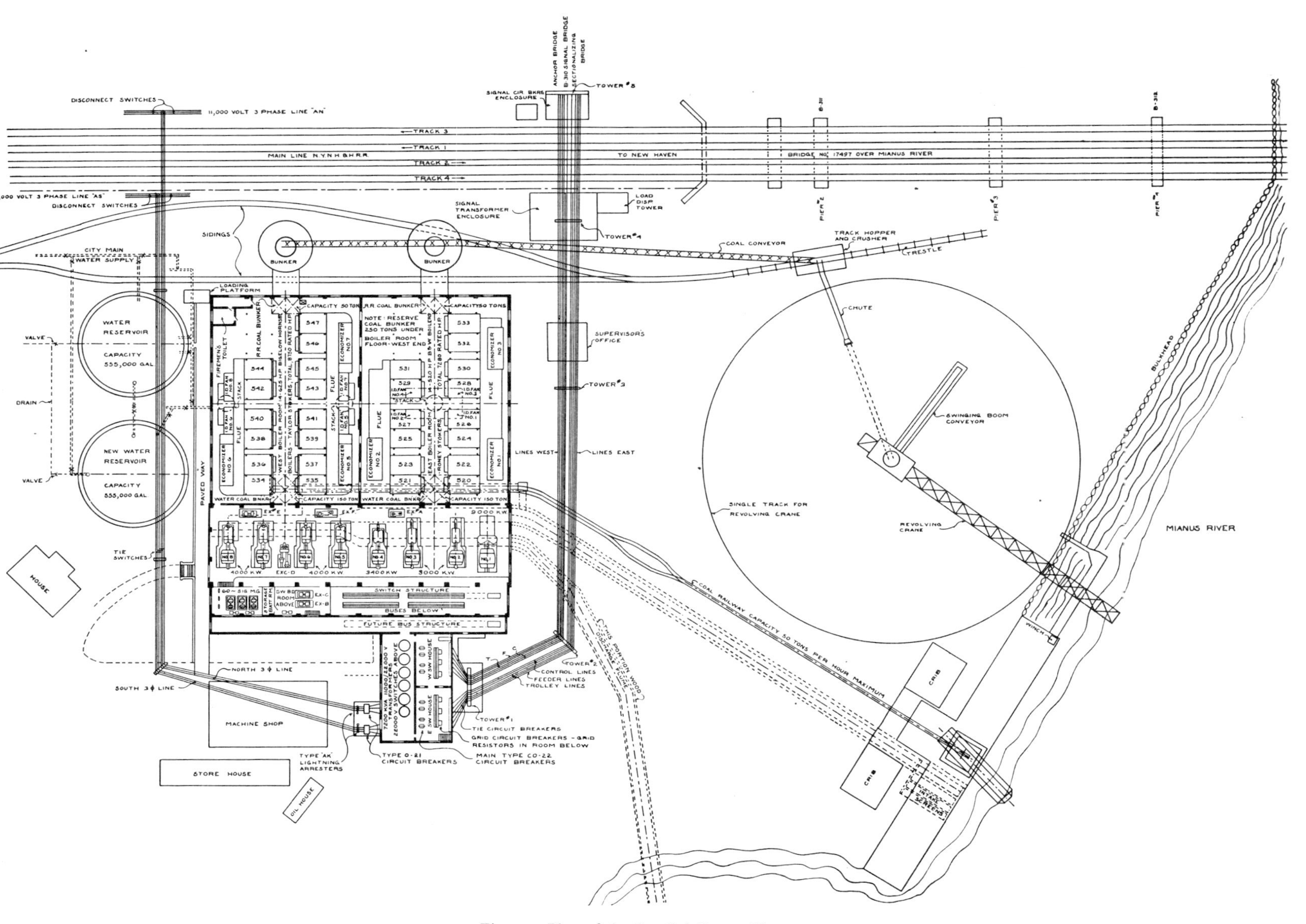

Fig. 22—Plan of the Cos Cob Power Plant

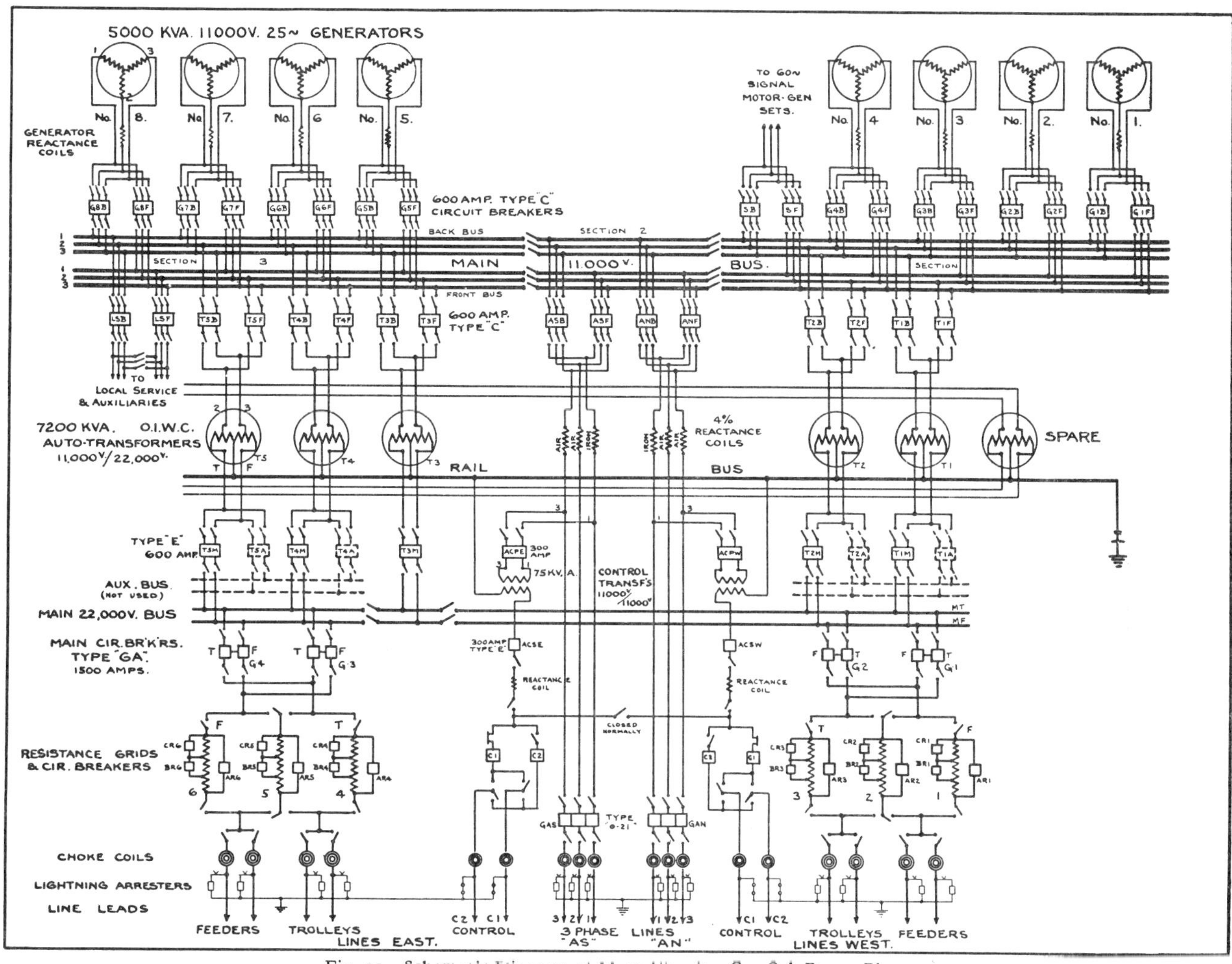

Fig. 23—Schematic Diagram of Main Circuits, Cos Cob Power Plant

contact system with auxiliary feeders; the return from the locomotives, wherever located, being through the rails. Virtually, one generator terminal was connected to the overhead system, the other to the rail and ground.

It was found that the inductive conditions could be improved and that advantages could also be secured from a transmission standpoint. With this in mind, a careful joint study of various neutralizing and transmission schemes was made in 1911 by the communication interests and the railroad, which resulted in the rearrangement of the circuits of both interests involved.

The existing railroad traction circuits were rearranged to form the present "three-wire system," wherein the rails form the neutral between "trolley" circuits and "feeder" circuits, having a potential difference of 22,000 volts, with "balancing" transformers located at proper intervals along the line. This scheme also afforded the advantage of 22,000-volt transmission for the pending extension from Stamford to New Haven, and required no change in the existing insulation.

At the Cos Cob power plant, the generators, instead of being directly connected to the line and ground or rail, were connected to quarter points in the windings of auto-transformers whose middle points were grounded, the terminals being connected to the "trolleys" and "feeders," respectively. This is shown in Fig. 20. There are six of these transformers, rated at 7200 kv-a. each.

Along the main line, balancing auto-transformers were located at sectionalizing bridges, with their terminals connected to the "trolleys" and "feeders," respectively; the middle points being connected to the rails and ground, which reduced the transmission voltage from 22,000 to 11,000 volts for use by locomotives. There are 24 of these transformers on the system, including the New York, Westchester & Boston Railway, seventeen of which are west of Stamford. These locations are shown in Figs. 20, 21 and 26.

The New Canaan Branch was equipped with "booster" transformers, such as shown in Fig. 30. These are one-to-one ratio transformers, which have their primaries connected across section breaks in the trolley and their secondaries connected across insulated joints in the rails immediately beneath. Their function is to keep the return current in the rail and prevent its return through the ground, as shown in Fig. 27. The communication circuits were suitably transposed and provided with drainage coils.

By these methods, the interference in the neighboring

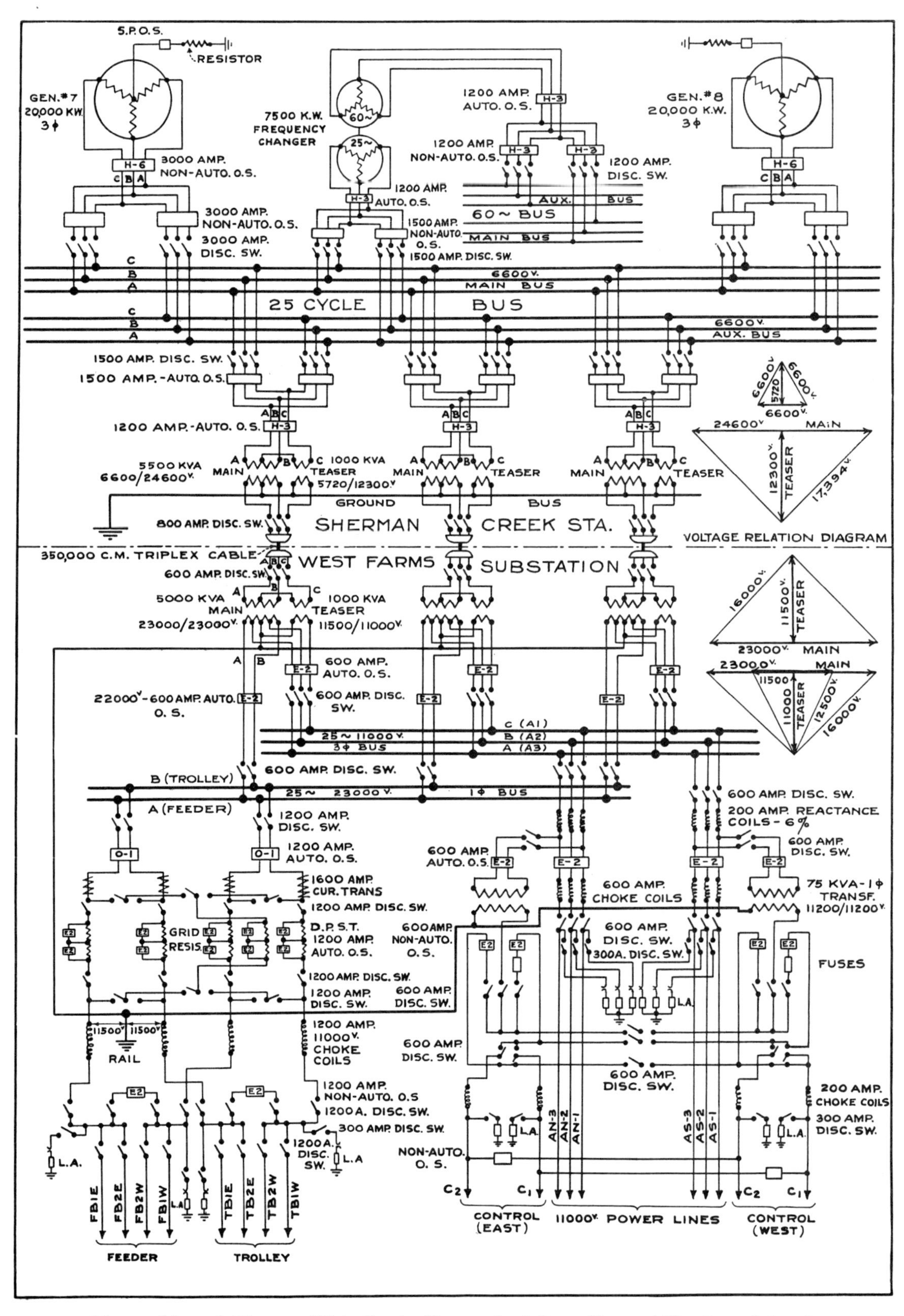

Fig. 24—Schematic Diagram of Main Circuits, Sherman Creek Power Plant and West Farms Substation

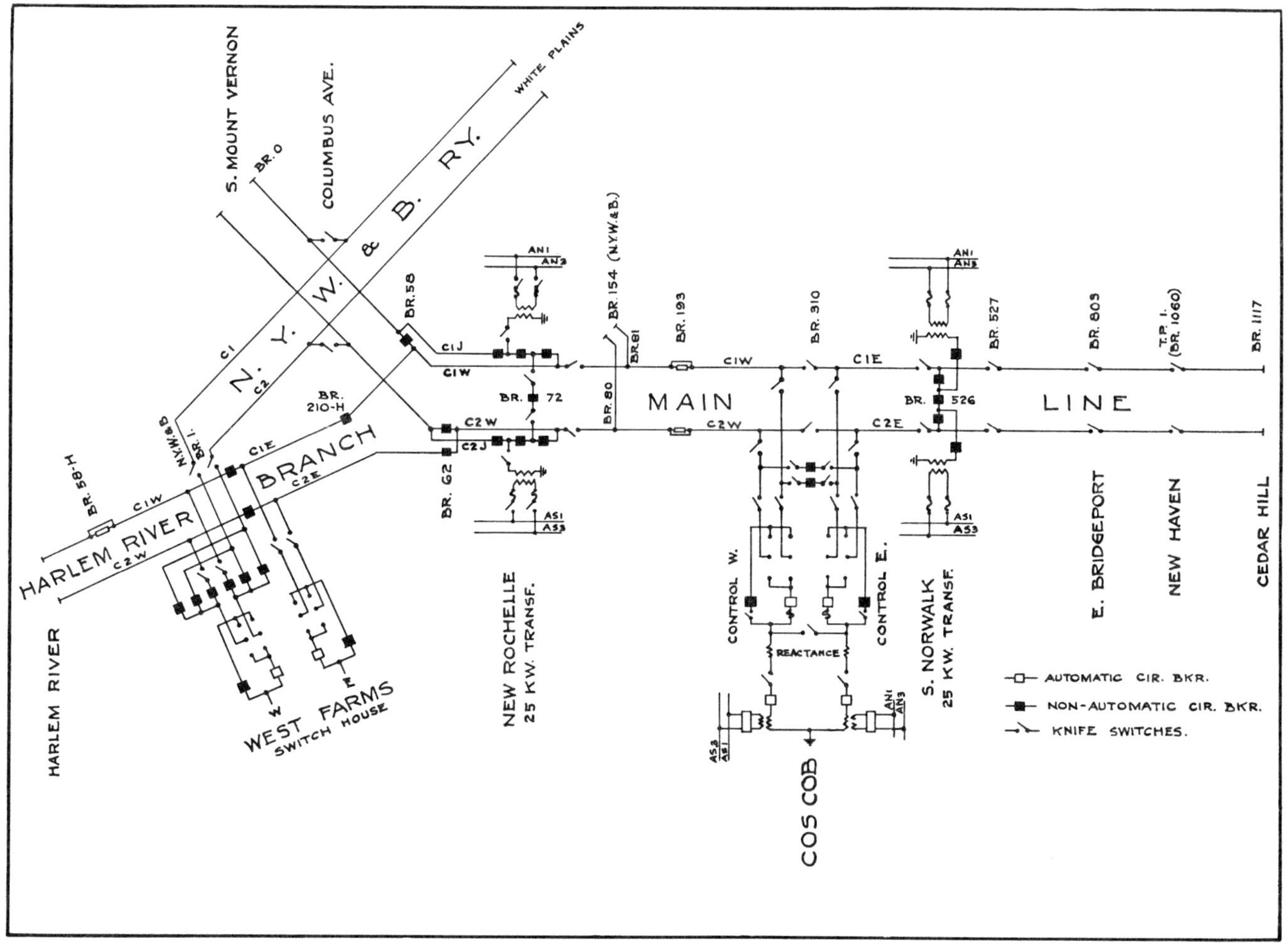

Fig. 25—Schematic Diagram of Line, Showing 11,000-Volt Control Circuits

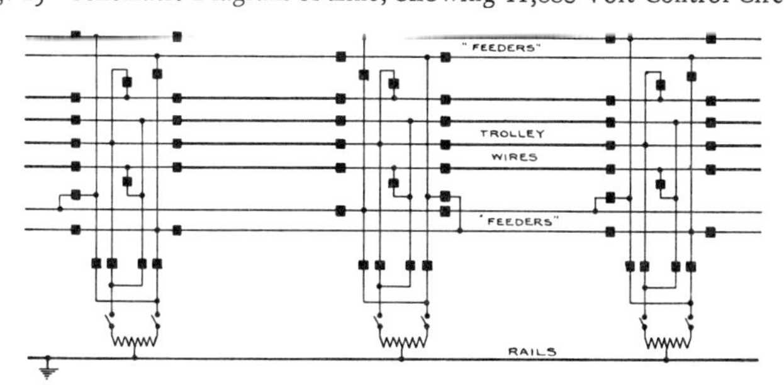

Fig. 26—Typical Sectionalizing Connections of Trolleys and Feeders, with 2000 Kv-a. Auto Transformers

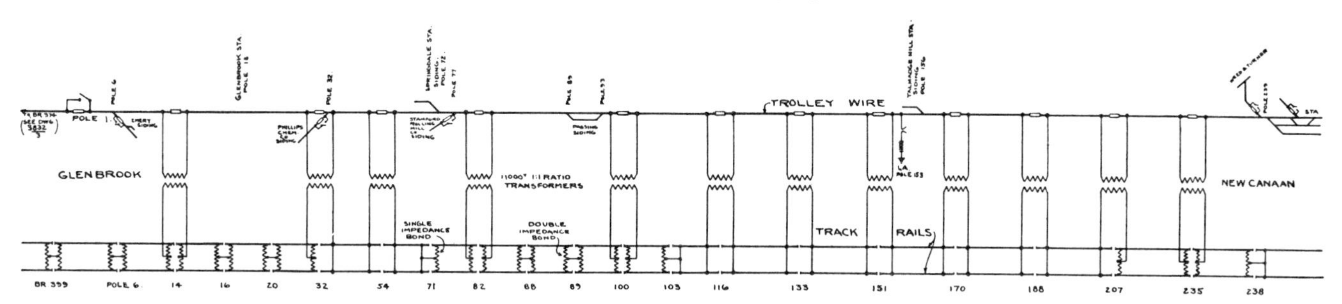

Fig. 27—Schematic Layout of New Canaan Branch, Showing Sectionalizing of Rail and Trolley and Booster Transformers
Impedance Bonds are Installed when there are Crossing Bell Track Circuits

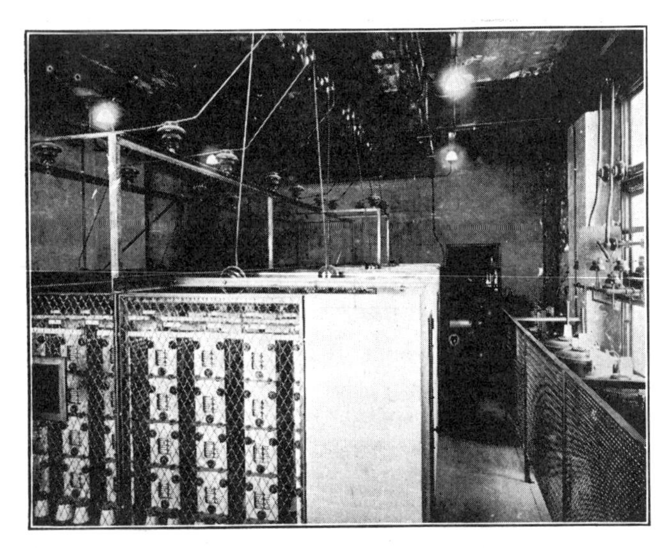

Fig. 28—The Resistance Used in Connection with the Current Limiting Equipment, Cos Cob Power Plant

communication circuits has been satisfactorily taken care of.

The railroad communication circuits are installed in a 45-pair lead cable throughout the electrified zone. This cable parallels the power circuit for about 72 miles, 56 of which are aerial and 16 underground, where the right-of-way is especially congested. This cable contains train-despatching and load-despatching circuits with selectors, as well as local and through telephone and telegraph circuits connecting with outside commercial switchboards; and communication over it leaves nothing to be desired.

In order to obtain selective operation on faulty sections on the overhead system, current-limiting resistances are installed at the Cos Cob power house and at the substation at West Farms, as shown in Fig. 28. In case of overload, these resistances, normally short-circuited, are first automatically cut in the circuit, after which an 11,000-volt control circuit is energized, and thus supplies tripping power to the bridge-type circuit breakers used for sectionalizing the trolley and feeder circuits along the line. The schematic diagrams (Figs. 20 and 25) illustrate this. At Cos Cob the resistances are arranged in two groups of three sets each, while at West Farms four sets are operated in two groups of two each. The supply to the trolley and feeder circuits, east and west, is under independent control.

A 12,850 kv-a. single-phase turbine-generator recently has been installed in the Cos Cob power house to provide the additional capacity necessary to meet the demands of increased traffic and to increase economy. This new unit is replacing one of the original 3750 kv-a. generators.

Power Supply Equipment

The following is a summary of the principal power supply equipment:

COS COB POWER HOUSE (SINGLE-PHASE RATING)

1—12,850 kv-a. Westinghouse turbine-generator
1— 3,750 " " " "
1— 3,750 " " " "
1— 4,160 " " " "
1— 5,000 " " " "

Fig. 29—Control Room at Cos Cob

1— 5,000 kv-a. Westinghouse turbine-generator
1— 5,000 " " " "
1— 5,000 " " " "
6— 7,200 kv-a. type O. I. W. C., Westinghouse 11,000-22,000-volt, single-phase, 25-cycle, auto-transformers.

WEST FARMS STATION

3— 5,000 kv-a. type O. I. S. C., Westinghouse 22,000-22,000-volt, single-phase, 25-cycle transformers.
3— 1,000 kv-a. type O. I. S. C. Westinghouse, 22,000-9,500-volt, single-phase, 25-cycle transformers.

SHERMAN CREEK POWER HOUSE

2—14,000 kv-a. Westinghouse turbine-generators.
1— 3,750 kv-a. Westinghouse frequency changer.

LINE BALANCING TRANSFORMERS

24— 2,000 kv-a. type O. I. S. C. Westinghouse, 22,000-11,000-volt, single-phase, 25-cycle, auto-transformers (including the N. Y., W. & B. Ry.).

Fig. 30—Typical Booster Transformer with Trolley Sectionalizing and Insulated Rail Joints, New Canaan Branch

Fig. 31—Signal Bridge, Showing Four Automatic Signals, Signal Transformers and Signal Relay Box. Note also Induction Bond Between Rails of Each Track. Telephone Cable at Right

CHAPTER IV

The Signal System

THE fundamental requirement of any automatic signal system is that a failure of any sort on account of interruption of power, stray foreign current, broken rail, defects of apparatus, etc., will result in signals immediately going to the "stop" position and that there will be no chance of a false "clear" indication which might result in a very serious accident. The design of all the apparatus involved has been made with this fundamental requirement in view.

The Functioning of the Automatic Block

Fundamentally, the automatic block system consists of a series of track circuits which control, at appropriate locations, automatic signals; one block control overlapping the next so as to provide advance information to the approaching engineman.

The track circuit (see Fig. 32) consists of the following: (1) Insulated joints in the rails to establish the limits of the track circuit; (2) the track transformer to supply current to the track circuit; (3) the track relay; (4) the reactor, the object of which is: first, to establish the proper phase relation in the track relay; and second, to create a choking effect when the track is occupied by a train. The track circuit (Fig. 32) is established as follows: From the track transformer "A," through a lead to one of the rails of the track circuit, from the rail to the relay coil "B," thence back to the other rail of the track circuit, thence back to the track transformer through the reactor "C."

At the entering end of the block is located a motor-driven signal "E." The circuit for the operation of the signal is as follows: From the bus bar at the location, through the contacts of a three-position or polarized line relay "F," hereinafter mentioned, thence to the signal mechanism, indicated as "E1," and thence back through the other contacts of the three-position line relay, and then back to the bus bar.

The signal mechanism is provided with a combination of contacts so arranged, that in the horizontal, or "stop" indication, the contacts are in a given position, but when the signal assumes the 45 degree or "caution" indication, or the 90 degree or "proceed" indication, the contacts are in another position. The position of these contacts controls the polarity of energy to a line circuit "L" at the far end of which is located the three-position line relay "F," mentioned as controlling the signal circuit. The object of the three-position line relay is to govern the signal indication as between zero, or "stop" indication; 45 degree, or "caution" indication; and 90 degree, or "proceed" indication.

The three-position line relay "F" is designed with two coils: one, "F," permanently energized from the local bus, and the other, "G," from the line circuit "L." The line circuit passes through the contacts of the track relay and, as outlined above, is subject to change in polarity. The functioning of the contacts for the control of the signal is as follows: When the line phase is de-energized by the opening of the track relay contacts, the moving member assumes a position by gravity, so that neither the front nor the back contacts are closed. This interrupts the circuit to the signal mechanism, which in turn, by gravity, assumes the most restrictive indication.

By energizing the line relay in one polarity, it will close two contacts through which the energy from the bus is transmitted to the signal mechanism and causes the signal to display the 45 degree aspect, or "caution." With the change of polarity of the line circuit, the moving member of the relay reverses its position closing three contacts to the signal mechanism, which, in turn, responds to the 90 degree or "proceed" indication.

Centrifugal Frequency Relay Developed

The heart of the automatic signal system is the centrifugal frequency track relay. The contacts of this type of

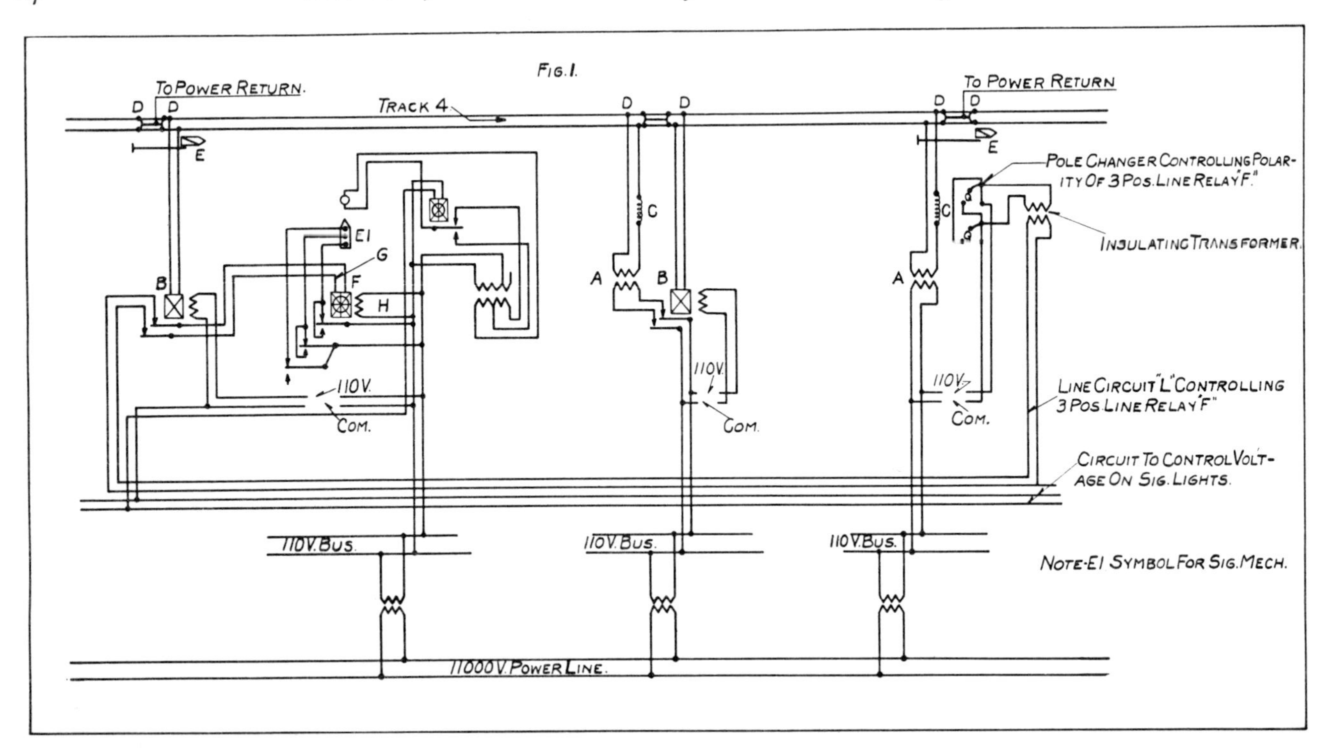

Fig. 32—Diagram of Track Circuits, Showing Three-Position Relays, Local Busses, Supply Circuits, Etc.

relay are closed by mechanical energy transmitted from a centrifuge which is lifted by the centrifugal force of revolving weighted arms, which operate in a manner similar to the ball governor of a steam engine. This revolving member is directly connected to a rotor of a two-phase induction motor. One phase of the motor is energized from the local bus at the signal location, while the other phase is energized by current from the track.

The selective feature of this relay is based upon the synchronous speed of the rotating member as driven by 25-cycle or 60-cycle current. The centrifugal force of the revolving weighted members, when operating by 25-cycle current, is not sufficient to lift the counterweight and close the contacts, but when traveling at a speed resultant from 60-cycle current, there is sufficient centrifugal force to cause the relay contacts to close. This relay not only has the advantage of being the safest type of relay that can be used, but also takes the major part of the required energy from a local bus and only a small part from the track. Therefore, it is comparatively economical in its consumption of energy.

Air-cooled transformers of good regulation properties with reactors connected in series between the track transformer and the track operate in conjunction with the track relays. The reactor is adjustable in order to create the proper phase relation in the two phases of the relay.

It is necessary to provide a means of carrying a propulsion current around the track insulators. The inductive

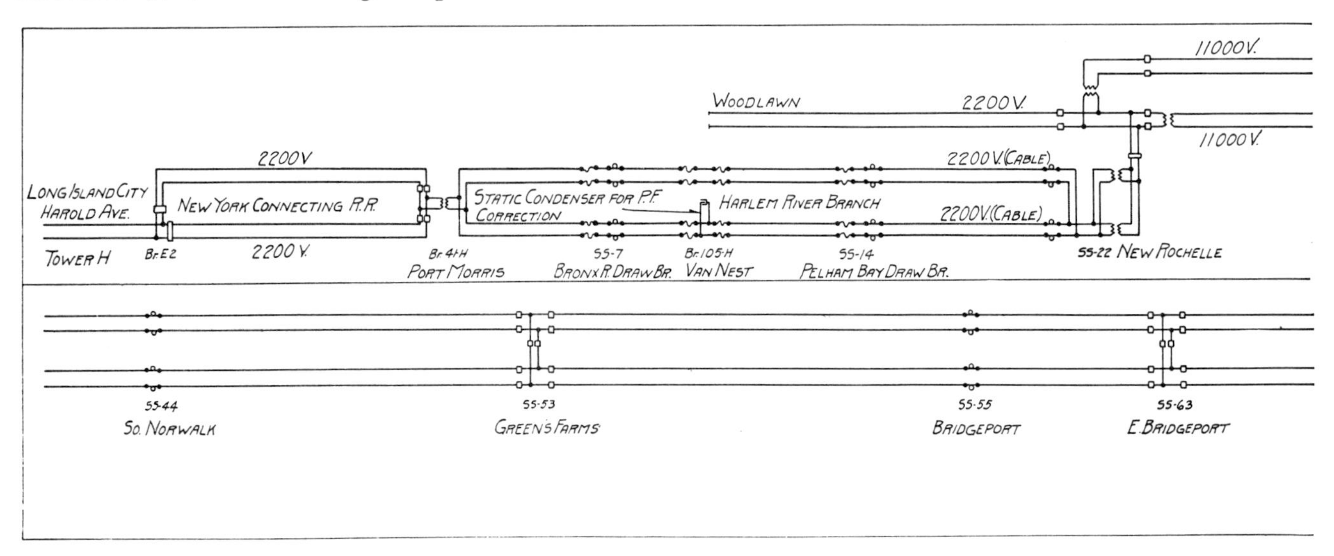

Fig. 33—Diagram of Line Showing Sectionalizing of Signal Supply Circuits

Fig. 34—Motor Generator Frequency Changer Sets at Cos Cob for Signal Power Supply

bond is similar to a large choke coil with an iron core, having three terminals, one from each end of the coil and one from the middle of the winding. The end terminals are connected to the rails, and the middle terminal is connected to a similar terminal of the bond in the next track circuit. The coil connected across the track offers considerable impedence to the 60-cycle signal current; but the 25-cycle propulsion current, entering both ends of the coil and passing out through the middle terminal to the next impedence bond, does not create any reactance in the coil; the only resistance to the propulsion current, therefore, is ohmic. The bond, as connected, has practically the same resistance as the regular rail joint bond.

Power System

The power for the operation of the entire signal system is furnished from motor-generators located in the power plant at Cos Cob. These are shown in Fig. 34. There are three 450-kv-a., 2300-volt, 60-cycle generators, driven by 25-cycle induction motors. Any two or all three units may be operated in parallel, furnishing current for the signal system, the demand being at present, about 700 kv-a. The 60-cycle current is taken single phase from the frequency changers at 2300 volts and transferred to 11,000 volts at an outdoor substation located at the power plant as illustrated in Fig. 35.

The signal transmission system consists of duplicate metallic circuit, 11,000-volt, single-phase lines extending east and west from Cos Cob. At the New Haven terminal, the circuit is reduced to 2300 volts for supply through the city, and at New Rochelle Junction, it is reduced to 2300 volts for the branch circuits extending to Woodlawn and

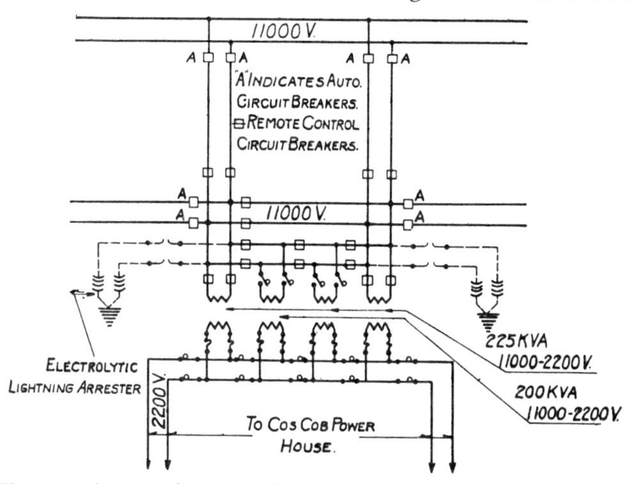

Fig. 35—Diagram of Open Air Substation at Cos Cob, Showing Connections for Supply Circuits

to Sunnyside Yard, over the New York Connecting Railroad. Facilities are provided at most of the interlocking towers for sectionalizing both of the transmission lines, which are located on the catenary structures on either side of the right-of-way, Fig. 33. The tower operators, under direction of the Load Despatcher, are in a position to cut in or out of service, any section of the transmission line. Between New Haven and South Mount Vernon, the signal load is connected to both transmission lines through automatic switches so arranged that the signal load is normally taken from one set of feeders. By the de-energizing of that set of feeders, the load automatically cuts over to the duplicate service, and also automatically restores to

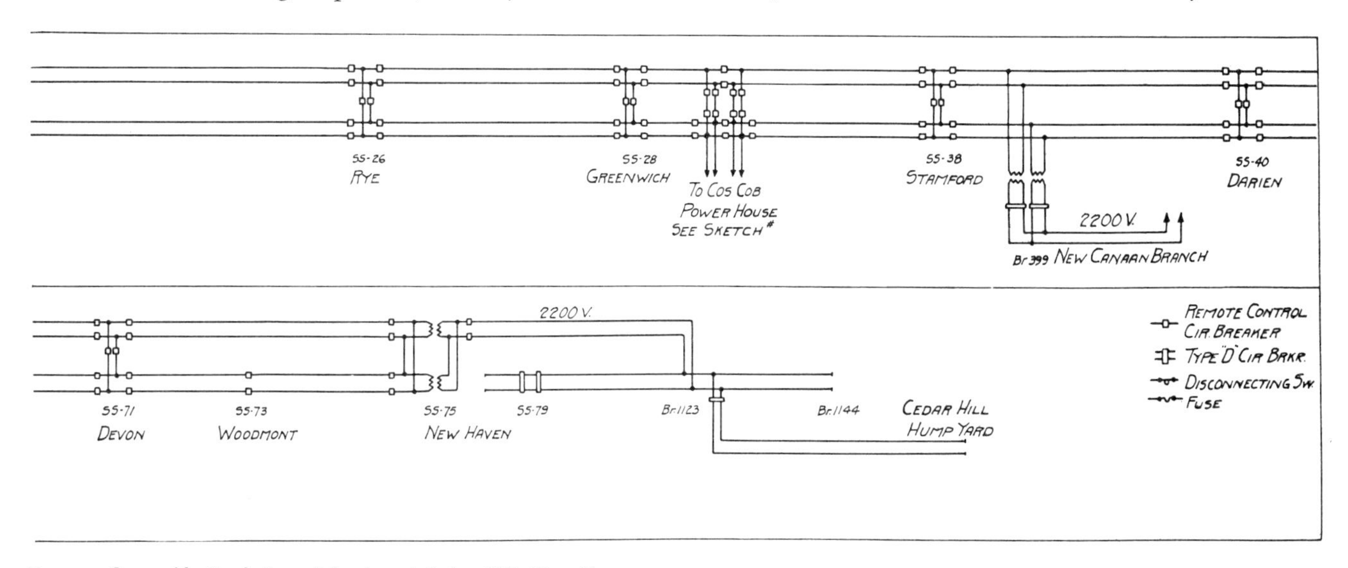

Between Sunnyside Yard, Long Island, and Cedar Hill, New Haven

the original source when that line is again energized. All of the oil circuit breakers for sectionalizing the supply lines are of the Westinghouse remote control type, and the system is thoroughly protected with electrolytic lightning arresters.

Signals

Practically all of the signals used in the electrified zone are of the motor-driven type with a "board" indicator, there being a few color light signals in the New Haven terminal. The motor signals function to the "proceed," or less restrictive indication, by means of a single-phase induction motor, and return to the more restrictive indication by force of gravity. These are illustrated in Fig. 31.

The motor-driven signals are equipped with high-powered semaphore lamps especially developed for this purpose and which burn day and night. The beam candle power of these lamps is approximately 7,000, when not transmitted through a colored roundel. The intensity of light is so great that the voltage is reduced one-half at sundown by means of a circuit operated by the nearest towerman and controlling a relay at each signal location.

Track Circuits

The maximum length of track circuit is 4000 feet, regardless of the length of the block. Where signals are more than 4000 feet apart, the track section is cut; that is, insulated rail joints are installed and the track circuits function through standard track relays around the cut section, as shown in the diagram, Fig. 32. Cross bonding for the propulsion power return is installed between impedence bond neutrals connecting all tracks at all signal locations, but not at cut sections.

The adjacent track circuit polarities are reversed so that, in case of the failure of an insulated rail joint, the current from the track transformer entering the relay of an adjacent track circuit will reverse the torque in the relay rotor. The mechanical design of the relay is such that the rotor cannot move in a reverse direction; and such a failure would cause the relay to drop, and the signal to go to a "stop" position. It is readily seen that if the polarity of adjacent track circuits were the same and an insulated rail joint were to fail, the energy from the adjacent track transformer would create a torque in the track relay which would allow the contacts to close as soon as a train had advanced far enough into the block to permit the current to reach an operating value and would result in a false "clear" indication.

Fig. 36—Sectionalizing Bridge at Greenwich, Showing Typical Tower and Sectionalizing of Traction System and Signal Feeders. Also Traction Auto-Transformers and Telephone Cable

Interlocking Stations

With the electrification of the railway and the reconstruction of the signal system, it became advantageous either to construct new signal towers, as shown in Fig. 36, or to relocate existing towers so as to establish a more efficient spacing of crossover points. The object of this was to establish, at the interlocking stations, universal crossovers so as to permit the directing of traffic and the changing of traffic from one track to another at proper distances; and also to provide for the proper sectionalization of the traction system. The interlocking stations are either all electric, electro-mechanical, or mechanical with electric control of signals. All signals operate electrically, and at all points approach locking, route locking and detector locking were planned.

The most extensive interlocking system is at Stamford where an all-electric type "F" system was installed. This system was first developed by Mr. W. F. Follett, Assistant Signal Engineer of the New Haven. From this point, energy is supplied through the type "F" bus for all signal units and track circuits on all four tracks and the New Canaan Branch, aggregating a single track mileage of approximately 20 miles. Crossovers are operated in the main line tracks as far as 3000 feet from the tower. The location of trains at this interlocking plant, as at all others, is given to the towerman by means of an illuminated track diagram. Switching and transferring traffic from one track to another over many of the more distant switches operated from the tower is conducted out of sight of the towerman.

One of the Most Extensive Signal Systems

The present signal system, furnished with power from Cos Cob, extends from Sunnyside Yard on the Pennsylvania System, to and including the classification yards at Cedar Hill, New Haven, with branches from New Rochelle Junction to Woodlawn, N. Y., and from Stamford to New Canaan, Conn.

The 700 kv-a. of energy from the generators at Cos Cob is distributed over 132 miles of power lines and 24 miles of power cables, along 89 miles of railroad, having 344 miles of signalled track, and is transformed by 1718 transformers ranging from 20 kv-a. to 225 kv-a. in connection with the operation of 940 track circuits, 1215 signal arms lighted by 1215 high-powered electric lights, and 133 electrically operated switch units. There are 36 interlocking stations having 1245 working levers; and in connection with which, there are 709 storage battery cells floated across 11 motor-generators, all of which are connected together by 175 miles of low voltage aerial cable. The consumption of energy for the signal units ranges from 700 watts for a switch motor down to .144 watts for a tower indicator. This is undoubtedly one of the most extensive single homogeneous signal systems in the world.

Fig. 37—The Latest Twelve Locomotives have Six Twin Motors and a Continuous Capacity of 2000 Hp.

CHAPTER V

Locomotives and Multiple-Unit Cars

WHEN purchasing the first electric locomotives, the initial specifications of the New Haven called for locomotives which must be able to haul passenger trains weighing 400 tons at a schedule speed of 26 miles an hour on runs averaging 2.2 miles between stops. Preference was given in the specifications for a double-unit locomotive to handle the maximum weight trains in order to provide greater motive power flexibility.

The contract for the first locomotives was awarded to the Westinghouse Electric and Manufacturing Company who bid upon single-phase alternating-current locomotives, which would also be capable of operating on direct current. All of the locomotives subsequently supplied to this railroad were built by the Westinghouse Company with the exception of one experimental locomotive.

Much Pioneer Work Involved

The design and construction of these first locomotives involved a vast amount of pioneer work and marks an epoch in the history of heavy traction electrification.

A brief consideration of some of the many problems, which had to be met, will give an idea of the magnitude of the undertaking. In the first place, the locomotives had to operate in a thoroughly satisfactory manner in both A-C. and D-C. zones. Provision had to be made for collecting direct current at 750 volts either from an under-running third rail placed on either side of the track, or from an overhead rail which was necessary at certain points where the complexity of the track work made the ordinary third rail impracticable.

On account of the clearances on the New Haven lines, the third-rail shoes had to be so constructed that they could be folded up out of the way when operating on the A-C. section, and the change had to be accomplished without delay when the locomotives were traveling at full speed.

Pantagraphs had to be designed which would successfully collect the necessary current at 11,000 volts from an overhead wire and at speeds up to 80 miles an hour. These pantagraphs not only had to operate on a wire varying in height from 16 to 22 feet above the rail, but had to be constructed so as to fold down snugly out of the way when operating in the close clearances of the Park Avenue Tunnel. All of the auxiliaries such as blower, compressors, lights, etc., had to be arranged so that they could be operated on direct- or alternating-current, under different voltage conditions.

An oil-burning steam boiler had to be provided on each passenger locomotive for the heating of the train.

The control system had to be arranged to give voltage control of the main motors when operating on A-C.; and series parallel and resistance control on D-C. The operation of the A-C. pantagraphs, D-C. pantagraphs, third-rail shoes, changeover switches, bell, sanders, etc., had to be arranged so that they could be operated by push buttons from either end of the cab.

These are some of the many problems which had to be worked out in building these locomotives; and when it is remembered that, in addition, the whole equipment had to be arranged so that a number of locomotives could be coupled together and controlled from one point by a single operator, some idea is gained of the magnitude of the work involved.

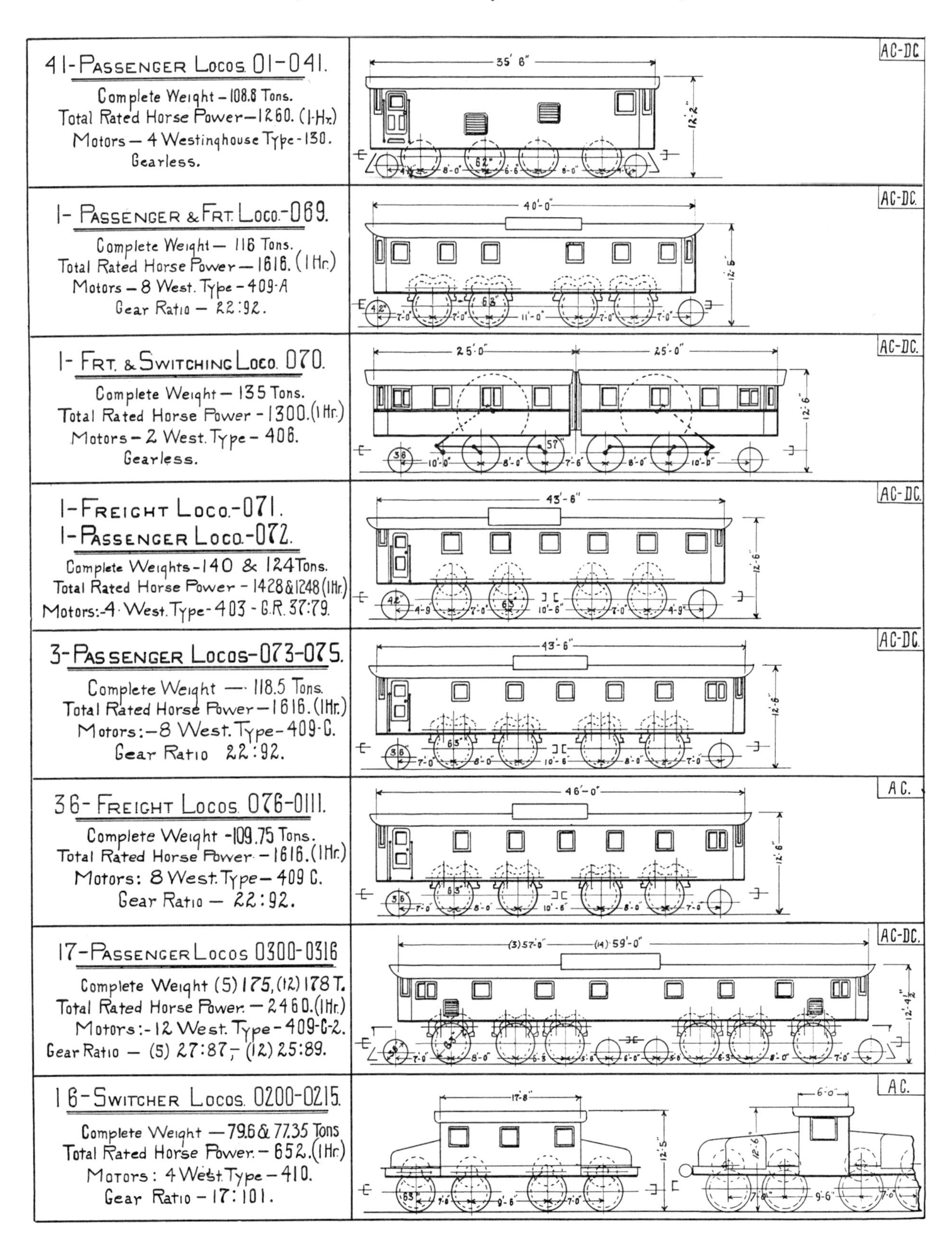

Fig. 38—Diagram of Various New Haven Locomotive Types

Fig. 39—Locomotives 01 to 041 are of This Type

Fig. 40—The 071 Locomotive, Articulated Truck Type

Fig. 41—The 069 Locomotive, Single Cab Type with Twin Motors and Experimental Under Frame

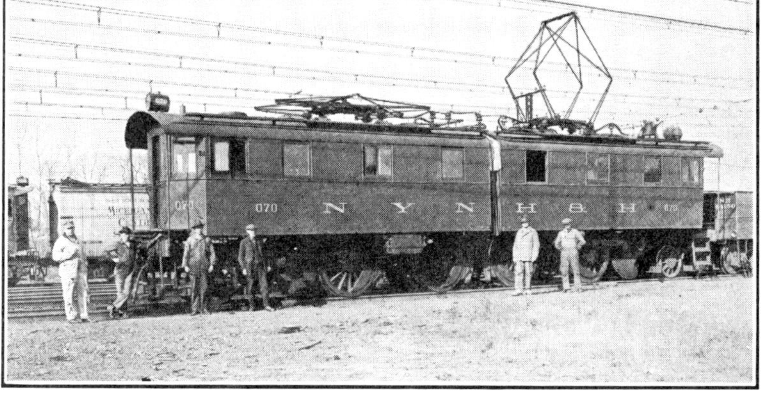

Fig. 42—The 070 Locomotive, Articulated Cab with Side Rod Drive

Fig. 43—Running Gear and Motors of the 071 Locomotive

First Lot Included 41 Locomotives

The first lot of locomotives built, Nos. 01 to 041, were of the swivel-truck type and were equipped with four 350-hp. Westinghouse type 130 gearless motors. This type is illustrated in Fig. 39. These locomotives had a continuous tractive effort of 5,200 lb., and a maximum tractive effort of 19,600 lb. When first placed in service, they used plain "bogey" trucks; but later on, guiding wheels were added to give greater stability at high speeds and to prevent "nosing." It is interesting to note that, even before the guiding wheels were added, the locomotives were operated on test at speeds as high as 89 miles an hour.

The armatures of the main motors are mounted on quills which surround the truck axles, with a liberal clearance all around. This is shown in Fig. 44. The ends of the quills terminate in discs which carry pins projecting outward and engaging with springs located in the pockets in the drive wheel centers. All of the driving torque, as well as the entire weight of the motors, is transmitted from the armatures through these springs. The main frames of the motor are split on the center line, and are supported through springs from a cradle resting on the main journal boxes. This method of motor mounting relieves the axle of dead weight and allows it to rise and fall, due to track inequalities, irrespective of the armature.

These locomotives, complete, weigh about 109 tons and now are operated singly in the lighter passenger trains or in pairs or threes on the heavy passenger trains. The first of these were placed in service in 1906, and the last of the first order for 35 was shipped in June, 1907.

Six months after the first of these locomotives had been placed in service, six more of the same type were ordered; and these all were delivered in June, 1908. The Westinghouse electro-pneumatic control system was used on these and all subsequent locomotives. The control was so arranged that it was possible to pass from the A-C. to the D-C. zone without stopping. Upon approaching the D-C. zone, the engineer lowers the third-rail shoes to the operating position by pressing a button. The control is thrown to the "off" position, the pantagraphs are lowered, the change-over switches thrown, and the control is thrown on again, so that only a momentary flicker of the lights marks the transition from A-C. to D-C. operation.

The fact that all of these locomotives still are in service, and have an average record of over a million and a quarter miles each, speaks well for the original design.

The increasing service demands, the increased weights of cars and trains, and the extension of the electrification soon made it evident that locomotives of greater capacity shortly would be required to handle the heavy through passenger and freight service. With a view to determining the most suitable form of locomotive to meet the future requirements, four different types of locomotives were built and tested in active service during the years 1910 and 1911.

An Articulated Truck Type

The experimental 071 was built first. It is of the articulated truck type, each truck comprising a "Rushton" truck and two pairs of driving axles. Gear and quill drive is employed with a single Westinghouse type 403 motor for each pair of drivers. The running gear, with the motors mounted in place, is shown in Figs. 40 and 43.

The 069 and 070 Locomotives

The next two experimental locomotives are known as the 069 and 070. These two locomotives were built at the same time, but are of widely different construction although of practically the same continuous tractive effort, namely approximately 10,840 lb. and 12,000 lb., respectively.

The 069 locomotive, shown in Fig. 41, is of the single cab type and is equipped with eight twin-geared Westinghouse type 409-A motors, while the 070, shown in Fig. 42, is of the articulated cab type and each half carries a single, large gearless Westinghouse type 406 motor with side-rod drive.

The motor equipment on the 069 is of particular interest as it was the first set of motors and drive of the type which was adopted a little later as standard for all future New Haven locomotives.

There are at present over 500 of these motors in operation on this line. The motors have a capacity of approximately 200 hp. each, at 300 volts, and are bolted together in pairs, each pair being geared to a single axle. The twin

Fig. 44—Gearless Motors Mounted on Quills Surrounding the Axle

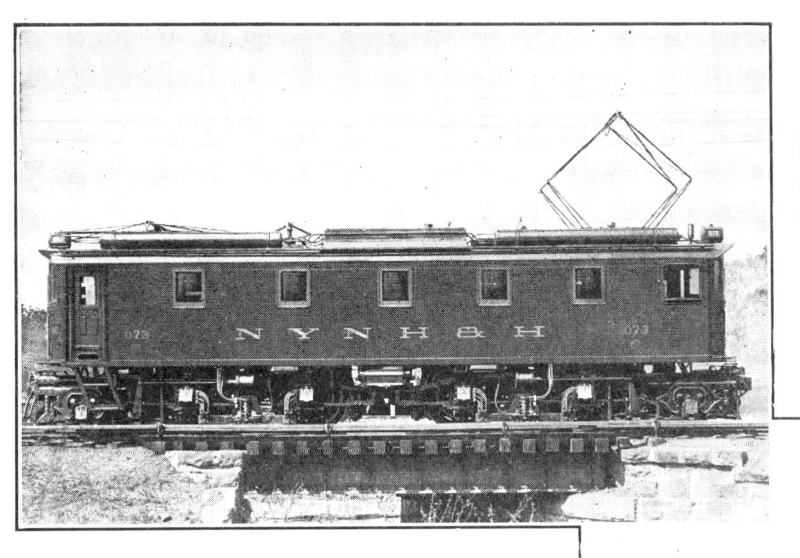

Fig. 45—There were 39 Locomotives of this Type Constructed and Three were Equipped for A-C.—D-C. Operation in Either Passenger or Freight Service

Fig. 46—Thirty-six of these Locomotives were Equipped for Freight Service on A-C. Operation Only

Fig. 47—There are 16 80-Ton Switcher Locomotives

Fig. 48—One Switcher is Provided with Control from One Position Only

Fig. 44—Standard Twin Motors with Pinions

motors, as they are called, are mounted above the axle and are connected by means of bearings to a quill which is concentric with the axle and connected at each end to the driving wheel through the medium of six helical springs. These springs are secured at one end to projecting arms on the quill and at the other end to pads on the driving wheel centers.

A single gear is mounted on one end of the quill, and a pinion on the shaft of each armature engages with this gear. The twin motor frames are mounted on the locomotive truck cross members, and proper alignment with the gear is maintained by means of the quill bearings which are securely bolted to the motor frames.

The running gear of locomotive 069 is of rather unusual construction. A deep structural steel girder forms each side and supports the entire cab weight. The two pairs of central drivers are located at eleven-foot centers and are spring-connected directly to the deep side girders. A truck having a single pony axle and a driving axle is provided at each end of the locomotive, and these trucks are connected to the locomotive frame by means of a radius bar arrangement which allows the side movement of the trucks. Weight is loaded on to the end trucks through spring pads mounted on cross members joined with side girders.

The 070 locomotive, as previously stated, is of the articulated cab type. The running gear of each half comprises a "Rushton" truck and two pairs of drivers mounted on the rigid frame of the locomotive. A jack shaft is mounted between the "Rushton" truck and the first pair of drivers, this shaft being carried in bearings on the main frame.

The single motor, which is used on each half, is mounted on the locomotive frame inside of the cab. The drive is by means of connecting rods from cranks on each end of the armature shaft to the jackshaft, and by side rods from the jackshaft to the drive wheels. This drive is the same as that which has proved so successful on the Pennsylvania Tunnel & Terminal locomotives. Both the 069 and 070 locomotives originally were equipped to operate on A-C. and D-C. although the D-C. control apparatus has since been removed from the 070 locomotive.

The 071 locomotive proved to be heavier than was necessary, and a second locomotive of this type was built of lighter construction, and was designated 072. The 069, 070, 071 and 072 locomotives, while marking a phase in the evolution of the most satisfactory all-around type of locomotive for this road, all have proved to be serviceable machines and still are in active operation.

Standard Type of Locomotive Selected

Service tests on these four locomotives were carried on for some time; and, as a result, it was decided that the 072 locomotive represented a mechanical construction which was admirably suited to service conditions, but that the twin motor construction on the 069 locomotive had certain inherent advantages as to lighter weight and greater ease of maintenance.

The logical step was to combine the mechanical features of the 072 locomotive with the motor equipment of the 069. This idea was carried out with the next 39 locomotives built, which were known as the 073 to 0111. The first three locomotives are equipped for operation on A-C. and D-C., as shown in Fig. 45, and the remainder for operation on A-C. only, as shown in Fig. 46.

The locomotives equipped for A-C.—D-C. operation weigh about 120 tons, and those for only A-C. operation weigh about 110 tons. The A-C.—D-C. locomotives have a continuous tractive effort of 10,840 lb. and a maximum tractive effort of 41,200 lb., while in the case of the A-C. locomotives, on account of lower gear ratio, these figures are 14,760 and 56,000 lb., respectively. These locomotives have been in service since 1913, and their operation has been extremely satisfactory.

In 1918, when locomotives of greater capacity were desired, the same type and the same main motors were used but an extra driving axle and two additional motors were added to each truck, and the locomotives were lengthened proportionately. These are the 0300 to 0304.

The latest locomotives, the 0305 to 0316, received during the latter part of 1923 and early in 1924, are shown in Fig. 37. They weigh approximately 178 tons complete, and their six twin motors give them a continuous capacity of 2,052 hp. They are similar to the 0300–0304 except for a slight change in gear ratio, and are capable of a maximum tractive effort of 52,500 lb. and a maximum speed of 70 miles an hour.

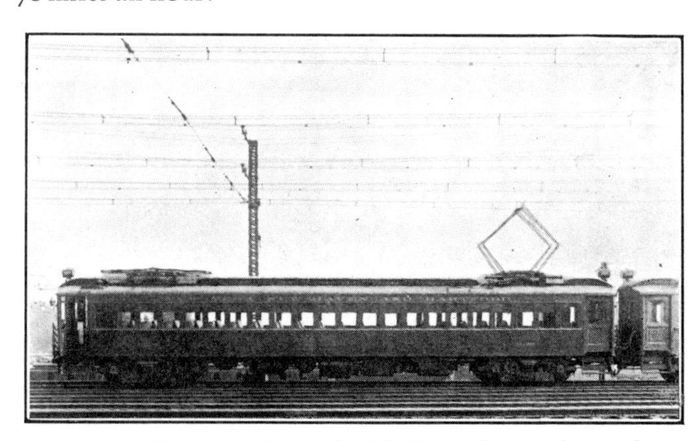

Fig. 50—Thirty-Five Multiple-Unit Motor Cars are Operated

ELECTRIC LOCOMOTIVE DATA

								Driving Wheels		Weights (in Pounds)		Dimensions	Wheel Base		Motors						Tractive Effort (in Pounds)		Horse Power of Locomotive with Forced Ventilation		Speed M.P.H.		Trailing Load in Tons Absolute Maximum Safe Rating			Max. Momentary Tr. Effort
	Locos	No. in Service	Year Placed in Service	Class of Service	Boilers for Train Heating	System of Traction	Wheel Order & Weight per Axle	No.	Diam.	Total	On Drivers	Length over All	Rigid	Total	No.	Type	H.P. 1 Hr. Rating with Forced Ventilation	Motor Volts (A.C.)	Method of Drive	Gear Ratio	1 Hr. Rating with Forced Ventilation	Continuous Rating with Forced Ventilation	1 Hr. Rating	Continuous Rating	Hour Rating	Maximum Safe	Passenger Service Express	Passenger Service Local	Freight Service	
MULT	01-041	41	1907	PASSENGER	YES	AC-DC	25610, 43370, 42820, 41940, 40820, 23480	8	62"	217,600	167,600	37'-7½"	8'-0"	30'-9"	4	W.130	*315	284	DIRECT	—	8,000	5,200	1,260	1,016	59.0	88	390	200	—	19,600
	069	1	1910	PASSENGER & FREIGHT	YES	AC-DC	28000, 43000, 43000, 45000, 45000, 28000	8	63"	232,000	176,000	46'-8"	11'-0"	39'-0"	8	W.409A	*202	330	GEARED	22:92	15,600	10,840	1,616	1,336	38.8	54	800	350	1500	41,200
	070	1	1910	FREIGHT & SWITCHING	NO	A.C.	38000, 47500, 47500, 48500, 48500, 40000	8	57"	270,000	192,000	52'-0"	8'-0"	43'-6"	2	W.406	650	310	SIDE ROD	—	16,800	12,000	1,300	1,100	29.0	61	800	350	1500	36,000
	071	1	1910	PASSENGER & FREIGHT	YES	AC-DC	45000, 47000, 47000, 48000, 48000, 45000	8	63"	280,000	190,000	48'-0"	7'-0"	38'-6"	4	W.403	*357	380	GEARED	34:79	10,400	10,120	1,428	1,432	51.5	67	800	350	1500	40,400
	072	1	1911	PASSENGER & FREIGHT	YES	AC-DC	32000, 46000, 46000, 46000, 46000, 32000	8	63"	248,000	184,000	48'-0"	7'-0"	38'-6"	4	W.403-B	*312	342	GEARED	34:79	10,400	10,120	1,248	1,240	45.0	67.0	800	350	1500	40,400
MULT.	073-075	3	1912-3	PASSENGER & FREIGHT	YES	AC-DC	28500, 45500, 45500, 45500, 45500, 28500	8	63"	239,000	182,000	50'-0"	8'-0"	40'-6"	8	W.409-C	*202	330	GEARED	22:92	15,600	10,840	1,616	1,336	38.8	62.5	800	350	1500	41,200
MULT.	076-0111	36	1912-3	FREIGHT	NO	A.C.	26900, 40700, 40700, 41800, 41800, 27600	8	63"	219,500	165,000	50'-0"	8'-0"	40'-6"	8	W.409-C	202	330	GEARED	17:97	21,200	14,760	1,616	1,336	28.5	45	—	—	2000	56,000
MULT.	0300-0304	5	1919	PASSENGER	YES	AC-DC	[illegible]	12	63"	350,000	233,000	69'-0"	14 3/10	59'-6"	12	W. 409-C2	*205	335	GEARED	27:87	18,000	12,540	2,460	2,040	51.0	70	900 (770 NORMAL)	420 (360 NORMAL)	—	47,520
MULT	0305-0316	12	1923-4	PASSENGER	YES	AC-DC	[illegible]	12	63"	356,000	240,000	68'-6"	14 3/10	59'-6"	12	W 409-C-2	209	349	GEARED	25:69	19260	13,080	2,508	2,052	48.5	70	900 (770 NORMAL)	420 (360 NORMAL)	—	52,500
	0200	1	1911	SWITCHING	NO	A.C.	39500, 39500, 39500, 39500	8	63"	158,000	158,000	37'-0"	7'-0"	23'-6"	4	W.410	125	190	GEARED	17:101	23,200	14,480	500	440	8.1	25	FLOAT SERVICE - WILL UNLOAD AT LOW TIDE FLOATS HOLDING 22 LOADED CARS (60 TONS AVERAGE WEIGHT) FOR PERIODS OF 10 MINUTES EACH IN EVERY HOUR. CLASSIFICATION SERVICE - WILL MAKE 60 MOVES PER HOUR, AVERAGE LENGTH RUN 350 FT. AV. WEIGHT OF TRAIN, 450 TONS			46,000
	0201-0214	14	1912	SWITCHING	NO	A.C.	39800, 39800, 39800, 39800	8	63"	159,200	159,200	37'-6"	7'-0"	23'-6"	4	W.410	163	228	GEARED	17:101	23,200	14,400	652	520	10.5	25				
	0215	1	1912	SWITCHING	NO	A.C.	38675, 38675, 38675, 38675	8	63"	154,700	154,700	37'-6"	7'-0"	23'-6"	4	W.410	163	228	GEARED	17:101	23,200	14,400	652	520	10.5	25				
	3-4	2	1915	SWITCHING (MAN. R.R. - N.H.)	NO	D.C.		8	36"	59,000	59,000	29'-5"	6'-0"	19'-0"	4	W. 101-K	°40	600 (D.C.)	GEARED	14:79	°7,300	°2,800	°160	°87	8.4	30				12500

NOTE:—"MULT." INDICATES LOCOMOTIVES WHICH ARE CAPABLE OF MULTIPLE UNIT OPERATION WITH EACH UNIT IN SAME GROUP

* A.C. RATING.

° INDICATES NATURAL VENTILATION.

Fig. 51—Comparative Data on Various New Haven Locomotive Types

In addition to the road locomotives, sixteen 80-ton switching locomotives were built and placed in service in 1912. The running gear of these locomotives consists of two articulated trucks with two driving axles in each truck. These locomotives are equipped for operation on A-C. only, as the yards are all in the A-C. zone. The motive equipment consists of four 187-hp. Westinghouse type 410 motors, each of which is geared to a single axle through a quill drive in the same manner as on the 072 type of locomotive. The entire weight is on the drivers.

These locomotives have a continuous tractive effort of 14,400 lb., and are capable of exerting a maximum tractive effort of 46,000 lb. Fifteen of these locomotives are arranged for double end control, as shown in Fig. 47, and one is provided with a high control cab and control from this one position only, as shown in Fig. 48. These locomotives are kept in service twenty-four hours a day with three eight-hour crew shifts, and average from 4300 to 4500 miles per month.

Multiple-Unit Cars

In addition to the 117 electric locomotives in service, 35 multiple-unit motor cars and 65 trailers are employed. The motor cars are equipped with four A-C.—D-C. motors. Westinghouse type 156 motors having a rating of 150 hp. were used on the first four cars equipped, but the later cars use Westinghouse type 409-D or type 412, 175 hp. motors. The control is the Westinghouse HB electro-pneumatic type. The energy for operating the magnet valves of the control and auxiliary apparatus is taken from a motor generator set and a storage battery at 32 volts.

The cars are equipped with pantagraphs and third-rail shoes of the same type as used on the locomotives. The main transformers, used for the purpose of stepping the 11,000 volts on the line down to the low voltage employed for the operation of the motors, are of the air blast type. The forced ventilation for both the motors and the transformers is supplied by a motor-driven blower and the air is carried from the blower to the apparatus through conduits built into the car body. The motors are axle-hung and are geared through spur gears of the flexible type. Both motor and trail cars are provided with independent main and auxiliary lighting systems, the main lighting circuits receiving power direct from the line on D-C. and from the main transformers on A-C.

The auxiliary, or center, circuit of lights is supplied with current from a 500-volt lighting transformer transformed at 32 volts when the car is operating in the A-C. zone, and from a storage battery when operating in the D-C. zone.

The changeover is made automatically throughout the train from the head end, which may be either a motor car or a trailer, when passing from one zone to the other. The motor cars are approximately 72 feet long, weigh 85½ tons, and seat 84 passengers. They are geared for a maximum speed of 55 miles an hour. The trail cars are of the same length and seating capacity as the motor cars and weigh approximately 51½ tons. These cars are operated in the ratio of one motor car to two trailers.

Fig. 52—Van Nest Electric Shops. Inspection Shed on Right and Special Building for Dipping Armatures at Left

CHAPTER VI

Equipment Maintenance Practice and Facilities

THE electrical equipment, as noted in Chapter V, at the present time consists of sixty-four A-C.—D-C. passenger locomotives, thirty-seven A-C. freight locomotives, sixteen A-C. switching locomotives, twenty-nine A-C.—D-C. and six A-C. multiple-unit motor cars, and sixty-five multiple-unit trailer cars, exclusive of the New York, Westchester & Boston Railway equipment.

The maintenance of these several classes of equipment is made more difficult not only on account of the variety of the equipment itself, but on account of the different classes of service and the fact that local conditions prevent concentrating the inspection and maintenance at a single point. Passenger equipment, including the multiple-unit cars, must meet the heavy demands for suburban service from several points, with allowance for inspection and with a very small percentage for shopping margin. The service is not such as to permit the freight and switcher locomotives to be handled at the same terminals as passenger equipment, and freight equipment frequently is substituted in passenger service, particularly in the summer season.

At present, the Van Nest electric repair shop is used as the point where heavy inspection and overhaul work on all classes, and regular periodic inspection of freight and switcher equipment, is accomplished. No train service is cared for at this point.

At Stamford, Connecticut, the passenger locomotives and multiple-unit cars are inspected and running repairs taken care of. Multiple-unit trains and certain local trains are made up at this point, and locomotives are also provided for passenger service, using Stamford as a terminal.

At Grand Central Terminal, New Haven, New Rochelle, Port Chester, and Oak Point, small maintenance forces make only light repairs and arrange for the assignment of locomotives and equipment to trains originating or terminating at these points.

All electrical equipment, with the exception of multiple-unit trailer cars, is inspected on a 2,500 mile basis, the trailer cars being inspected every three months. Passenger locomotives are given a general overhaul on a 200,000 mile basis, freight and switching locomotives on a 100,000 mile basis.

Van Nest Electric Repair Shop

On account of the several points at which maintenance work is done, interchangeability of wearing parts is of great importance, and many special tools and fixtures have been built at Van Nest shop to insure this interchangeability. The general practice is to remove a defective unit from equipment at any point where it is reported, and forward the defective part to the Van Nest shop. It is there repaired and returned for service in a car especially arranged for the transportation of motors and the ordinary supplies.

Most of this repair work on small parts is done in what is known as the assembly department, located on a balcony, as shown in Fig. 55; this department also makes repairs on all switching equipment and similar parts removed from locomotives being overhauled. Here also small parts

Fig. 53—Shop Floor at Van Nest—Heavy Machine Tool Bay in Foreground and Erecting Bay at Left

are manufactured, and third-rail equipment is tested, and bearings babbitted. Under the assembly balcony is located special air-brake testing apparatus.

The heavy repair work is done on the main floor. This includes the repairs to cabs and trucks, the rewinding of armatures, turning of commutators, turning and replacement of wheels, welding on all classes of equipment, and other work that is necessary at the time of general overhaul. Armatures are dipped in an adjoining fire-proof building and are baked in ovens at the rear of the heavy machine tool bay. The armatures are handled by cranes and are lowered through doors in the top of the oven.

When the Van Nest shops first were opened, armatures were dipped in a tank in the floor of the shop adjacent to the bakeoven. Due to the fire hazard of having this inflammable material inside the shop, a separate building, entirely fire proof, has been constructed as a dip house. The base is concrete with brick walls, a tile roof, and wire glass windows, and all doors and shutters are of steel. The dipping tank contains 2300 gallons of varnish and is provided with an air inlet at the bottom for agitating purposes. It is also equipped for either manual or automatic emptying by gravity into a distant underground tank in

Fig. 54—Heavy Machine Tool Bay with Baking Oven at the Rear

case of fire. Material to be dipped is handled by an air-operated crane and there is no electricity inside the building.

The Van Nest shop equipment includes also facilities for electric welding, but this is not as extensively used as the oxyacetylene method. This work is greatly facilitated by the location on the premises of any oxyacetylene generating plant. Pipes conveying both the oxygen and the acetylene are carried throughout the shop and outlets are provided at numerous points. The generating plant has a capacity of 200 cu. ft. of acetylene gas an hour.

Maintenance of Multiple-Unit Equipment

The New Haven multiple-unit equipment is designed to operate in trains of a ratio of at least one motor car to each two trailers. As the changeover from A-C. to D-C., or vice versa, is made at approximately 30 miles an hour, all apparatus affected must be operated electrically or electro-pneumatically and controlled from the engineman's control station.

Trailer cars are inspected every three months at Stamford, the work usually being done out of doors. A new inspection shop is now under construction at Stamford, which will make it possible to do this work under cover. This shop will be long enough to have a six-car train completely

Fig 55—Assembly Department Located on Balcony

under cover for inspection with another track for repairs on individual cars or locomotives. Here the electrical equipment is inspected, cleaned and lubricated.

Trailer cars are sent to Van Nest shops when requiring paint or heavy repairs, but usually the painting is the limiting feature for shopping. At the time of shopping, all equipment is thoroughly gone over, being removed from the car if necessary, and put in good condition for road service. The trucks are overhauled, and any other mechanical features on the car requiring repairs are taken care of.

In connection with motor car maintenance, the A-C. oil circuit breaker inspection every six months is made at the time the oil is changed due to change of season, during April and October. The main motors are gone over at each inspection, thoroughly cleaned and worn parts renewed. Lubrication is supplied to the armature and axle bearings in accordance with a chart showing proper depth of oil to be placed in the wells.

Fig. 56—Erecting Bay where Heavy Repairs are Made

The switch groups are opened and all parts inspected, new parts being supplied if necessary. Other switching equipment is looked over in the same way to make sure that all terminals and contacts are in proper condition.

Forced ventilation is used on these cars for the transformers and main motors, the blower motor being practically the same as the compressor motor. These motors are inspected in a manner similar to the main motors, renewing brushes, and lubricating, and cleaning. The later type blower motors are equipped with ball bearings and are lubricated with grease.

The grid resistors used on this equipment are heavy and it is essential that they be well maintained. An insulation test of 1,000 volts is applied, after inspection is complete, to all main circuit apparatus, including separate tests on resistors and switch groups which are mounted on insulated bolts. The car is tested out for proper sequence of switches and action of the auxiliary motors before releasing from the shops.

At every third inspection, the car is lifted from the truck center bearings and these parts are inspected and lubricated. After each inspection, a form is filled out showing the parts renewed, the general condition of main and auxiliary motors and just what other work was done. This is kept on file for checking at future inspections.

Periodically, motor cars are sent to the Van Nest shops for painting, and at that time the trucks are removed and dismantled, the motors also being dismantled, if necessary, for turning of commutators or other work. While the car is at Van Nest shop, the electrical equipment is gone over more thoroughly than is possible at the ordinary inspection, and the various relays and magnets operating changeovers, third-rail shoes, trolleys and other devices are removed, cleaned, and tested for proper operation.

At the time the car is being painted, it is checked for mechanical defects, proper operation of doors, windows and all air brake devices. The brake valves, distributing valves and brake cylinders are inspected and cleaned at this time and a hydrostatic test is applied to the various air reservoirs.

Locomotive Maintenance

Light Inspection

All locomotives are inspected on a basis of 2500 miles of service. On account of train service using Stamford as a terminal, the passenger locomotives are inspected here during a day and night period, except Sunday. Freight and switching equipment, however, is inspected at Van Nest shop, two locomotives being inspected each working day within an eight-hour period.

At the end of trips, all passenger locomotives are looked over and any defects that may have been reported by the crew are corrected. These usually are light repairs, such as the replacement of a trolley shoe, but may cover switch group repairs, control repairs, renewal of brake shoes and defective piping. During the winter season, it is also necessary to fill oil and water tanks to take care of the train heating. Three passenger locomotives are inspected during the day, and two at night, usually of the smaller type.

The smaller locomotives, which are the oldest in service and use the gearless motors, have two classes of inspection, known as "light" and "heavy." The heavy inspection is each third inspection for a given locomotive, and at this time, the work done at light inspection is taken care of, but a longer time is allowed for this inspection and certain other things are done as well. The main motors are thoroughly cleaned, the center bearings are lubricated, and several other items designated as "heavy inspection" items are taken care of. Locomotives for heavy inspection are so selected that one of the smaller type chosen for daylight inspection will be due for this work.

The main motors are inspected in a manner similar to those on motor cars, and brushes and brushholders are renewed as may be necessary, and the bearings lubricated. The switch groups are examined and all the usual wearing parts are changed as required.

The locomotive is given a complete mechanical inspection, brake shoes examined and renewed, journal boxes and other bearings packed and lubricated. On locomotives using the geared-type motors with quill drive, the

Fig. 57—Front End of Erecting Bay, Showing Multiple-Unit Car Undergoing Repairs

quill height is checked and if necessary the motor is adjusted to keep the quill properly concentric with the axle. The air-brake apparatus is adjusted and tested to see that it functions properly and such other work done as may be due for periodic inspection.

The pantagraph trolleys are tested in operation, the shoes examined for wear, and the spring tension measured both going up and coming down. The high-tension cable connecting the two trolleys, together with the supporting insulators for the cables, hose, and trolleys themselves are given a megger test periodically. This test tends to eliminate defective parts, such as cracked porcelains, before they bring about a failure in service.

Each locomotive is equipped with two motor-driven blowers and compressors. These motors are cleaned, brushes and brushholders examined and renewed, and bearings lubricated. Ball bearings are used on the blower motors and are grease lubricated. At the completion of the inspection, the sequence of switches and the action of the auxiliary motors are tested.

The inspection on the freight and switching locomotives is handled in the same general manner as that of the passenger equipment, a number of items being cared for on a periodic basis. As these locomotives use a large volume of cooling air for the electrical equipment, the roof, main motors and other equipment are cleaned with compressed air before the locomotives are placed in the shop. The draft gear also requires more repairs on this class of equipment than on the passenger locomotives.

As Van Nest shop is the point where all major repairs are taken care of, in case any parts are found defective, they are changed if possible while the locomotive is undergoing inspection. In case it is found that such items as wheels or main motors must be changed, the inspection is completed with the exception of the part to be changed, and the locomotive is then taken to another portion of the shop where such work is done.

As the freight and switching locomotives are designed for A-C. service only, the number of pieces of control apparatus are few in number and are relatively easy to keep in repair. Therefore, there is no heavy inspection of such parts, defects usually being found at the regular inspection, and additional work is not done except at periods of general overhaul.

General Overhaul Work

Passenger locomotives are given a general overhaul on

Fig. 58—Inspection Shed at Van Nest. Tracks at Extreme Right are Used for Painting

Fig. 59—Inspection Shed at Stamford

a basis of 200,000 miles, while freight and switcher locomotives are overhauled on a basis of every 100,000 miles.

At the time of overhauling, the locomotive cab is lifted from the trucks and supported in such a way that work can be done both inside and outside of the cab. All riveted parts are checked; and, if found to be loose, are again riveted properly and the trucks are completely dismantled and rebuilt with new parts or such repaired portions as can be used again. The cab sheets are replaced if found to be rusted through, and the cab is painted inside and outside.

Air tanks are removed and given a hydrostatic hammer test and the date of such test stencilled on each tank. The working parts of the air-brake equipment are removed and repaired, cleaned and tested, and again replaced. The dates applied to the overhaul period now become the basis for future periodic inspection.

The motors that have been taken from the trucks are dismantled and have new bearings applied, fields and armatures cleaned, and the commutators turned if necessary.

On the small passenger locomotives using the gearless motors, all parts are worked very close to their rating or above it; and on this class, the switching equipment is entirely removed from the cab, and is dismantled, new insulation applied, if necessary, and again rebuilt. In order to do this work on this class of locomotives, a spare set of switch groups is maintained, so that as the old equipment is removed, a new set can be put in place almost immediately without waiting for the time necessary to overhaul it.

On the later types of passenger locomotives, and on the freight and switching equipment, it has not been found necessary to completely remove all apparatus from the cab. The working units themselves, such as the switches, cylinders and valve magnets, reversers, and auxiliary magnets are removed and either overhauled and again installed for service, or replaced by similar ones which have been repaired and put in first-class condition.

Each overhauled locomotive, after the equipment is all assembled again, and the parts connected for service, or after motor changes at Van Nest shops, is given a short test under its own power, including the individual operation of main and auxiliary motors, to see that it functions properly, both on A-C. and D-C. for passenger, and on only A-C. for other equipment. After releasing from general overhaul, a locomotive is restricted in service for a few days to see that hot bearings and other defects do not develop, after which it is released for general service.

Fig. 60—One of the New Locomotives with Washington to Boston Express Train Crossing Hell Gate Bridge

CHAPTER VII

Service and Results of Operation

Passenger Service

THE New York, New Haven & Hartford Railroad has the most dense passenger traffic of any standard road engaged in interstate traffic, and part of its main line is a four-track railroad carrying the heaviest passenger traffic of any road of its length in the country. Of every ten or eleven passengers riding on standard railroads in this country, one rides on the New Haven. The extent of transportation service offered by the New Haven may readily be visualized as being equivalent to the transfer of 18 passengers and of 28 tons of freight every minute between New York and Boston.

The New Haven brings to its passenger terminals at New York about 23,000 passengers daily; and its subsidiary, the New York, Westchester & Boston, delivers about 7000 daily at its various Bronx stations.

Most of the New Haven's passenger business is handled at Grand Central Terminal, jointly used with the New York Central Railroad. The present terminal was opened in 1913, and passenger traffic was handled without interruption throughout the ten year construction period. The Terminal handles normally some 376 revenue trains daily, in and out, 149 New Haven and 227 New York Central, in addition to 191 light engine and transfer movements to and from Mott Haven Yard. The New Haven, in 1916, handled 11,400,000 passengers at Grand Central, and 17,600,000 in 1923, an increase of 54½ per cent. In March, 1924, the New Haven carried 1,442,209 passengers into and out of the Grand Central Terminal.

The interlocking signal machine used in the lower level of the Grand Central Terminal, where most of the suburban trains of both roads are handled, is the largest ever built. This is in a four-story building with separate interlocking plants on the upper and lower levels.

Pennsylvania Terminal is used by the through trains operated by the New Haven and Pennsylvania Railroads between Boston, New York, Philadelphia, Baltimore, Washington and Pittsburgh, with additional through service to Bar Harbor and the White Mountains and to Florida and other Southern points during the respective seasons. This terminal is the largest "through" station in the world; it is connected by means of two tunnels with the Pennsylvania main line on the Jersey Shore, and by four tunnels with the Long Island Shore. New Haven trains enter over the New York Connecting Railroad and Hell Gate Bridge and through the East River tunnel. The building proper covers an area of eight acres, the entire area occupied being twenty-eight acres with sixteen miles of track.

The handling of passenger travel presents many difficulties not reflected in averages. The greater part of the ordinary day's suburban movement (or nearly 33 per cent of the entire number of trains each day) into Grand Central Terminal, is handled during two morning hours, and in a corresponding period outbound in the late afternoon. In addition, there are large, and constantly growing, movements in and out of New York over the various holidays, and to and from the resorts in Maine and other sections of New England during the summer. Many of the through trains operate under these conditions in six or eight sections.

As an example of a peak movement, the New Haven carried to the Yale-Harvard football game at New Haven, in 1922, nearly 57,000 people, most of whom were from

Fig 61—Grand Central Terminal Area Before Electrification

Fig 62—Grand Central Terminal Area After Electrification

Fig 64—The New Haven Also Uses the Pennsylvania Terminal—the Largest "Through" Station in the World

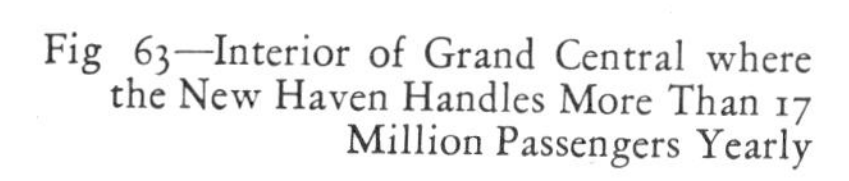

Fig 63—Interior of Grand Central where the New Haven Handles More Than 17 Million Passengers Yearly

Fig 65—Baldwin-Westinghouse, 110-Ton, 1260-Hp. Locomotives in Main Line Express Passenger Service

Fig 66—The Newest Locomotives of the New Haven are Operating in Passenger Service

Fig 68—Baldwin-Westinghouse, 110-Ton, 1616-Hp. Locomotives in Main Line Freight Service. Hunt's Point

Fig 67—Forty-Three Float Loads are Handled Daily from the River Floats at the Oak Point Yards

Fig. 69—The Switching Locomotives Operate 24 Hours a Day, Making 4500 Miles Monthly. Note Four Electric Switchers in Operation

New York. They were hauled on 22 special trains from Grand Central and one special train from the Pennsylvania, over the Hell Gate Bridge route, all in addition to the regular service.

In March, 1924, the total mileage made by electric locomotives in passenger service was nearly 306,000. This resulted from an average daily movement of 171 miles for each serviceable locomotive, or 152 miles for all locomotives, including those in the shop, assigned to this class of service. During the same month, motor cars made over 96,000 miles with a daily average of 103 miles for all cars in service.

Freight Service

The New Haven receives daily an average of 704 freight cars through the New York gateway, 434 coming from the Pennsylvania and 36 from the Long Island (both by way of the New York Connecting Railroad), 122 from the Lehigh Valley, and 92 from the Central Railroad of New Jersey over the all-float route. These are practically all loaded cars. A corresponding number are moved west, of which about sixty per cent are empty. At Oak Point there is an all-rail interchange with the New York Central; here, too, the four Brooklyn contract terminals, Bush Dock, New York Dock, Jay Street and Brooklyn Eastern District, interchange with the New Haven, using their own floating equipment.

The Harlem River freight house in the Bronx handles freight to and from points above 59th Street. The Produce House at Harlem River is occupied by various dealers handling potatoes, apples, cabbages and other perishable farm products. This class of freight traffic is especially exacting in its requirements. Other freight stations operated by the New Haven within the city limits are at West Farms, Van Nest, Westchester, Baychester and Woodlawn. At Fresh Pond Junction, a freight station is operated jointly with the Long Island Railroad.

New Rochelle Junction, the point where the Harlem Branch leaves the main line, is, considering its restrictions, the busiest traffic junction on the system, or probably on any road within a hundred miles of New York Harbor. In a normal 24-hour period, 261 trains with 3483 cars pass over this junction. Of this number, 50 trains, including 2307 cars, are freight. As many as 1708 freight cars have been run east from Harlem River in 24 hours; all of these cars, along with those westbound, moved over the junction crossovers. There were 37,334 cars moved east during a 33-day period in the spring of 1923—a sustained average of 1131 a day. During six hours on June 23, 1923, 16 trains, totaling 813 cars, were started east from Oak Point.

Since March, 1924, all freight between Oak Point and Cedar Hill (New Haven) has been handled with electric locomotives. It is not an unusual performance for an electric locomotive in this service to make the round trip from Oak Point to Cedar Hill (69 miles) and return in six hours—often permitting two such round trips per day for these engines. During March, 1924, 156,052,095 ton miles of freight were moved by electric locomotives, and the average train speed was 17.43 miles per hour. The average locomotive mileage per day for all locomotives in this service, including those in the shop, was 117.49 and for serviceable locomotives, 140.

Switching Service

Switching service in the yards at Oak Point, Harlem River, Westchester and Stamford is handled by electric switching locomotives. There are 16 of these locomotives used in this service. The three most important yards are those located at Oak Point, Harlem River and Westchester.

The number of locomotives assigned to the Oak Point

Fig. 70—Hell Gate Bridge which Connects New England with the South and West

yard varies from five to seven according to the amount of traffic to be handled. When five engines are assigned to the yard, four are used in what is known as "float yard" operation, consisting of unloading eastbound cars from the floats and loading westbound cars on the floats. These four electric engines also perform whatever switching is necessary with the westbound cars. The fifth locomotive takes the eastbound cars from the float yard to the classification yard and makes up the eastbound trains. One switching locomotive is regularly used as a helper engine on the 1.5 per cent grade of the Hell Gate Bridge for westbound freight trains to Bay Ridge.

A representative month includes an average of forty-three float loads handled daily at the Oak Point yard. This would give a total of 15,000 tons for each electric switcher in twenty-four hours, on a basis of 900 tons for eastbound float loads and 600 tons for the westbound. Approximately 75 per cent of the eastbound cars received go to the classification yard, the balance being taken to the Westchester yards. This means that the engine working in the classification yard handles a daily tonnage of something like 29,000. Considerable miscellaneous switching and some movement of westbound cars is done by the four locomotives working in the float yard in addition to loading and unloading the floats.

Several of the locomotives in switching service have monthly mileage records of 4300 to 4500. The locomotives are kept in service 24 hours by using three eight-hour crew shifts and a number of the engines have made continuous service records of 24 hours per day for 30 days.

COMPLETE electric operation of the electrified division has been the goal toward which the New York, New Haven and Hartford Railroad has been working since 1906. This goal has now been achieved and in the foregoing chapters an endeavor has been made to draw a picture of the present electrification. The intent has been to present a story which would be of interest to both the non-technical reader, interested either in electrification in a popular way or in the New Haven Road, and the engineer who is interested in the technical aspect of this great undertaking. To do this, it has been necessary to go somewhat into the historical, to eliminate many of the technical details, and to cover the subject in a rather general way. It is hoped that the outline given will prove of interest to both classes of readers.

Transportation Hints

Special Publication 1711
October, 1924

Westinghouse Electric & Manufacturing Co.
East Pittsburgh, Penna.

CONTENTS

Modernize and Merchandise Electric Railway Service

Rail Transportation Should Be Up to New Standards

The American people, all said and done, will not be satisfied with anything but the best of service. They are willing to be liberal where that service satisfies them, and they like luxury. Standards of living have changed a great deal in the last five or six years and transportation ought to be changing with them. Consequently, I think that our track ought to be better maintained, our speed ought to be materially increased, and the exterior of our cars ought to be more attractive and the interior more luxurious. All this will tend to sell the thing that we have to sell, and that is transportation.

By BRITTON I. BUDD
President American Electric Railway Association

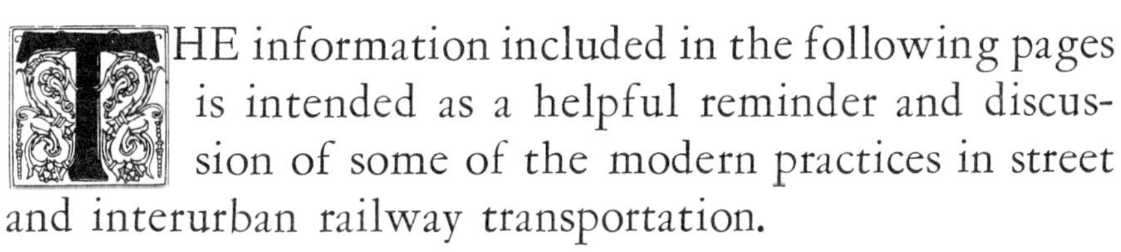

THE information included in the following pages is intended as a helpful reminder and discussion of some of the modern practices in street and interurban railway transportation.

It is hoped that this summarization of modern transportation practice will be of similar benefit to the transportation engineer as have our operating and maintenance publications of past years been to the equipment engineer.

Transportation in a Busy City.

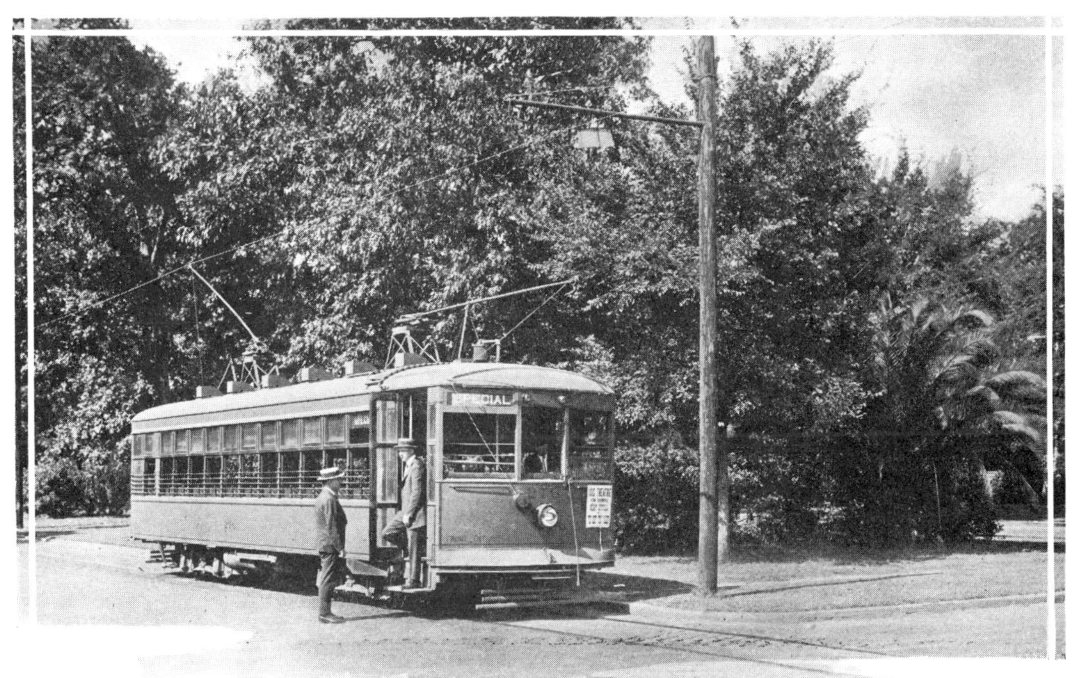

Chapter I

One-Man Operation

THE use of a one-man crew is not new—it was common with horse cars and horse buses, some of which employed a coin tube into which the passenger could put his fare which then rolled by gravity to the driver. Others relied on a pay as you-enter plan with which, in order to avoid delay, the car was started as soon as the passenger was aboard, and a change shutter in the door allowed the fare to be paid after he was inside.

The early trolley car was successfully operated with but one man, though less frequently than with two men. Sometimes the car would be operated by two men in the busier section and by one at the outer end, the conductor changing from car to car at the passing point.

One-man operation soon became very rare. Labor costs were less, the industry was flourishing, and with the cars as then built, delay, inconvenience and loss of fares were involved. It was, moreover, an advertisement of poverty not especially desired. It survived in isolated cases such as the branch of the Intramural Railway, at the St. Louis Exposition in 1904, which was operated by one man.

The Safety Car an Economical Development

Changed economic conditions, however, brought out the modern safety car making it entirely practical to again use only one man. The various necessary elements appeared from various sources—air brakes, fare boxes, etc., and modified car designs, such as pay-as-you-enter and other prepayment cars. The combination of these with new devices of safety and convenience made it possible for one-man operation to reappear, not apologetically but as the very essence of the single truck safety car, the adoption of which, beginning about 1916, spread all over the country during and after the War. It was impelled by the pressure of necessity—the natural and general opposition from the public was weakened or neutralized by its simultaneous success all over the country, by more frequent service, and by carefully prepared publicity.

Platform Expense a Large Item

The natural step followed—why not a larger car? So the double truck one-man safety car appeared, and is in wide use.

The great advantage of the one-man car is, of course, the saving in platform expense. At present wage scales, this is a very large proportion of the total operating cost, especially in city service. Since it is usual to pay the one man somewhat more than each of two men, the wage reduction per car hour amounts to perhaps 45 per cent.

There are minor advantages as well. It is often found that one-man service appeals chiefly to the younger, more enterprising men, and that these men take a certain pride in having entire charge. There is no divided responsibility as regards making the schedule or other details of operation. It is probably true also, that one man can maintain schedules better than can a motorman if burdened with a slow conductor.

There are, of course, precautions to be observed. Fare

collection at crowded loading points often requires an additional man to avoid delays. Some companies use successfully a street collector who is on hand at points and times of heavy loading. He opens the rear door so that the car loads in two streams instead of one. Where an enclosure is possible as at parks, some factories, and similar places, fares may be collected when the passenger enters the enclosure, and he may then enter through either front or rear door with no delay. This is probably the most rapid method.

"Pay-enter" operation when inward-bound toward the crowded district, and "pay-leave" when outward bound, also is a successful solution when the heavy loading district is near the end of the line.

Many cars are built for either one-man or two-man operation, the latter for times of heavy traffic. To avoid confusion and irritation to the passenger, care should be taken to mark very conspicuously on the front of the car which entrance should be used. Some roads in introducing front entrance cars, paint a brilliant patch of color on the front end to catch the patron's eye. The same idea can be embodied in a conspicuous, removable sign on the dash.

The Railroad Crossing Problem

Much discussion has centered about one-man operation at railroad crossings. There being no conductor it is conceivable that a single operator might get out, see a clear track, return to his car, start up, and be struck by a fast train coming from around a curve. Actual practise indicates, however, that such accidents do not occur. Occasionally there may be a main line crossing which requires automatic signals or a watchman, but in the vast majority of cases the question solves itself. High speed main line crossings in city streets are becoming more and more rare and are protected by the railway against street traffic by watchmen or gates. Crossings in the outskirts are usually more or less open to sight so that an adequate view may be had from the vestibule. And the practice of "flagging across" is a more or less illusory safeguard, since observation all over the country indicates that it is often purely perfunctory—the motorman proceeds before getting the signal— the conductor waves "come ahead" before looking down the track. At all events, during several years' experience, the great number of companies using one-man cars have not, in practice, found crossings a serious difficulty.

Perhaps the greatest problem is not actual operation with one man, but its introduction. The employees, and the public usually meet it with a dubious eye. The platform men fear the loss of their jobs and the imposition of an unreasonable amount of work. The public is likely to sympathize with the men and apprehend that its safety and convenience are to be sacrificed for the company's profit.

The men should be shown the facts; that each man will actually profit by the new service. The pay may be higher, the duties will be more responsible and more interesting, it will be more of a good man's job. Most men's self respect welcomes a job that calls for reasonable alertness and activity. As for the fear of discharge, on many properties the normal labor turnover will take care of that—the road simply ceases to hire new men for a time and before long the usual rate of resignations and discharge have done the work. Also, the new cars frequently operate with reduced headway so that the number of trainmen is not halved.

Cost of Training Men Reduced

It is recognized that a very definite expense is attached to employing and training every new man—the one-man car cuts this expense almost in half. It may reasonably be assumed that operation of a one-man car reduces the likelihood of a man's resigning, as the work is more profitable and interesting.

Many roads allow the new one-man runs to be "bid for" by a list of the better employes. It is interesting to note that young men take more readily to the change than the older men, and that conductors learn more readily than motormen the duties of the operator.

If the cars are new, that has a strong effect in mollifying both employes and public. If possible, the first one-man cars should be new and attractive. Later when one-man operation is taken as a matter of course, old cars may, if desired, be changed over.

Safety and convenience devices are essential. Merely to close up the rear door of an old car and call it "one-man" is courting trouble. It has been tried. To forget the interests and safety of the public and the employe is to forfeit their confidence as well as the expected profit.

May Also be Applied to Interurbans

Most one-man operation has been in city service, but it is used successfully on interurbans as well. Its advantages there are less urgent, however, and no doubt for many lines it need not be considered. Platform expense is a much smaller proportion of the total outgo on an interurban, and a saving in this item is less important.

Many roads carry baggage and express, which often requires the strength of two men. If orders are received by telephone, a degree of safety is lost if two men are not available to check each other. In case of accident or mishap to car or motorman in the city help is at hand but when miles out in the country it might be a very serious matter. Moreover, some roads feel that it is more important to maintain a certain degree of "pride of service"—an element of the luxury idea, and that a conductor is a part of the same scheme which includes plush seats, elegant finish, high speed and perhaps heavy smooth-riding cars. It is the same idea that reaches a high point in parlor, dining, and sleeping car services.

For the shorter roads where luxury does not appear so desirable, and where heavy baggage handling is not essential, one-man cars can be and are used very successfully.

In brief the one-man car is a definite step forward in the industry. It is in line with the universal tendency to use more machinery, fewer men, and to pay those men better. Such a tendency is for the best ultimate interest of company, employe, and public, and those who recognize and employ such innovations earliest are those who profit by them most.

Chapter II

Multiple-Unit Operation

VERY early in the development of the electric railway it became evident that train operation of passenger cars was essential to the succesful handling of large masses of people. The first multiple-unit problem was that of converting elevated lines from steam to electric operation, requiring the creation of a complete new system, and the engineering fundamentals derived at that time still form the basis for the application of multiple-unit equipment. The advantages of multiple-unit operation were so obvious that no one gave serious consideration to the operation of single cars under these conditions. Since that time multiple-unit control has been applied to all classes of electric railways such as rapid transit, interurban and city service.

Local conditions have governed the selection as to the length of train, the loading facilities provided, and the headway maintained. Interurban problems with their high speeds seem to solve themselves by requiring a fairly heavy car, and although normal service may call for single car operation, there are usually holiday conditions or other similar circumstances which justify the provision for multiple control of as many as 3- or 4-car trains. Trailer operation is used very little except for special service such as parlor cars or dining cars.

Rapid transit started with trains of four or five cars, about equally divided between motor cars and trailers. The tendency on new equipments is to use more motor cars and fewer trailers, and to increase the length of trains to 8 or 10 cars. In some cases station platforms are being lengthened and car equipment apparatus revised to accommodate longer trains and to reduce the platform labor expense. The minimum practicable headway seems to have been reached in the New York subways, where trains are about one minute apart and the stops are 15 to 30 seconds in duration. This means moving passengers at the rate of about 75,000 per hour on one track, and the general impression is that the physical limit has been reached. Further growth means a duplication of tracks, or the creation of an entirely new system of transportation.

Train Operation Solves the Problem

In general it can be said that in city service the following conditions must be considered in making a selection between single car operation, either with or without a trailer and, two or three car train operation. In analyzing these conditions it must be borne in mind that our communities whether large or small are all developing and expanding. Traffic density increases from year to year and conditions which now appear insignificant may change in a relatively short time.

(1) It is probable that the movement of more than 3000 passengers per hour in one direction for a distance of more than one mile through business districts, with a reasonable run beyond this point in less congested areas, would justify more than single car operation.

(2) When it becomes necessary to operate single cars on a headway of less than two minutes multiple operation justifies itself by saving in loading time and in the reduced interval during which cross traffic is blocked at intersections.

(3) In a growing community where the single car is reaching the limit of its efficiency, multiple-unit operation can be applied, and still maintain the same type of car and the same size of motor equipment; thus standardizing the new equipment with existing cars and with the least possible disturbance in operating and maintenance practice.

(4) When rush hour trailer operation has increased

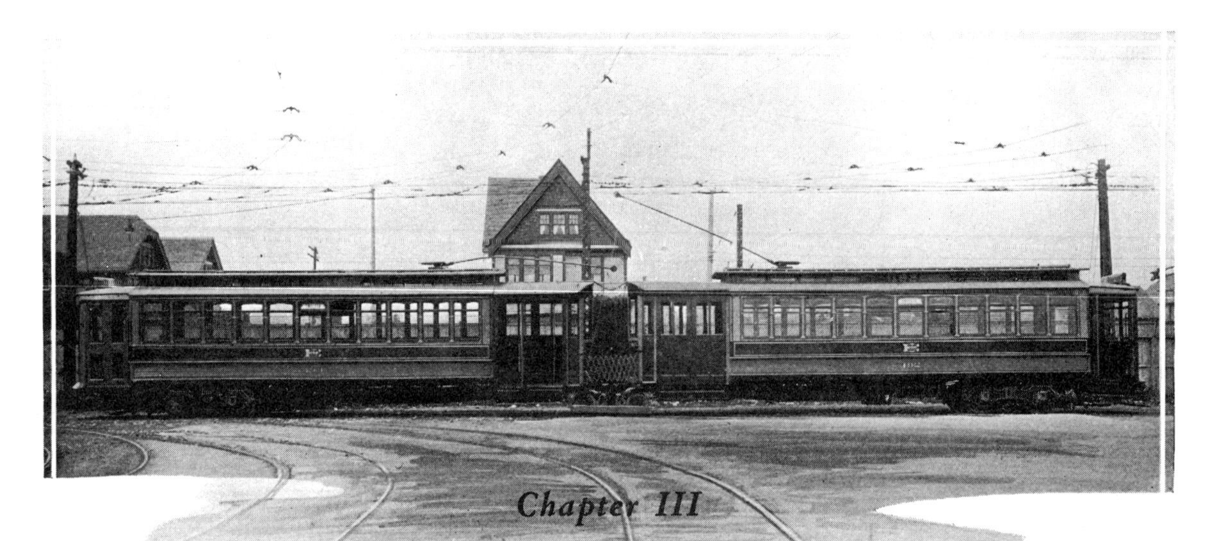

Chapter III

Articulated and Permanently Coupled Cars

AN INTERESTING phase of railway transportation is the increasing use of articulated or permanently couple train units. Beginning with the first trials in Boston in 1912, an increasing number of train units have been placed in service in America for handling passenger traffic under varying operating conditions. While the idea may appear new here in America, the train unit has solidly established itself on some European steam railroads in service varying from heavy suburban to deluxe main line trains on long runs.

The increasing number of trial installations indicates that train units have desirable characteristics for certain kinds of traffic, and give economies in first cost and operation. Traffic must be dense enough to give a fairly constant load, or the unit in all-day service becomes uneconomical in power consumption as it cannot be uncoupled and operated in smaller sections to meet varying loads as can multiple-unit trains. The unit is adapted to handle rush-hour traffic or peak traffic from definite loading centers on runs where unloading points are comparatively few. Where trains of more than two cars are required, the articulated unit has some advantages in first cost and weight saving over a corresponding number of multiple-unit cars.

Advantages of Articulated Trains

From the standpoint of operation the articulated train is an improvement over the usual motor car and trailer in the following respects:

1. Better distribution of passenger load because of articulation and passageway between cars.
2. Adaptability for various methods of fare collection.
3. Economy of platform labor, depending principally upon loading arrangements.
4. No loss of headway or slowing down of schedule as occurs with increased loading on motors when trailer is added to motor car.
5. Better distribution of motive power on unit resulting in smoother acceleration, less weight transfer which in turn decreases wheel slippage, and in general decreases wear and tear on motors, gears and trucks.
6. No time lost as in coupling on trail cars, or making runs to storage yards to pick up trailers.

On the other hand, the equipment for an articulated train unit costs more than for motor car and trailer; it uses much more power if operated for light loads when ordinarily the trailer is dropped; should have higher maintenance because of upkeep of interconnections, electrical and mechanical, between cars; and in general is a much less flexible unit; hence decidedly limited in application for varying conditions which frequently occur on the same property. This latter objection is a deciding factor in many cases and the articulated train cannot be expected to do the work of multiple-unit cars in such service.

Articulated Cars Not Flexible

When the articulated, or permanent coupled, train unit is compared to the usual multiple-unit train, the most noticeable feature is lack of flexibility for economical and efficient all-day service of varying density. The overall economies of multiple-unit cars are offset to some extent by their higher first cost, which approximates twenty percent more for corresponding new articulated train units of the type used by the City of Detroit, but higher first cost is usually more than balanced by the power economies in all-day service on a given line. It must be remembered that the difference in first cost is affected by various factors such as single or double end operation; use of rebuilt old cars; or fewer multiple units because trouble or inspection retires to the shop only one car instead of a train.

Fig. 1. Multiple-Unit Operation on the Pacific Electric Railway.

to a point requiring three-car trains or all day two-car service, trailers may be motorized, thus utilizing existing rolling stock for a greater percentage of the operating period.

(5) The economic use of platform labor when trains are being increased in length is an item which cannot be overlooked. Efficient loading and unloading facilities are essential to the economical use of trains. In street loading there is a tendency for passengers to crowd the leading car and many ingenious schemes have been evolved to divide the load and decrease the time of stopping to as nearly as possible that of a single car. Convenient signalling and a certain amount of safety interlocking in connection with the doors is essential. Under normal conditions it will be possible to operate two or three cars with three men, four cars with four men and in case of rapid transit with adequate platform facilities a greater ratio between the number of cars and the number of guards can be realized.

(6) Where mixed single car and two-car service occupies the same track, a trailer is sure to slow up the service, whereas a two- or three-car train of all motor cars can maintain the same schedule as a single car.

(7) With multiple-unit equipment trains can be built up from one to three cars, thus increasing the available seats 200 percent without changing the headway or the schedule speeds.

(8) Train delays are reduced to a minimum when all cars are motor cars capable of independent movement.

(9) Aside from the direct traffic advantages involving the efficient handling of a large number of passengers, there is a distinct gain throughout the entire system in having all equipments complete motor car units, capable of indiscriminate train operation. The time involved in adding or dropping cars is reduced to a minimum. There is no need for yard pushers in distributing cars on storage tracks. The flexibility of rolling stock as compared with trailers or permanently coupled units is obvious.

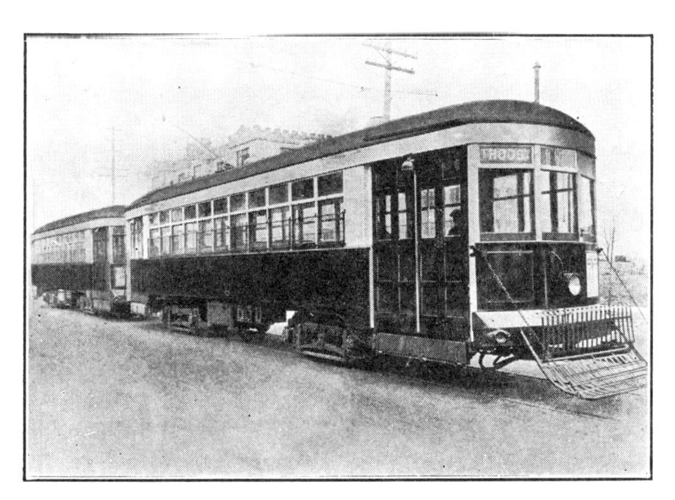

Fig. 2. Two-Car Train on the Kansas City Railways.

Fig. 3. Lackawanna & Wyoming Valley Railway Multiple-Unit Train.

Multiple-unit trains are particularly desirable for trunk line, through service where small cars or buses act as feeders. The electric railway industry is making great progress in the coordination of car and bus service and the public is gradually seeing the accuracy of the idea that all transportation service in each city should be furnished by one responsible agency.

Students of transportation agree that the rail car will continue to handle the greatest part of urban passenger transportation for many years. One of the most promising means of increasing the efficiency of rail service is multiple-unit train operation.

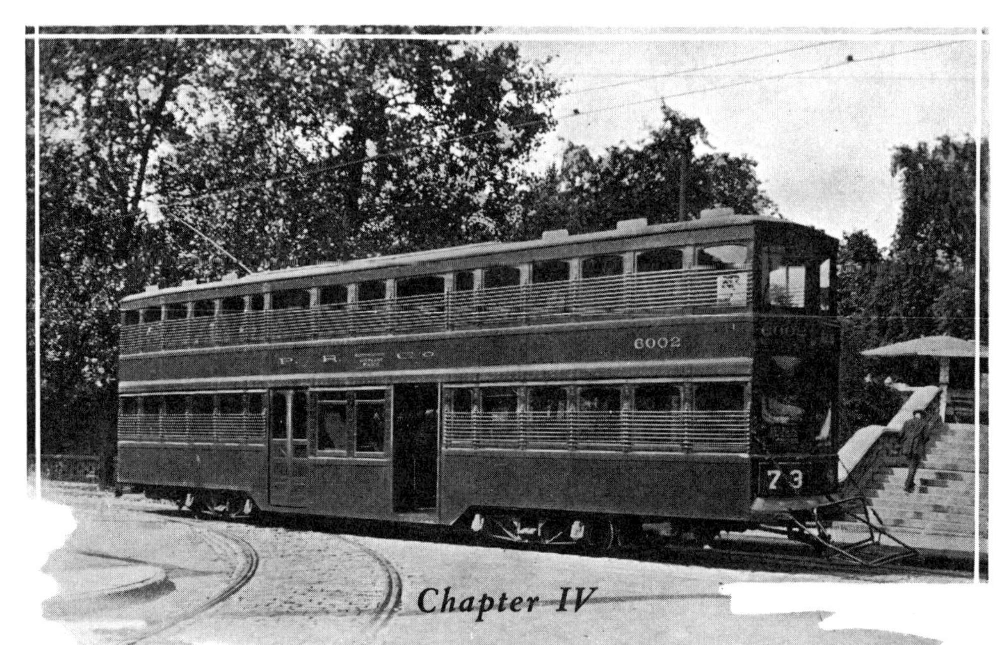

Chapter IV

Double Deck Trolley Cars

HE double deck type of trolley car is used almost universally in the British Empire, and to some extent in other countries. In this country however, the single deck car is practically universal, and only in a few instances are double deck cars used. There are probably many locations where double deck cars could be used to a good advantage and they might serve to increase the travel and result in larger receipts and reduced operating costs.

The double deck design has probably been perfected to a greater extent in England than in any other country. Considerable attention has been paid to the body design with a view to obtaining maximum strength and rigidity with minimum weight of material. Only the best air dried lumber is used and the joints are fine examples of expert wood working. Many believe that in this country sufficient attention has not been paid to these points, with the result that double deck car bodies are unnecessarily cumbersome and heavy, thereby resulting in increased operating expenses.

Double Deck Motor Coach Popular

A large percentage of motor coaches are of the double deck type and these are especially popular during the summer months. A ride on the upper deck furnishes recreation for thousands of residents in New York and Chicago. At the same time the inside of the coach is also used by those who are unable to obtain access to the upper deck. In general it may also be said that for a trip of some length the upper deck is utilized, but for short trips passengers do not go to the trouble of ascending the stairs and as a result there is a tendency for both decks to receive equal patronage. If the double deck is desirable for motor coaches why not for trolley cars?

On the British Tramways double deck cars have always been popular as a means of sight-seeing and enjoying smoking privileges. It is true that the maximum speeds are much lower than in this country, but due to the fewer stops the schedule speeds are not much less. One of the most objectionable features is the duration of stops due to loading and unloading, and with the same number of passengers distributed in a motor car and trailer, the duration of stops would probably be less. The weight of the motor car and trailer would be greater than the double deck car so that the acceleration of the double deck car should be faster with the same motor equipment. As a result there need not be much difference in schedule speed. The great advantages of double deck cars are economy of wheel space and low platform labor cost per passenger.

It would seem that the maintenance of the double deck type of car would be much lower than that of a motor car and trailer. The advantages of flexibility, however, would of course be with the motor car and trailer, as the trailer would not be used except when the traffic demanded.

It is believed that no attempt should be made to use double deck cars where the streets are narrow. There should be sufficient room to operate a car having such a width that transverse seats can be used rather than longitudinal seats, in order to furnish the maximum of comfort.

The Pittsburgh Railways Company is one of the few properties in this country which has thoroughly investigated the possibilities of double deck cars. The cars in Pittsburgh are handicapped to a certain extent by restricted clearances, but a number of these cars are used on a line which has two fairly well defined loading and unloading districts, one near each end of the line.

The articulated, or permanently coupled, train units used for operation on city streets have largely been equipped with drum type control while the articulated trains for subway service utilize remote control of the multiple-unit type. The drum control limits the length of trains to one unit of two or three cars which is an objection where longer trains are necessary either for regular or peak traffic conditions. Although remote control on the articulated train permits longer trains of two or more units without a corresponding increase in platform expense, the train unit cannot equal the multiple-unit car in flexibility for train make-up and adaptability for service.

Most two-car articulated train units have the center truck under the articulation and consequently must use small diameter wheels on it. This arrangement leaves only the end trucks to carry motors. It is possible to use trucks with the same wheel diameter throughout, as is done on the Detroit Municipal Railways, but no motors are placed on the center trucks principally because removal of these trucks is much more difficult than removal of the end trucks. Also in case motors were placed on the center trucks the control would be somewhat complicated.

Multiple Units Have Motive Power Advantage

The usual multiple-unit car has four motors, two on each truck, like the permanently coupled subway trains, and thus gives much better distribution of motive power throughout the train. Such distribution of motors is particularly advantageous for negotiating grades or for handling the train when one motor fails. Since the motors are usually grouped in pairs for the sake of control simplicity and to prevent abuse of motors, a motor failure on a four-motor two-car articulated unit will require cutting out of circuit fifty percent of the motive power of the train, while a similar failure on a two-car, eight-motor multiple-unit train will require cutting out only twenty-five percent of the motive power. The difference is more marked in favor of the multiple-unit car, if a three-car articulated unit like that of the Detroit Municipal Railways is compared with a three-car multiple-unit train.

The inflexibility of the articulated train, especially if equipped with drum controllers, is particularly disadvantageous when control trouble occurs. A control failure usually ties up the entire unit, or if motors can be cut out, does not leave enough power to move the unit without slipping the wheels. The disabled unit must then be pushed along by the next train, which not only places double duty on the motors of the next train, but also places the motorman in the center of the train where he cannot obtain a view of the street ahead and must rely on signals from the disabled unit. Obviously, this would interfere much with schedules during rush hours when real need of the train unit exists. Failure of control on a multiple-unit train can be taken care of usually by cutting out the control on the car giving trouble, or by cutting out a pair of motors in case of motor trouble, and leaving the other car in operating condition. In case the motors on the remaining car are unable to move the train, the train can be readily coupled to the next one that comes along and both trains operated as a unit from the head end where the motorman's view is unobstructed.

The accompanying tabulation gives briefly some of the characteristics of train units in use on both steam and electric lines. Perhaps the most noticeable feature is the wide variation in weight per seat, ranging from 510 to 536 pounds in the trains built new for surface car operation, to practically double that for the new subway trains, and to about four times as much for the "deluxe" steam road service where the run is over four hours' duration. The articulated train is especially fitted for rush hour service where the standing load is from one and one-half to three and one-half times the seated load. Based on rush-hour loads the weight per passenger ranges from about 225 pounds in the new Twin City Rapid Transit cars to 300 to 325 pounds for the subway cars and to 425 to 500 pounds for the rebuilt city cars.

TYPE	USER		SERVICE	OVERALL LENGTH	WEIGHT	SEATING CAPACITY	WT. PER SEAT	MOTORS	WHEEL DIAMETER	CONTROL	OPERATION	CURRENT COLLECTOR	UNIT MADE FROM-	DATE IN SERVICE
ARTICULATED	BOSTON ELEVATED RWY.		CITY SURFACE CAR LINE	61'-9¼"	39000#	104	750#	4-30HP.	30"	K-28	DOUBLE END	WHEEL TROLLEY	REBUILT OLD CARS	1912
ARTICULATED	BROOKLYN-MANHATTAN TRANSIT CORPORATION		CITY SURFACE CAR LINE	63'-10"	72000	71	1013	4-40HP	30" 22	K-28	DOUBLE END	WHEEL TROLLEY	REBUILT OLD CARS	1924
ARTICULATED	MILWAUKEE ELEC. RWY. & LT. CO.	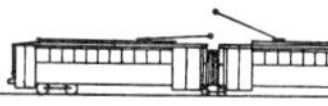	CITY SURFACE CAR LINE	89'-2½"	75000	102	735	4-65HP	34" 22"	K-35	DOUBLE END	WHEEL TROLLEY	REBUILT OLD CARS	1920
ARTICULATED	DETROIT MUNICIPAL RWY	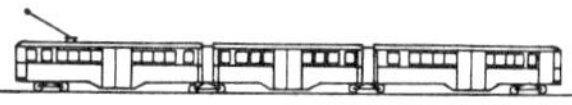	CITY SURFACE CAR LINE	122'-8"	75000	140	536	4-60HP	28"	K-35	SINGLE END	WHEEL TROLLEY	BUILT NEW	1924
ARTICULATED	NEW YORK RAPID TRANSIT CORPORATION	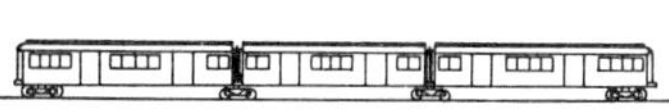	SUBWAY	136'-9"	171800	160	1072	4-190HP APPROX.	34" 30"	MULTIPLE UNIT	DOUBLE END	THIRD RAIL SHOE	BUILDING	
ARTICULATED	GREAT NORTHERN RWY OF ENGLAND	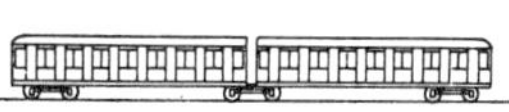	SUBURBAN	102'-0" APPROX.	100000	128	781	— FOR STEAM RAILROAD SERVICE						
ARTICULATED	GREAT NORTHERN RWY. OF ENGLAND	 THIRD CLASS / THIRD CLASS / KITCHEN / FIRST CLASS / FIRST CLASS	MAIN LINE	246'-1"	264700	128	2068	— FOR STEAM RAILROAD SERVICE						
PERMANENT COUPLED	BOSTON ELEVATED RWY.		SUBWAY	94'-6"	88800	88	1009	8-40HP	26"	MULTIPLE UNIT	DOUBLE END	THIRD RAIL SHOE	BUILT NEW	1923
PERMANENT COUPLED	TWIN CITY RAPID TRANSIT CO.		CITY SURFACE CAR LINE	94'-0"	53000	104	510	8-25HP	26"	K 43	SINGLE END	WHEEL TROLLEY	BUILT NEW	1922

Fig. 4. Some examples of Articulated and Permanently Coupled Cars.

Chapter V

The Trolley Bus

THE trolley bus, being a rubber tired vehicle, may, in general, be used wherever such a vehicle is applicable. It has inherent characteristics which make it a high grade transportation facility. Though it is especially adaptable in certain particular fields, it is well worthy of consideration in many fields where bus service is desired.

Theoretically there is a certain field in which it will operate most economically. This field includes a certain range of headways and if headways longer than these are to be maintained, the motor coach is more economical, while if they are shorter, use of the rail car will be justified. While the limits of this field may be rather definitely established, other factors many times determine which type of transportation should be used.

The extended use of the automobile has resulted in a demand for rubber tired service. There was some question at first as to the advisability of extensively introducing the use of the automobile into the field of transportation, as an aid to, or to take the place of, the well esablished rail lines. The continued demand has resulted in the introduction of the motor coach into service for which it was at first considered unfit. In many cases it is a direct competitor of the rail car. It has been found more costly to operate than safety cars, but in many cases enjoys a higher fare than the rail car. It is possible that time may prove the motor coach to be less economical than rail cars in fields where both are used, and thus result in its use being restricted to only those fields to which it is economically adapted. It is very probable, however, that there will be sufficient patronage, willing to pay a higher fare, to result in an even more extended use of the motor coach. This will be true to a greater extent as the use of the private automobile becomes relatively less, due to increasing traffic congestion, and thus of less convenience or advantage to the person using his private car for shopping and business.

Four Distinct Bus Applications

In any field suitable for motor coach operation, with the exception of that in which very long headways are maintained, either the gas or trolley bus may be considered applicable. There are four rather distinct cases in which bus operation is used:—

First, in districts where traffic is not great enough to warrant the laying of track. This district may vary from a new real estate development with few inhabitants, in which transportation facilities are desirable to attract settlers and enhance the value of the property, to a well settled side street district, where transportation is desirable to give patrons better service by permitting them to ride to the trunk surface line instead of walking—and possibly being picked up by a friend driving a private automobile. In either of these districts, rail service may follow.

It is not always possible to determine in a new district what will be the best location for a rail line when the population has increased. In such a case the motor coach has over the trolley bus the advantage of greater flexibility, as the route can be changed at any time. The trolley bus requires the installation of overhead which limits its operation to a definite route. The very fact that this overhead must be installed is of great advantage to those purchasing land in the new sections, as there is assurance of more or less permanent transportation facilities, which enhances the value of the real estate. The overhead construction also paves the way for future rail lines in that relatively slight modifications are required to adapt it to rail car use.

The second case in which buses may be used is in a residential district where rail service is opposed due to noise and to the laying of tracks. In many such sections the main streets are classed as boulevards, and tracks are undesirable. Traffic often moves at high speed which makes it dangerous for patrons to go to the center of the street to board a car. With a bus, this necessity is eliminated, as the patron is received at and carried to the curb. From this standpoint, the trolley bus and motor coach are on a par. From the standpoint of silent operation, the trolley bus has the advantage. As far as the use of rubber tires is concerned, the two types of buses are equal, but the trolley bus has the added advantage of none of the noises of the gas engine and the elimination of gear shifting. The lack of gasoline and oil odors in the trolley bus is also a considerable advantage.

A third case in which a bus may be applied, is where the streets are so narrow that double track or even turnouts could not be used conveniently. The loop system could be used, in which the cars would operate inbound and outbound on different streets, but this increases the cost of track and overhead. With a bus, the narrowness of the street is not such a vital factor. The bus is not limited to a definite path, and is therefore not necessarily held back due to parked or slow moving vehicles. Also, should a unit be damaged or for some other reason be put out of commission, the result would be much less serious with a bus than with a rail car. The former could be drawn over to the curb or even around a corner into a side street, and thus cause very little delay or inconvenience to other traffic.

The fourth case is where transportation service is needed or demanded, and where rail service may even be justified, but its installation may be temporarily inadvisable for economic reasons. In such cases the bus fits in very well, as it affords immediate service with a minimum investment. The motor coach would have the advantage of the least initial investment. The trolley bus would require a somewhat greater investment due to the need for overhead, but this expenditure would not be lost as it would be required at such time as rail service was inaugurated. The setting up of permanent overhead incident to trolley bus operation, would give the residents a feeling of assurance that rail service would follow, and might be sufficiently advantageous so that the capital investment necessary for the installation of rail service could be delayed until it was necessary or could be more conveniently handled.

Seven Trolley Bus Installations in America

There are at present seven installations of trolley buses in this country and in Canada. These are at Windsor and Toronto, Ont., Rochester, N. Y., Staten Island, N. Y., Petersburg, Va., Baltimore, Md., and Philadelphia, Pa. In these installations a total of 47 buses are used. On at least two of these properties, more buses and extensions of

Fig. 5. The Street Car Type of Trolley Bus.

route have been considered as a result of the satisfactory performance of the original equipments.

The service handled by these buses may be divided into three classes; feeder, crosstown and independent. In some cases the traffic handled is rather light, while in other cases it is quite heavy. All of the present installations are more or less experimental. The variety of service into which the trolley bus has been introduced, affords it an opportunity to prove itself satisfactory or unsatisfactory in several different types of service.

The installations at Windsor, Toronto, Petersburg, Philadelphia and Baltimore, are principally of the feeder class. In these cities the trolley bus route is an extension of existing rail lines, or extends laterally from such lines.

In Petersburg, the route extends from the end of a rail line, through a new residential district with no other transportation facilities. As a result of the performance of the trial installation, the extension of the route into the business section was considered. This plan included the abandonment of some track.

In Philadelphia, the route extends out from a rail line, along the edge of a rather closely built up residential section, to a manufacturing center. Heavy rush-hour loads of workingmen are handled, as many as 50 passengers being carried over a considerable part of the route at times.

In Rochester, the trolley bus is used in what may be classed as crosstown service, as it ties together five main surface lines. The route is in a residential district and the loads carried during the day are relatively light, but at rush hours capacity loads of fifty or more passengers are hauled. It is interesting to note, as an evidence of ability to render good service, that during the first six weeks of operation ninety thousand passengers were handled, with a total of five buses.

On Staten Island the routes may be classed as independent for although they connect with rail lines, they are really main arteries in their sections. The routes are through sparsely settled districts, and assure transportation facilities to people desiring to establish their residence in those parts of the Island.

The accompanying table showing schedule speeds recently observed on some of the routes, indicates the serv-

Fig. 6. Motor Coach Operation in Baltimore.

its other advantages invite continual consideration of this form of drive.

For light transportation work the advisability of using gas electric drive is open to some question. The enthusiasts for this type of vehicle point to the high losses in the transmission gears of the direct gas drive as offsetting the additional weight and losses of the electric transmission, this however, is not borne out by the facts.

The gears used in modern automobiles are cut and ground with great precision so that the losses in the transmission are comparatively small. Actual tests have shown that the losses between the engine shaft and the rear axle on an automobile run from five to ten percent, the latter figure being attained when running in low gear. For the greater portion of the time the transmission gears are not in use so that an average gear loss of eight percent with a bevel gear drive and sliding gear transmission is probably a fair average figure. With worm gear the losses will be higher, but it must be remembered that if an electric transmission is employed the final worm or bevel gear drive losses will be common to both the mechanical drive and the electric so that the actual difference in the gear losses in the two cases will be only that due to the change gears.

To balance against the above, we have in the case of the electric transmission, the losses in the generator, motor, wiring and control. Very good design will be required for the generator and motors to obtain an average efficiency of 80 percent of each machine over the operating range, so that the overall efficiency of the two machines will be about 64 percent without including wiring losses.

This means that if we substitute a generator and electric motor drive for the mechanical transmission, we are adding approximately 36 percent average loss in the electrical units to offset a saving of perhaps 8 percent by the elimination of mechanical transmission. If this is the case, what do we gain in the electric transmission to make it even worthy of consideration? The principal gains are greater flexibility, elimination of the change gears and clutch, the ability to run the engine at a more nearly constant load, and the possibility of readily driving all of the wheels without complicated mechanism.

Gas Electric Drive with Storage Battery

The electric transmission has in some cases been further complicated by the addition of a storage battery. The advocates of this combination argue that, as practically every vehicle now propelled by an internal combustion engine uses a storage battery for ignition, lighting and starting, the expansion of this battery into one large enough to provide part of the energy to drive the machine does not entail a great hardship. They argue that by providing a battery of this sort in connection with an electrical transmission it is possible to run the engine continuously at nearly constant load and at or near its most efficient point, the excess energy not required for propulsion of the vehicle being stored in the battery. These arguments are based upon engineering facts, but must be balanced against increased weight and lower efficiency. The battery will weigh approximately 100 lb. per horsepower hour, and it must be remembered that all of the energy which goes through the battery will be subject to an average loss of approximately 30 per cent.

Tracing one horsepower back from the engine shaft through an electric transmission and battery, the following conditions would exist:—.80 hp. would be delivered by the generator to the battery; .56 hp. would be delivered by the battery to the motor; and .47 hp. or less than half the original power at the engine, would be delivered at the motor armature shaft.

The foregoing figures are based on average conditions and will of course be subject to variation above and below these amounts.

In general the battery increases the total weight and decreases the efficiency of transmission, and it is a question of considerable doubt whether the advantages offset these facts. The electric transmission, however, eliminates gear changing and gives a flexibility far greater than is obtained by the gear method so that it gives very desirable conditions if they can be obtained without too great complication, weight increase or loss of efficiency.

The most promising form of electric transmission is that wherein only a portion of the energy is transmitted electrically and where the major portion is transmitted directly by mechanical means. Several systems of this sort have been developed and they have the advantage of retaining the flexibility of the "straight" electric transmission but at the same time, giving better efficiency and reducing the size and weight of the electrical units required.

Cost of Power

The cost of power is not the governing factor in the cost of operation of a motor vehicle but it is sufficiently large to make it an important item, as for instance in the case of a heavy motor coach it may easily run from one-eighth to one-sixth of the total cost of operation and may therefore make the difference between profit and loss.

The cost of power varies with the price and composition of the fuel, both of which vary considerably in commercial practice, so that the cost of a horsepower hour produced by a gas engine can be stated only upon the basis of certain assumptions.

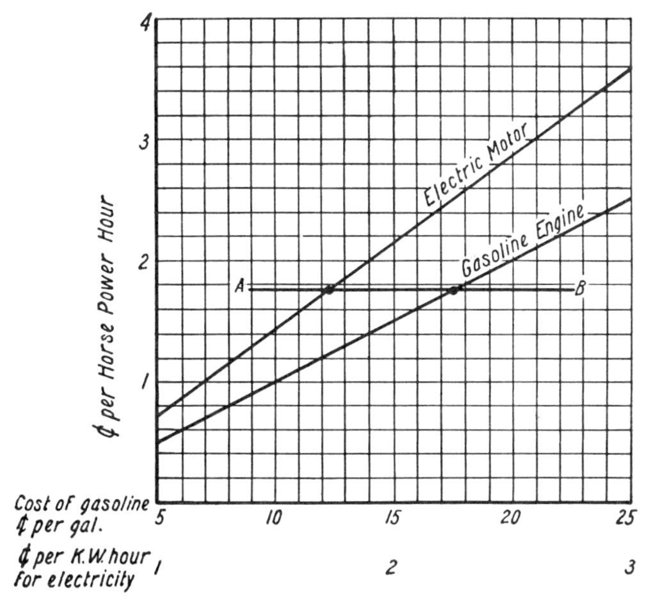

Fig. 7. Approximate Cost Per Hp. with Gasoline Engine and Electric Motor.

In the first place, the internal combustion engine used for traction work obtains its energy primarily from oil products, and the electric motor as used on a traction car or trolley bus gets its energy from either gas, coal or water power, back at the power house.

The usual commercial type of automobile engine spits about 30 percent of the energy of the fuel out of the exhaust, radiates about 52 percent in heat and kindly delivers about 18 percent in work, a considerable portion of which is used in friction.

As far as the electric motor itself is concerned, it is a pretty efficient machine and will deliver in useful work about 80 percent of the electrical energy which is put into it. For a fair comparison with the gas engine, however, we must follow the electric current back to its original energy source. This means that in addition to the losses in the oil, gas, or steam engine back at the power house, we have the added losses in the generator, line, and motor, before we reach the motor shaft.

With these additional losses it would seem at first thought, a foregone conclusion that electric power would be more expensive. It has been found in actual practice, however, that where power is generated in large central stations, cheaper fuel, greater efficiency, and lower maintenance and other operating economies will frequently offset the losses in the generators, line, and motor and enable the power to be delivered at a figure which compares favorably with the cost of that obtained directly from a small gas engine unit.

Figure 7 shows the approximate cost per horsepower hour for power obtained with a gasoline engine and with an electric motor, with varying costs of gasoline and electric power.

A fair average cost for electric power delivered at a vehicle is 1¾ cents per kw.-hr. and the line A-B which cuts this point on the curves (Figure 7) shows that gasoline would have to be selling at 18 cents a gallon for the cost of power produced by the gas engine to be as low as that produced by the electric motors.

To get the whole picture of the relative cost of propelling a vehicle by a gasoline engine or an electric motor fed from an overhead line, it is necessary to go a step further for the reason that due to the difference in transmission, and the fact that the gas engine is running all the time, it requires more actual energy per ton mile for the gas driven vehicle than it does for the electric driven. Re-plotting our curves on the basis of conservative data obtained from a number of rubber-tired vehicles of the two types in actual practice, we would have the results shown in Fig. 8.

It will be seen from these curves that with electric power at 1¾ cents per kw.-hr. gasoline fuel would have to be down to about 5½ cents per gallon in order for the cost of power per ton mile to be as low with gas engine drive as with electric motor drive.

High Maintenance

The maintenance in a gas engine is unfortunately rather high, due mainly to the reciprocating parts, the large sliding surfaces, and the fact that the power is being generated by a series of intermittent blows, instead of a uniform torque.

The use of a number of cylinders irons out the torque variations as far as the power delivered at the engine shaft is concerned, but it must be remembered that each individual cylinder and its operating parts is still subjected to the same racking strain.

In conclusion may be stated the rather trite fact that the internal combustion engine is with us to stay for a long time, and that if a simple automatic type of transmission, either electrical or mechanical, can be developed to give it the proper operating characteristics for traction work without gear shifting, and greater fuel economy can be attained, it is going to prove a very strong factor in the light traction field.

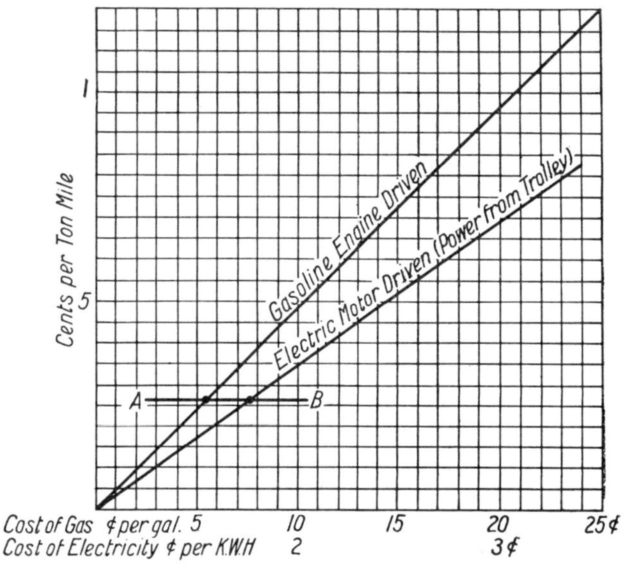

Fig. 8. Relative Cost of Power Per Ton Mile to Drive a Rubber Tired Bus.

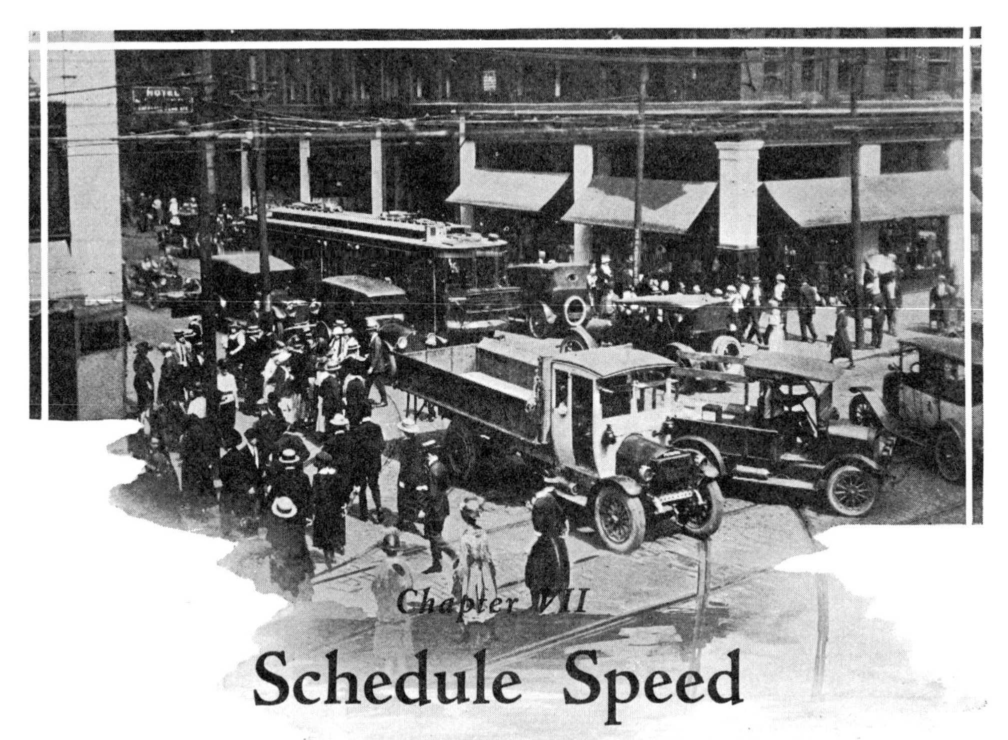

Chapter VII

Schedule Speed

SCHEDULE speed involves both running time and stop time, and is technically defined as the distance in miles between two terminals divided by the time in hours, *including stop time,* required to cover the distance. It is different from average speed in that stop time is included. The transportation industry is performing increased service to the public when more transportation is furnished without jeopardizing safe operation or causing inconvenience through excessive stop elimination or uncomfortably rapid acceleration. Faster passenger transportation, and by this is meant a shortening of the schedule time, is one of the services appreciated most by the public, and their appreciation is usually demonstrated by increased riding. It is acknowledged that faster schedule speed and frequency of service are major factors in developing riding.

Increased Schedule Speed More Economical

It is fortunate that increased schedule speed in electric car service leads to economical operation. It should be realized that earnings depend on car miles while expenses are closely proportional to car hours; in other words, faster schedules require that each car hour produce more car miles, thereby effecting a relative reduction in operating expenses. Low schedule speeds mean less attractive service, larger investments in equipment and resulting higher operating costs. The running time required by a street car is of vital importance in the quality of service and its cost. Schedule speed has a more potent influence upon service and operating expenses than any other single item of railway operation. It has been stated that at least 60 percent of an electric railway's maintenance and operating costs vary directly with the number of car hours operated. This is not surprising when it is known that conducting transportation alone amounts to nearly 45 percent of the total cost of operation. Fig. 9 shows graphically the effect schedule speed has on platform wages and fixed charges on rolling stock.

Shortening Stop Time

An increase in operating speed is not always necessary to gain a speeding up of schedules. Shortening the stop time effects an increase in schedule speed just as directly as does an increase in operating speed, and in city service it is the stop time that usually causes the slow schedules.

While loading time forms an appreciable percentage of total stop time in city operation, the fact that delays due to other causes form a surprisingly large percentage of total stop time, should not be overlooked. Causes of such delays include railroad crossings, waiting at switches (on single track lines), equipment failure, track failure, traffic congestion, draw bridges, slippery rails, etc. Many of these delays can, with proper attention, be appreciably reduced if not fully eliminated.

It is not always possible to eliminate railroad crossings, though the tendency is in this direction as a matter of safety as well as a means for increasing speed. Waiting at switches is a matter for the traffic department to investigate and remedy. Track and equipment failures are largely eliminated by proper maintenance. Traffic may often be regulated by proper cooperation with city officials to relieve congested areas by more stringent parking regulations, one-way traffic, elimination of left-hand turns, preference to electric cars at street intersections, semaphore control of traffic, use of electric switches, etc. Traffic regulation is essential in eliminating motionless traffic and to accommodate the mass of people.

Possibly the selective stop arrangement referred to in a later chapter effects the greatest savings in stop time on lines where the "stop at every block" policy is in operation. In congested areas, multiple berthing and the use of street collectors have accomplished gratifying results in reducing stop time. The pre-sale of tickets, use of tokens, and loading stations show a distinct improvement in loading of cars. These subjects of fare collection will be covered in more detail in the following chapter.

The design of the car should be closely studied and those features incorporated which will assist in loading under the particular conditions of service to be met. Wide aisles, wide doors, various types of entrances and exits, location of the fare box, operation of the doors, lighting, etc., all have their effect upon the loading time and deserve serious consideration. Usually such features can be incorporated in the car when under construction with very little additional cost.

Car Routing Important

The routing of cars has considerable bearing on schedules. Wherever through routing is feasible, better schedules will justify its adoption. Arrangement of schedules demanding fast time and rigidly enforced by inspectors is necessary to the proper maintaining of fast schedules. A further refinement in schedules calling for inbound trips in the afternoon rush hours to be made in less time than the outbound trips, and vice versa in the morning serves as a direct and satisfactory scheme of increasing the schedules on the light trips. Cars having equipment with wide difference in speed characteristics should not be mixed on the same routes.

Complete cooperation between crews, passengers and railway officials is essential to the best service and may be attained by a program of education for the crews and public as to prompt means of loading and unloading, payment and collection of fares, issuing of transfers, etc. Fast accelerating and braking rates should be used consistent with comfort and safety. There should be no lagging of the car when approaching street intersections as the opportunity to go ahead at the next traffic signal may be lost. Courteous treatment of passengers by the employes helps materially to aid the service. The conductors and the motormen are the salesmen of the railway industry and should spare no effort to improve the service.

The equipment, including rolling stock, power machines, track and overhead must be maintained in good condition if reliable schedules are to be maintained. With the growing service, an investigation may reveal the fact that a change in gear ratio would provide improved service. A change in motive equipment may be desirable in order to increase the power per unit of weight and to obtain better speed characteristics for the particular service involved.

In interurban service especially, low voltage has a marked effect upon speed. Low voltage conditions may be due to insufficient substation capacity, insufficient feeder capacity, or both. In some instances if it is not feasible to increase the substation and feeder capacity, better line voltage may be obtained by the use of lighter cars and modern equipment which not only increase the speeds due to better voltage and better speed characteristics, but also serve to improve the service because of their better appearance, improved lighting and additional comfort.

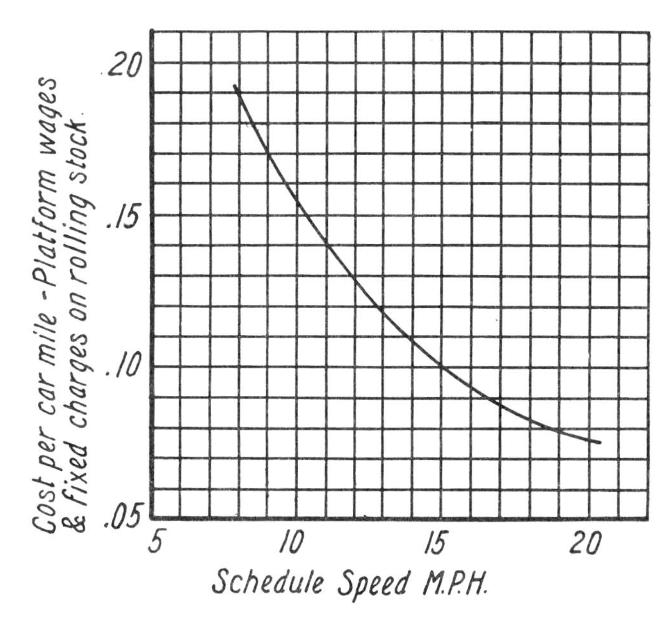

Fig. 9. Effect of Schedule Speed on Operating Cost.

Higher Speed on Interurbans

On interurban runs fast schedules must be attained by higher speed as stops and delay time form a smaller percentage of the total time than in city service. However, waiting at switches and railroad crossings should be investigated in an endeavor to eliminate such delays. Private right-of-way, double track and automatic signals are becoming more and more necessary with the demand for increased speed. Equipment failures should be minimized and track conditions improved to conform to high-speed operation. Considerable time is often consumed in getting in and out of urban centers, and it is only by keeping off the more congested streets and conforming in every way with the city service that the time can be shortened. It is believed that two-man operation is essential to long, high-speed interurban runs. In snow-bound territory, emphasis should be put on the need of proper snow fighting equipment and its prompt use.

High schedule speeds are necessary to meet the demands of the traveling public, to meet competition and to approach more economical operation. The schedule speed should be as high as safe operation will allow. There are numerous factors which affect schedule speed and it is only by concentrated study of these factors and by steps taken to speed up the service that the service can be made to cope with the ultimate demand. The need of adequate transportation is becoming greater every day. Higher schedule speeds form one very practical means of meeting the increasing demand.

Chapter VIII

Methods of Fare Collection

IN THE preceding chapter it was stated that the reduction of stop time was the most important item in increasing schedule speed. In the same manner it may be said that methods of fare collection more than anything else effect the duration of stop time. The rapid passenger interchange on rapid transit lines is due largely to the payment of fares before access to the station platform is obtained. The benefits of prepayment areas cannot be obtained to the same degree on surface lines because a prepayment area cannot be made available at every stop and the cars must be designed for loading of passengers from the street level.

Prepayment Areas

Under certain conditions prepayment areas can be used on surface lines with very satisfactory results. In some of the largest cities, notably Philadelphia and Boston, subways are used in the down town districts for surface cars and passengers pay their fares before going to the loading platform just as is the case with rapid transit lines. At other points prepayment areas are obtained in conjunction with rapid transit lines of which the surface car lines are feeders. The most extended use of prepayment areas is in conjunction with parks and factories where large passenger loads are obtained at certain hours of the day. These areas are usually arranged so that passengers can pass directly through them except under rush-hour conditions. The expense of fare collectors is thus incurred only during the rush hours when the heavy traffic justifies it.

Street Collectors

The use of street collectors is a variation of the prepayment area. The time of fare collection on the car is undoubtedly decreased when street collectors are employed. The volume of traffic which can be handled readily is somewhat less with street collectors than with prepayment areas.

Street collectors are of particular advantage when cars with two doors are operated by one man. It is generally agreed that in the non-congested districts of most surface railway lines there is no necessity for two-man crews.

In the congested districts, particularly during rush hours, the operation of the car by one man with the increased duties of fare collection, increases materially the length of stop and seriously reduces schedule speed. If street collectors are provided with means for opening the rear door of a one-man car, passengers can load at the rear at the same time as they are loading at the front, the street collector obtaining the fares of those at the rear door and the regular car operator collecting the fares of those that board at the front. Conditions are thus entirely comparable to two-man car operation.

Pay-As-You-Enter Collection

The pay-as-you-enter system of fare collection placed the conductor where he could most efficiently superintend the operation of the car, and it reduced the lost fares to a minimum. Duration of stop was necessarily increased and many of the changes in car design since the introduction of the pay-as-you-enter system of fare collection have been brought about in an effort to reduce the time necessary for collection of fares. With the pay-as-you-enter system all passengers after boarding at the rear should move to the front of the car as far as possible and leave at the front door. If passengers followed this practice, there would be little opportunity for improvement of the end entrance design of car. However, many passengers insist on crowding at the rear of the car, filling the platform and interfering with those who desire to move to the front. The distance between doors being great, the average passenger must move a considerable distance in the car.

Pay-As-You-Pass Collection

The pay-as-you-pass system of fare collection was introduced with center door cars with wide front doors. The front half of the car ordinarily has longitudinal seats and the rear end of the car has transverse seats. Passengers board at the front door and the entire front half of the car is, in effect, a loading platform. Large numbers of people can thus be taken on the car at heavy loading points with a minimum stop time. There is a natural desire on the part of the passengers to sit on the transverse seats and a number of them pass to the rear of the car immediately, paying their fares as they pass the conductor who is located at the center of the car. The passengers remaining in the front end of the car pay their fares as they pass the conductor on leaving the car at the center doors. In these cars the two center doors are sometimes located together and the conductor is in front of both doors. In other cars, the center doors are separated by a panel, the conductor standing between the two center doors. The latter type of car is believed to be better adapted to systems of fare collection other than the pay-as-you-pass system.

Pay-Enter, Pay-Leave System

The system of fare collection which provides for pay-as-you-enter on inbound trips and pay-as-you-leave on outbound trips has met with general approval during the last few years. Almost any type of car except the car with the conductor located in front of the two center doors can be used readily with this system of fare collection. The main advantage of the system is the fact that fares are always collected in the outlying districts where passengers are boarding or leaving the car a few at a time and many of the fares are collected while the car is in motion. Such portion of the standstill time as is actually necessary for fare collection takes place in the outlying sections instead of in the congested districts where the effect of all delays is cumulative and where it is especially desirable to move the cars as quickly as possible.

Zone Fares

Zone fares have been little used on city systems in this country. Two fare zones can be handled by a combination of pay-enter in the first zone and pay-leave in the second zone. Where zones overlap, some form of ticket must be used for identification. Where there are more than two zones, some form of identification must be provided or fares must be collected by hand in each zone. Zone fares are especially difficult to handle with one-man operation and there are several railways now operating without any form of identification check. It was found that the use of checks required too much of the car operator's time and that the loss of fares was very small provided the cars were not crowded.

Tickets and Tokens

The five cent fare was so universal prior to 1915 that it required an enormous amount of education to get the public to agree to any change in rate. The next step in our monetary system was the ten cent piece which was too great a step from the previous fare. For this reason it was necessary for many roads to go to 6, 7, 8, or 9-cent fares all of which required the use of more than one coin and hence complicated making change and made it more difficult for the operator to check the amount of fare paid when fare boxes were used.

Fig. 10. Street Collectors Reduce Stopping Time and Speed Up Schedules.

To overcome this trouble it was necessary to use some form of tickets or tokens which could be sold in small quantities. In order to promote the sale of these tokens they are generally sold at "bargain prices" which are less than the single fare.

This system has become almost universal today and has so many advantages over the one coin cash payment that there is little chance of going back to the former practice. Other than the advantage of making change and watching fares dropped in the box the bargain price for a small number of rides appeals to the average rider and when the tokens are once purchased he does not look upon them as the same amount of money. In other words the average rider when once having made the purchase of tokens is inclined to use them more freely than money. This results in increasing the traffic and particularly off peak and short-haul riders.

Weekly Pass

The use of the weekly pass also facilitates fare collection. Where a large number of passengers are pass-users, they can file quickly past the conductor showing him their passes and no delay is required for the making of change or the dropping of the fares in the fare box. Whether the weekly pass system is a benefit or detriment to the street railway company, depends on a large number of variables and is difficult to predetermine. It at least tends to increase the use of the street cars and to reduce the time of passenger interchange.

Chapter IX

Safety Zones

ONE of the arguments advanced for motor coach service is the fact that the passengers can board the coach at the curb and do not have to brave the dangers of going to the middle of the street as is the case with a street car. The dangers are probably over-estimated by the bus enthusiast and in any event the average person does a great deal more crossing of streets in walking from one place to another than is necessary in going to or from a street car. In spite of numerous laws regulating the use of motor vehicles and designed to protect the large number of people using street cars there is an accident hazard connected with waiting for cars on a busy street. For this reason the use of safety zones performs a two-fold benefit. They provide "refuges" for the street railway passengers and they permit automobiles to pass street cars loading and unloading passengers without danger to the passengers and thus tend to reduce traffic congestion.

Three Important Types

There are various kinds of safety zones. One of the most usual kind consists only of painted lines or similar marking on the surface of the street. Another type is formed by stanchions either with or without chains. A third type comprises a raised platform. The first type involves the least expense but it is more difficult to keep automobiles from running through the safety zone. The second type is somewhat more expensive and keeps automobile traffic out of the zone fairly effectually. These two types of safety zones are suitable for use on heavy traffic streets where safety of street car passengers and mass of vehicle traffic warrant the setting aside of a part of the street area. The raised loading platform type is most effective in protecting passengers from the vehicular traffic. This type has the further advantage that loading and unloading time is reduced by the shorter step from the car to the loading platform than from the car to the street level. Heavy trunk line traffic justifies the use of raised loading platforms.

One great advantage of safety zones from the general street traffic standpoint is the fact that other street traffic is not held up while a car is loading or unloading as automobile traffic is permitted to pass the safety zone without interruption. To permit the free passage of traffic, it is sometimes necessary to widen the streets a few feet so that the number of lines of traffic which can run on the street is not reduced by the safety zone. In some cities automobiles are permitted to run on the car track at points where safety zones are located, while in other cities this practice is prohibited. In some cities traffic is permitted to run through the safety zone when it is unoccupied by passengers. It is almost always necessary to prohibit parking at safety zones and for a reasonable distance from the safety zone to prevent traffic restriction at those points.

Safety zones must be located after due consideration has been given to density of street car traffic, density of automobile traffic and width of street. The extensive use of safety zones in the larger cities of the country furnishes ample proof of their effectiveness.

Fig. 11. Safety Zones not only Protect Passengers but Decrease Traffic Delays.

Chapter X

Multiple Berthing

AMONG the means which have been adopted to speed up service in congested sections is "multiple berthing." With the multiple berthing plan of operation, one or more cars arriving at a regular stopping place where another car has already stopped, take their position directly behind the first car, discharge and receive passengers, and then proceed without making a second stop.

Such practice is employed in congested areas for the ultimate purpose of increasing schedule speed by reducing stop time. Obviously, if cars having come to a dead stop can discharge and receive their load and then move forward without coming to a second stop, there has been accomplished a doubling or tripling of the stopping area, and the total stop time of the following cars is appreciably reduced as compared to the practice of holding the cars with closed doors until the position of the first car is reached. Cutting down of stop time on streets where vehicular traffic is heavy not only enables the cars in question to make better schedules, but by their moving out of the way assists materially to allow other cars and traffic to proceed. For practical reasons, zones are usually limited to about three cars.

The objection has sometimes been raised that the multiple berthing system is annoying to passengers in that they do not know where to wait for the car which they wish to board. On surface lines spaces are rarely provided for more than three or four cars and these should be plainly marked "first car", "second car", etc. The prospective passenger usually has a view of the approaching car and if a car or train stops at the loading point and he sees the car which he wishes to board following, he can quickly go to the exact stopping point designated for the next car. Experience has shown that no great inconvenience is caused passengers or that any particular degree of alertness is required to avoid confusion.

However, confusion is sometimes caused at certain pre-payment loading stations where provision is made for stopping some eight or ten cars. Under these conditions the distance which one must walk from the first car to the last car is so great that the particular car one wishes to board may arrive at and leave the last position before the average person can reach the point where the car stops. The principle of multiple berthing should not be extended to the point where it causes the average car rider inconvenience, as otherwise he will seek other means of transportation.

Safety zones as described in the preceding chapter are practically essential to the proper employment of double berthing in order to keep the space cleared for the second car to approach without interference from vehicular traffic. They further assure the safety of passengers who have to hurry for a second or third car in the zone.

Multiple berthing is employed in many cities and has proved very successful in facilitating the movement of cars and in decreasing congestion. The practice is sufficiently old to prove its worth and there should be no hesitancy in adopting the scheme at points of heavy traffic and close headways. It is a scheme which will appeal to the city traffic department, to the car rider and to other vehicular traffic. Further, it is a scheme which entails very little investment or supervision.

The accompanying illustration shows the scheme of double berthing in operation on the lines of the Pittsburgh Railways, which was one of the first properties in the country to adopt the plan.

Chapter XI

Routing Problems

THROUGH ROUTING

HE rapid increase in the amount of traffic on city streets in the past few years has made the question of congestion in downtown business areas a serious problem for street railway operators. Of course, wider streets assist in relieving the situation but changes of this character involve the expenditure of large sums of money and are seldom made. Regulation of motor traffic has assisted in improving conditions to some extent but it has not been possible to keep pace with the rapid increase in traffic.

During the morning and afternoon rush hour periods delays which occur in congested areas make it extremely difficult for street railway companies to operate cars on schedule. For this reason it is essential that every means be employed to avoid operating methods which will tend to increase congestion. On some properties studies of routing are being continually made in an effort to relieve congestion, but in many cities there is considerable opportunity for improvement in operating methods.

Helps Solve Congestion Problem

One of the simplest means of reducing congestion and car mileage in business sections is through-routing of lines. Fig. 12 shows a typical condition existing in a city where all lines make loops in the downtown business area. For the purpose of illustration only two lines have been shown, line A and line B. These two lines connect residential districts with the downtown area and make practically the same loop in the latter section. Cars on line A operate over Second Avenue from A Street and Second Avenue, to B Street, to Fourth Avenue, to E Street, to Second Avenue, and back to Second Avenue and A Street, covering a distance of twelve blocks. Operation on line B is from I Street and Third Avenue on Third Avenue to E Street, to Second Avenue, to C Street, to Fourth Avenue, to E Street, to Third Avenue and I Street a distance of sixteen blocks. This is a total of 28 blocks in the business district for the two lines.

Fig. 13 shows a plan for combining lines A and B and operating them as a single through route. The new line would originate in one residential section, pass through the congested city district and terminate in another residential section. The proposed arrangement would not a' versely affect the service since the line passes dire through the business area.

On the through route all cars traverse 18 city ' and there will be a direct saving of ten blocks operation . the congested section where the possibility of delay is greatest. Schedule speeds in this section will not be more than five to six miles per hour which means 1.25 minutes to traverse the average city block. This gives a total saving of 12.5 minutes on each trip for the two lines when they are combined. Line A is 8.25 miles long (round trip), is operated on a five-minute headway with 55 minutes running time and requires 11 cars. Line B is 10.5 miles long (round trip), is operated on a five-minute headway with 70 minutes running time and requires 14 cars. The combined running time for the two lines is 125 minutes and with a saving of 10 minutes by through-routing the running time will be 115 minutes. A five-minute service can be provided with 23 cars which compares with 25 cars for lines A and B when they are operated separately. This is a saving of two cars and will reduce operating expenses 8 percent, a real economy in the operation of the two lines.

Factors to Consider

When combining lines to make through routes there are several factors which should be given careful consideration:

(a) Connect lines that have the same amount of traffic.

(b) Join lines of approximately the same length and that serve similar sections of the city.

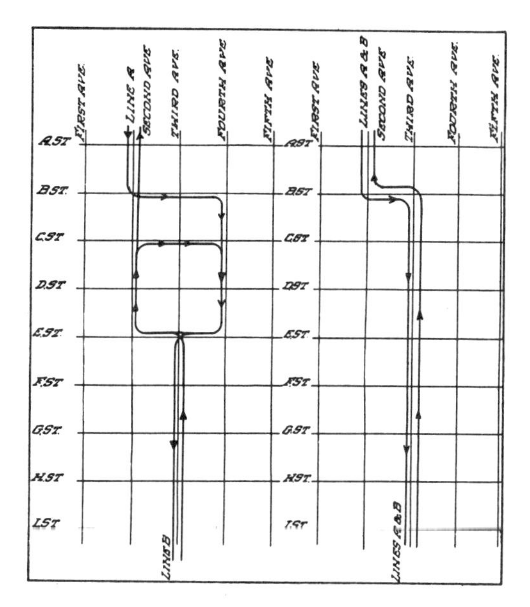

Fig. 12. City lines with Loops in Congested Area.

Fig. 13. Showing Effect of Through Routing on Operation Illustrated in Fig. 12.

(c) If practicable select lines that serve the same class of patrons.

(d) Consider type of car used, schedules and grades.

Naturally lines having the same headway must be connected unless it is satisfactory to modify the headway of one line. Since traffic determines headway the former will be consistent if the latter is considered. If lines of approximately the same length are joined the distribution of load will be improved since the location of the business districts is the determining factor influencing the traffic. Schedules are less difficult to maintain when heavy loading is not near one end of a line.

Consideration should be given to the sections of the city served and the class of pasengers handled. For instance it would not be desirable to operate the same equipment through a mill district and an exclusive residential section as it would be difficult to keep the cars in proper

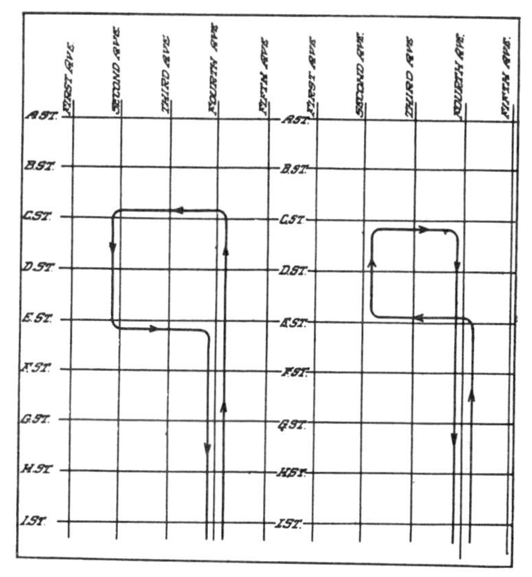

Fig. 14. City Line with Three Left Turns in Congested Area.

Fig. 15. City Line Shown in Fig. 14 with One Left Turn in Congested Area.

condition for the latter service. Also in joining lines the type of car, grades and schedules must be given consideration. There are always some sections of a city that should be served with the most modern cars and profile and schedules occasionally limit the use of certain classes of equipment.

In some cases all of the factors discussed will be in accord for combining two lines in a through route but the length of the lines may make such operation undesirable. This is particularly true in cities which cover a large territory. In this case a line which makes a down-town loop may be eight to ten miles long (one way) and it would not be a good plan to combine two lines of this length.

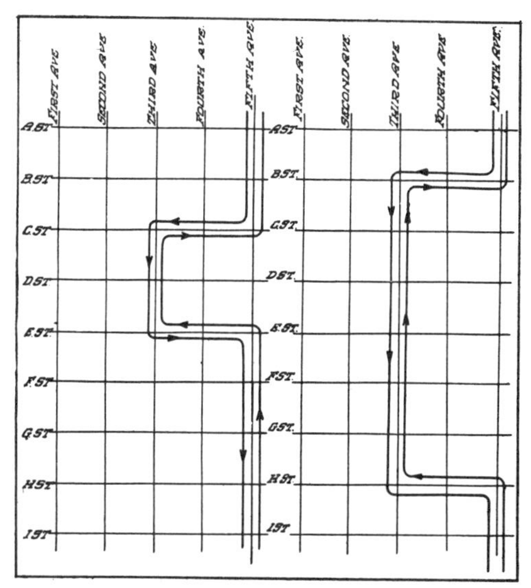

Fig. 16. City Line Routed to Pass Through Business Center.

Fig. 17. Change in Routing Shown by Fig. 16 to Remove Turns from Central Part of Congested Area.

A twenty-mile line (one way distance) in city service would subject a comparatively large number of cars to the influence of local delays. For instance, assume the line to be operating on a five-minute headway with 9.25 miles per hour schedule speed. The one way running time would be 2 hours and 10 minutes or 4 hours and 20 minutes for the round trip and 52 cars would be required to fill the schedule. A delay on any part of the line would influence the operation of 52 cars while in case of separate operation of the lines it is not likely that more than half of the cars would be affected.

Other objections to the operation of long lines are difficulties in arranging proper crew hours, possible increase in dead mileage and long pull-ins in case of equipment trouble.

RE-ROUTING

Left-Hand Turns

Left-hand turns are a source of congestion and should be avoided when possible, particularly in the downtown area. A left-hand turn stops the movement of traffic in both directions at street intersections and naturally cars making such turns are subject to delay as well as being the cause of delay to other traffic. Sometimes it is necessary to have a few lines that make loops in the downtown districts and when such conditions exist it is essential that left-hand turns be avoided when possible.

Fig. 14 shows the route of a line which makes three left-hand turns and one right-hand turn in operating around a city loop. The route of the line is along Fourth Avenue to C Street, to Second Avenue, to E Street, to Fourth Avenue. Operation can be improved considerably by a change in routing as is shown by Fig. 15. In this case the line follows Fourth Avenue to E Street, then on E Street to Second Avenue, to C Street, and return to Fourth Avenue. This arrangement results in one left-hand turn and three right-hand turns.

The one objection to reducing the number of left-hand turns in this manner is the fact that the line crosses itself at Fourth Avenue and E Street (Fig. 15). If this is a busy corner there will be a tendency to increase congestion at this point so it is desirable to locate the turn and crossing where traffic is light whenever possible. For instance, Fifth Avenue and E Street may be a better location for the left-hand turn and crossing.

Care in routing cars through the center of the congested district to avoid turns will assist in speeding up service. This is illustrated by Figs. 16 and 17. The center of the congested district is along Third Avenue from E Street to C Street. In Fig. 16 the route of the car line shown is along Fifth Avenue to E Street, to C Street, and then to Fifth Avenue. Right- and left-hand turns are made at E Street and Third Avenue, and C Street and Third Avenue, which are two of the busy intersections. Fig. 17 shows a change in the routing of the line to locate the turns at corners where traffic is less dense. The route is on Fifth Avenue to H Street, to Third Avenue, to B Street, to Fifth Avenue. The same downtown section is served but turns are made at corners which are less congested.

Parallel Lines

The downtown districts of most cities have several parallel street car lines and frequently some of the tracks

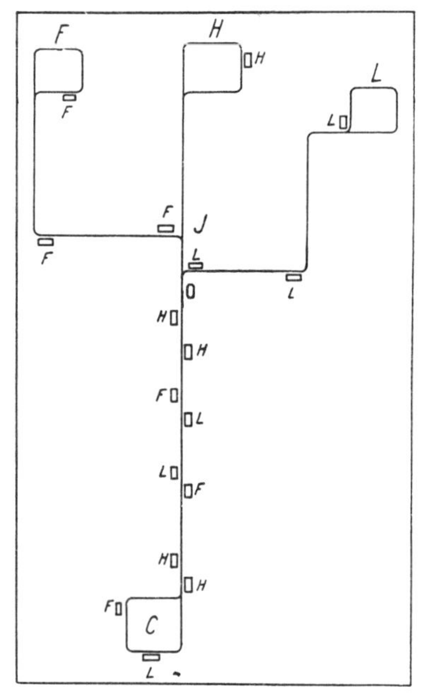

Fig. 18. Split Routes.

are seriously overloaded. A track plan on which the number of cars that pass over each track is marked, particularly for the rush-hour service, will show at a glance which tracks are handling the maximum number of cars. Changes in routing can be made to eliminate dead mileage and divide cars equally between the available routes through the congested area. Such a plan was recently proposed and put in effect in a midwestern city and the result was an annual saving of more than $60,000.

It sometimes happens that a large number of cars are operated to one central point which naturally promotes congestion. In one city operating under conditions of this nature where cars are routed to a central square the situation has been improved by having several loops and confining all cars entering the square by one street to the use of one loop. This prevents cars on one line from interfering with those of another line and also greatly facilitates loading. Another means of relieving congestion in the square that has proved very effective is the looping back of cars on the edge of the district.

SPLIT ROUTES

In many cities the natural trend of traffic is between a central business zone and the surrounding manufacturing and residential districts. Of course, there is also among these outlying communities a certain amount of travel which has no desire to pass through the heart of the city.

In fact such patrons have every incentive to travel by the most direct route. Where this latter class of traffic is very heavy it may be served best by belt or cross lines intelligently located and wisely operated with limited transfers to and from the routes connecting with the business district. However, where the majority of the travel centers at and radiates from one section of the city, a different method of operation, which has been christened "split routes", may prove most profitable.

Referring to Fig. 18, C represents the common center to and from which there is the major traffic. F, H, and L are outlying terminals with little interchange. Obviously, there are two methods of serving this territory. First; a through line CH with transfers to a cross line F J O L. Second; three through lines, one from C to each terminal, F, H and L with transfers between any pair of terminals. This latter is the "split route" method.

(1) Lines, CF, CH and CL, of approximately equal length and equal traffic density between O and each outlying terminal.

(2) Traffic density between C and O equal to or greater than the total traffic between O and all outlying terminals, F, H and L.

(3) Hourly, daily and seasonal traffic variations such that the number of cars required to secure the maximum revenue from the CO portion of the line will be neither much too small nor too great for the most profitable operation of the sections OF, OH and OL.

(4) Absence of local conditions (such as railroad grade crossings) which may interfere with maintenance of schedules on one line more than on another.

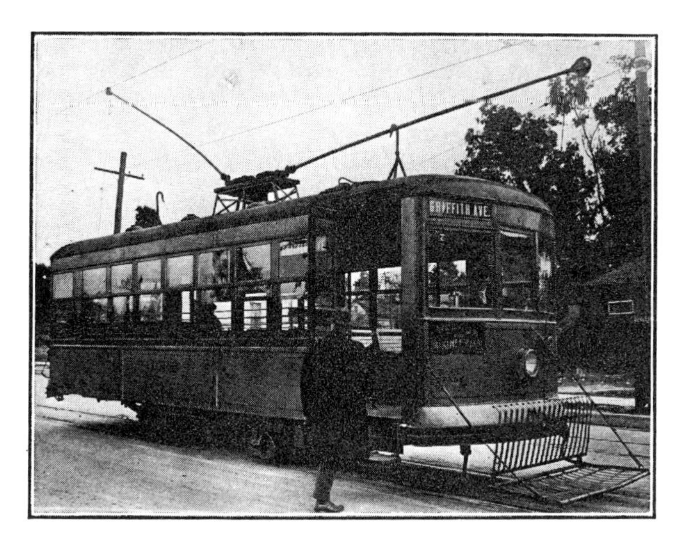

Fig. 19. Safety Cars may be Used in Split Route Operation.

(5) Character of traffic such that one type of car is most suitable for all lines having a common section, CO.

Fig. 19. Safety Cars may be Used in Split-Route Operation.

The maintenance of uniform headway between C and O, as well as between O and F, H and L, is essential to the best operation. Therefore, where the several lines are of unequal length or require different headways, it may be difficult to arrange for uniform headway along CO. However, careful study will usually develop a means of securing this result. In such a study, it is important to avoid long layovers at any terminal, as it is only a rolling car that gathers fares.

The wide application of the safety car has demonstrated thoroughly the value of short headway in producing revenue. At the same time many of our cities have such extreme traffic congestion that train operation becomes imperative. For some of these cases a combination of multiple-unit trains and "split routes" may be the logical solution. For example, assume that CO has become so congested that trains are essential, although there remains ample room for single car operation on OF, OH and OL. The solution may be found in operating trains of three one-man-two-man cars between C and O, composed of one car destined for each of the terminals, F, H and L. Each car may be operated between O and its outlying terminal by one man, while a second man may be desirable on the head car when in train.

As another instance, the traffic on OH may be double that on OF or OL. The solution may be found in two-car trains; one train running solid between C and H and the next train being split at O, one car going to F and the other to L.

It is impractical to outline, and in fact impossible to preconceive, all of the variations in and combinations of traffic conditions. The main thought to be brought out, is that careful analysis of city traffic problems with the principle of "split routes" in mind may be profitable, particularly when considered in conjunction with multiple-unit train operation.

Chapter XII

The Selective Stop

IT HAS already been outlined how the proper methods of fare collection and the use of safety zones and multiple berthing can perhaps accomplish more than anything else to increase schedule speeds as it is the stop time that offers the greatest possibilities of reduction. Aside from the methods of rerouting it is believed that the selective stop, more commonly called the skip stop, offers the greatest possibilities of reducing running time. The term selective stop would seem to be a more fortunate designation as it is a positive rather than a negative term and does not convey the idea of taking something away from the customer. The selective or skip stop idea came into prominence during the war when it became imperative for railway companies to adopt every possible means of reducing expenses and increasing earnings. It has proved to be not only an economic method of operation but to offer a very material improvement in service to the public.

Selection of Stops

In the inauguration of selective stop operation on any line, stops should be selected to provide the best service to the car-riding public. An excessive number of stops increases the time of the trip, at the same time making the journey of the passenger tiresome. On the part of the railway company it requires an excessive number of cars to fill the schedule of a given line, increases the crew expense and also the energy consumption and maintenance cost. An insufficient number of stops increases the walking distance between stops, unnecessarily inconveniencing the patrons of the railway. It loses part of the short-haul traffic which is considered profitable business. A moderate number of stops determined by proper investigation and study in selection is just as necessary as to select the proper number of cars to fill the schedule on a given line.

The number and location of stops should be such as to attract the maximum patronage and insure a minimum operating cost. In other words the stops should be of sufficient number and so located that no business will be lost to the railway company and the energy, crew and maintenance costs will be a minimum.

In one city where the selective stop has been in operation for over ten years, the officials of the railway company not only believe the selective stop to be advantageous at present but are planning to extend its scope. At present with the selective stop system the service on one of its lines in the non-congested section, is as follows:

Stops per mile—6.48
Schedule speed—10.6 miles per hour
Duration of stops—5.6 seconds
Energy consumption—2.06 kw.-hr. per car mile.

It is proposed to reduce the number of stops in which case the service will be approximately as follows:

Stops per mile—5.87
Schedule speed—10.85 miles per hour
Duration of stops—6.2 seconds
Estimated energy consumption—1.99 kw.-hr. per car mile.

It will be seen by comparing these tabulations that even

with an increase in the schedule speed there is a small reduction in energy consumption per car. There is also a further energy reduction possible in case a fewer number of cars per line can be used to fill the schedule. This is brought about by the fact that shorter running time of the cars in the future service would on some lines result in a time saving sufficient to eliminate one or more cars from the schedule.

Fig. 21 shows graphically the schedule speed with a given number of stops per mile, from which it is readily seen that the speed which can be maintained changes rapidly with the increase or decrease of number of stops per mile.

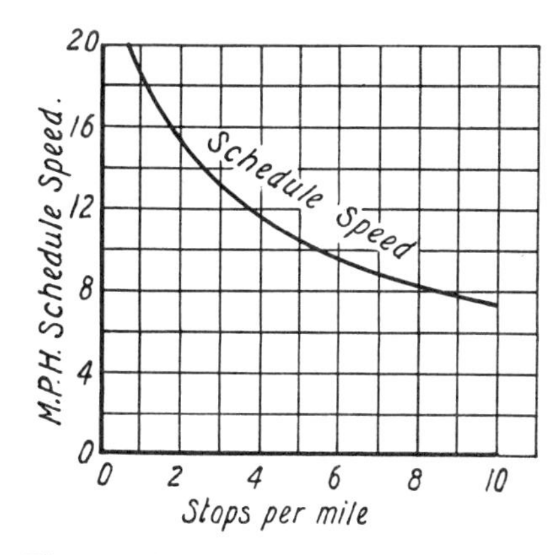

Fig. 21. Operating Characteristics, 5000 Lb. Passenger Load—33000 Lb. Car, 4-55 Hp. 500 Volt Motors.

Stop Location

The past training of the public with regard to proper location of car stops has in many instances been very bad. A large percentage of the people expect the car to stop at every corner and yet make good time on the run. This, of course, is impossible. Then again some prominent resident or politician or even some group of individuals may insist on having a stop or even a number of stops located to suit them, when as a matter of fact a better arrangement could be made which would serve more advantageously a greater percentage of the people of that district.

Fig. 22 shows a section of line slightly less than two miles in length which with the selective stop system has 17 scheduled stops in each direction. This is approximately 10 stops less than would exist if stops were scheduled at every cross street. With 17 stops as shown the number of scheduled stops is 8.55 per mile while the actual number of stops shown by test is from 6½ to 7 stops per mile, as not all of the scheduled stops are made in any one trip during non-rush-hour service. This is a relatively low number of stops per mile for city service.

It will be noticed that some of the stops are near side stops while others are far side stops. For instance, on the inbound trip the stop at 32nd Street is a far side stop while the one at 30th Street is a near side stop. Furthermore, there are a number of stops in the center of the square such as occur between 52nd and 53rd Streets. It is evident that a study of the stop location has revealed to the railway company that it is not advisable to adopt either the far side stop or the near side stop but rather to place the stop at the most advantageous location based on observation. It will be noticed, furthermore, that the stops on the outbound trip are frequently located at streets at which no stop is made on the inbound trip so that the passenger that has to walk one square for his car in the morning will be discharged at a point nearest to his residence on the outbound trip in the evening. Usually it is neither convenient nor advisable to reduce the number of stops in the congested section although there are exceptions.

Automobile and truck traffic has an important bearing on stop locations. At crossings where buildings are close to the sidewalk the fact that the view of the motorman is thus obstructed always requires slowing down for reasons of safety. In other words it may even be necessary to stop at a cross street before crossing on account of the danger of collision with moving vehicles. In this instance it is just as well to take a few seconds longer at this stop and load passengers. On the other hand if a stop should be located at a busy corner and loading operations could be better performed at the far side, the better policy is to use a far side stop. The elimination of a stop at a cross street does not necessarily mean that a car may cross this street at full speed without danger of collision with fast moving vehicles. Part of the advantage of the selective stop is lost when the car must slow down to a safe speed before crossing such a street.

With the selective stop system it may be advisable and economical to change the location of stops in certain districts in summer and winter seasons. These instances, however, would be very few on city lines. On interurban lines or for instance lines leading to parks, a change in location of schedule stops may be advisable.

Advantages of Selective Stop

Some of the benefits derived from the use of the selective stop system are:

1. Minimum time required for the passenger to reach destination.
2. Maximum comfort in transit due to fewer accelerating and decelerating periods.
3. Minimum investment cost and fixed charge expense for equipment.
4. Minimum operating expense.

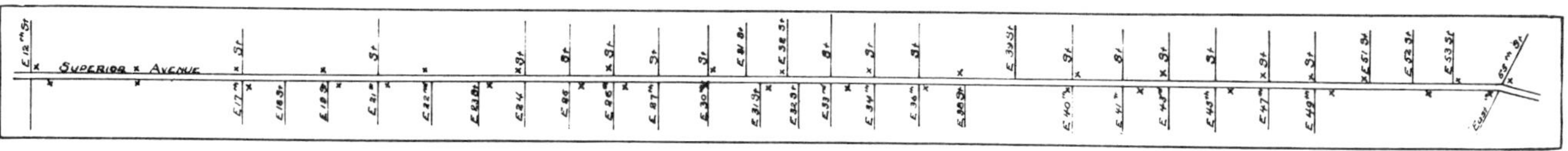

Fig. 22. Stop Locations on Part of a Line Operating on the Selective Stop Principle.

The first two advantages are for the benefit of the public while the last two are for the benefit of the railway. In cities having both subway or elevated lines and surface car lines a large percentage of the people prefer to ride on the higher speed subway or elevated lines in order to reach their destination more quickly even though it requires trudging up or down long stairways. The layout of car lines in our cities and traffic conditions make it practically impossible to operate advantageously surface express cars or trains between large population centers and at high speed. This type of operation eventually requires private right-of-way so that for surface systems one of the greatest reliefs to be hoped for is that of a reduction of the number of stops in order that reasonably high schedule speeds may be maintained.

Results of Selective Stop Operation

In the year 1918 it was estimated that the total yearly saving by the introduction of the selective stop system in cities of the United States having a population of 25,000 or over would be $30,000,000 a year and that the schedule speed would be increased 10 to 12 percent while the power saving would range between 8 and 16 percent. Since that time a sufficient number of companies have retained the selective stop system to enable a poll to be made which shows that the introduction of this system was fully justified.

Of 40 large street railway companies using the selective stop during the World War, 17 companies have continued it. Six of these companies have increased the use of the selective stop on their systems, 5 have decreased it and 6 have made no change in this installation since the war.

The companies reporting their experience with the selective stop show an increase in schedule speed from 7 percent to 24.4 percent or an average increase of approximately 17½ percent. An average figure for speed without the selective stop is approximately 8½ miles an hour, while the use of the selective stop makes possible a speed of 9½ to 11 miles an hour. The decrease in energy consumption per car mile by the use of the selective stop is from 4½ to 10 percent, an average figure being about 5½ percent.

Cleveland had the honor of being the first city to utilize the selective stop, then called the skip stop, on an extensive scale. This was in the year 1912 and since that time it has been retained to advantage. This was brought about largely by the fact that the people themselves voted in favor of instituting such service. The fact that the vote was in favor of this system was due largely to proper placing of the facts before the public by the railway management. The selective stop system has been one of the means of retaining a relatively low fare in Cleveland as compared with other cities. Pittsburgh has consistently followed the practice of eliminating useless stops and this work has been carried out in an unobstructive manner without incurring the displeasure of the public.

Public Opinion

In one city, opposition was directed against the selective stop, based upon the argument that the number of accidents were increased by its use. This particular incident required considerable effort and expense on the part of the railway company to present adequately and properly its side of the case to the public in order to allay the criticism. In another city the selective stop was opposed because the passengers were compelled to walk one square farther than formerly during the rainy weather. The railway company met this opposition by substituting a compromise service. Two classes of cars were put in operation—motor cars and motor car and trailer trains. The single cars made all stops whereas motor cars hauling trailers passed up a number of the stops in order to maintain their place in the schedule. This operation developed the fact that many riders prefer to take the motor car-trailer train rather than the single car in order to have a more comfortable journey.

The question of public relations has always been and always will be an important one. There will always be persons or groups of persons antagonistic to the railway, and conversely there always will be groups friendly to the railways. The larger this latter group the easier it will be to maintain pleasant public relations. This requires constant reassurance that the company is rendering to the public the highest grade service it has at its command.

The leading citizens of any community are, as a rule, progressive and open to reason on matters pertaining to improvement of the railway service. Focusing their attention upon the fact that with a selective stop system patrons can make a trip in less time and with more comfortable riding conditions than under the "stop at every corner" plan, will do much to bring about the desired attitude of mind on the part of the public. Realization that the selective stop system is the result of a careful study of railway operation should lead such citizens to favor its adoption.

Conclusion

The benefits to be derived from the selective stop system are such as to warrant its careful consideration by all progressive railway systems. When it is necessary to initiate the service by publicity channels it is felt that it can be best performed under the term "selective stop" rather than under the old term "skip stop". To compete more satisfactorily with gas propelled vehicles, it is imperative that the schedule speed be raised and that riding be made more comfortable. At the same time the railways have an opportunity of reducing operating expenses with the possibility of rendering materially improved service. Past experience has proven that the use of the selective stop system is advantageous to the car-riding public and beneficial to the railway.

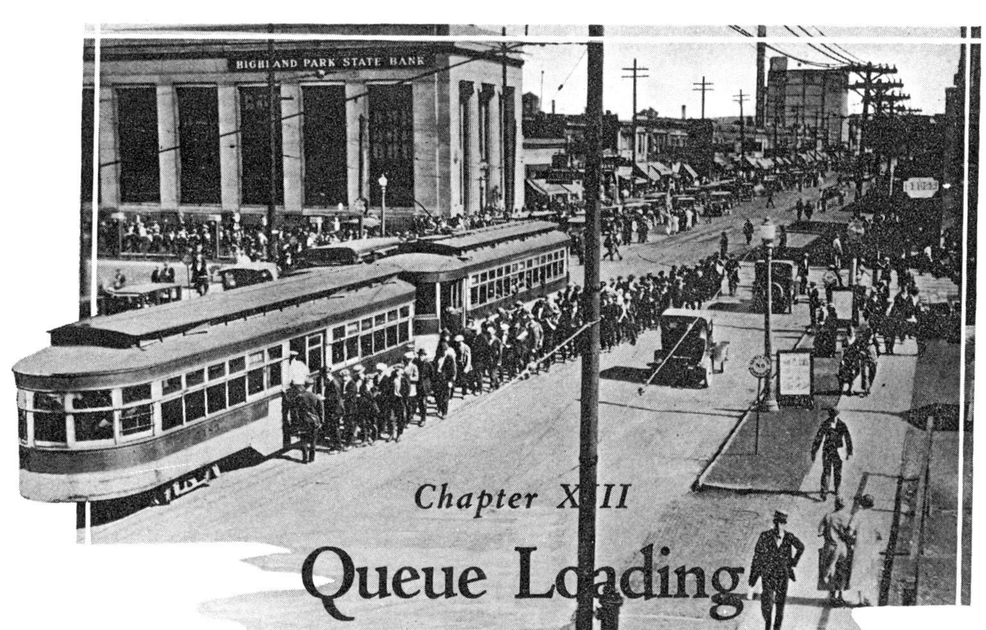

Chapter XIII

Queue Loading

HE definition of the word "queue" is a file or line of persons waiting in the order of their arrival. "Queue loading" as applied to the transportation industry means that the patrons of the road line up in the order of their arrival at designated car stops and board the car in an orderly manner. This scheme, to expedite the handling of large crowds at congested centers, is one of the more or less modern ideas now occupying the minds of transportation officials.

Advantages

The scheme of queue loading has many advantages accruing for both the people and the railway company.

The advantages to the car rider are:

(1) *Saves Time*—A systematic and orderly movement of the boarding passengers permits of their more rapid handling.

(2) *Dangers Avoided*—An orderly line-up eliminates disputes and dangers of a personal nature occasionally encountered with a crowd trying to board the cars.

(3) *Protects the Weak*—Women, children and old men have an equal chance with the more rugged and robust men in boarding cars and securing a seat.

The advantages to the railway company are:

(1) *Saving of Time*—Faster loading permits running a greater number of cars within a specified time, as time is saved by each car getting its quota of the load.

(2) *Saving of Money*—Systematic and orderly loading of passengers results in the collection of all fares, as passengers are prevented from entering the cars through the windows and from riding on the outside or crowding past the fare box.

(3) *Saving of Equipment*—It is a well known fact that considerable damage has been done to the windows, doors, etc., of cars at heavy loading points by passengers forcing their way uncontrolled into the cars through the doors and sometimes through the windows. All of this is eliminated by the queue formation of loading.

Feasibility of Scheme

There is plenty of evidence that this method of handling large crowds is not theoretical but entirely practical as such practice is followed at theater box office windows, baseball park ticket windows, business windows of large banking institutions, railway ticket office windows, bread line and soup kitchen locations, fire drills in schools and industries.

One very striking example of what can be accomplished in connection with the practical application of the queue method of loading in this country is at the Ford Motor Works, located on Woodward Ave., Detroit, as shown in the accompanying illustration. At this point, between 30,000 and 40,000 employees of the Ford plant are successfully handled by this method of loading between the hours of 3:30 and 4:45 P. M. each working day.

How to Accomplish

To inaugurate queue loading successfully the following points should be given some consideration:

Co-Operation—It will be necessary to get the cooperation of the patrons, the city officials and the merchants and manufacturers.

Education—Present established customs will have to be broken down and the people educated and trained to see the benefits of this new scheme of loading. At loading centers, proper signs or markers will be required.

Publicity—Some well directed advertising will be essential in which an appeal should be made to the people's sense of fair dealing, by citing examples of queue formation and the advantages to be obtained.

Legislation—Following the example of some of the foreign cities, local councils might be persuaded to cooperate by passing favorable city ordinances.

Chapter XIV

The Traffic Problem

WITHIN the last few years American cities have suddenly found their streets so choked that passage is both slow and dangerous. Fire apparatus can make no haste through the jam of traffic, and, arriving, finds a barricade of motor cars between itself and the blaze. Street cars containing 70 percent of the people crawl through and accumulate in the jam. Automobiles with the other 30 percent wander slowly about, seeking a place to stop. Business men and publicists from coast to coast clamour for a ready remedy, something new, something expensive—arcaded streets, sidewalks in the air or underground subways.

Meanwhile the cause of all this difficulty is so plain that people dislike to mention it. A small percent of the populace are storing their automobiles in the streets, free and without charge eight hours a day. Another small percentage leave their cars in the street while they do errands or shop. Perhaps these people who park are our leading citizens—the presumption is that many of them are buying things from our leading storekeepers. Hence the delicacy of the subject.

Street Cars Carry Seventy Percent

Traffic counts have shown that 70 percent or more of the people traveling in the congested district are in street cars, which occupy only 1.5 percent of the available space. Of the remaining 30 percent carried in automobiles, some go through, some park in garages; it is only a part, even of this 30 percent, which fills the streets, imperils life and property and wastes incalculable time.

Of all the remedies proposed absolute abolition of parking in the congested district is by far the most successful, the most reasonable, and the least expensive. Allowing parking "not to exceed" a certain time has been widely tried and is usually a failure from impossibility of enforcement. But absolutely no parking is easy of enforcement and when tried has immediately produced a great improvement.

The street car rider must be shown that the fight is in his behalf. The irritated motorist must be shown that the clogged streets which make movement so difficult, slow and dangerous are the fault not of the street car but of his parked brethren. The store-keeper must be persuaded that a line of parked cars in front of his store does not always mean a crowd of wealthy customers within. It is the street car rider, not the motorist, who is the chief patron and support of the merchant and any hindrance to the street car is a hindrance to his buying.

Merchants Depend on Railways

A very recent check on all the entrances to the largest stores in New York City brought to light figures which indicate the dependency of merchants upon the electric railways as a means of transporting customers.

The figures, which are equally applicable to any large city, covered a period of only one hour during a busy shopping period. They showed arrival of shoppers during that hour as follows:

Store	By Auto or Taxi	Otherwise
B. Altman & Co.	116	1362
Stern Brothers	59	1706
Arnold, Constable & Co.	12	511
Lord & Taylor	98	1032
Franklin Simon & Co.	58	1636
Best & Co.	22	3500
Bonwit Teller Co.	78	754
John Wanamaker	102	2298
James McCreary & Co.	30	3264
R. H. Macy & Co.	92	9302
Saks & Co.	30	1860
Gimbel Brothers	90	5066
Hearn & Sons	26	2148
Totals	813	34,439

What is to become of the motorists forbidden to park? One may respond: What becomes of the motorist now who (in large numbers) finds all parking pre-empted? He leaves his car outside the congested district. It must not be forgotten that the total street area in the congested district of any city is far too small to contain all the cars that might desire to park there.

The Russell Sage Foundation Regional Plan Committee (New York) estimated that the enormously expensive removal of surface lines from four avenues, substitution of subways for two elevated roads, and widening of three avenues, would increase traffic capacity 67 percent, while 50 percent increase in capacity could be gained merely by abolishing parking.

What Some Cities Are Doing

Syracuse has prohibited parking in the business district. Traffic was immediately relieved and flows freely. The retail merchants, however, are as yet dubious as to the effect on them.

Buffalo "keeps everything rolling" and finds its streets are wide enough.

St. Louis finds that when parking ceases, automobiles move more rapidly and freely, and street cars reduce their time in the district from 33 minutes to 23 minutes.

The New York State Conference of Mayors advocates no parking except perhaps on streets wide enough to leave ample space for traffic.

Detroit merchants and owners of real estate are converted to the idea. The change increases schedule speeds nearly 2 m. p. h. and reduces accidents 10 percent.

Modifications of the plan also are beneficial. Parking may be prohibited on streets occupied by car lines—that helps the trolley and the great majority who ride in it.

In Philadelphia where a million people enter and leave an area of two square miles, parking is forbidden on the left side of one-way streets, leaving a clear channel for motor traffic. Automobiles are kept off the car tracks by sufficient police. To the right of the car track, a clear space is maintained 100 feet back of each car stop. This arrangement permits a clear flow of automobiles on the left, a perfectly clear track for street cars and entire safety for passengers getting on and off. Cars load three at once. Time in the congested district is reduced from 40 minutes to 33 minutes, and reduction in accidents is expected.

Some forms of parking are worse than others. Angle parking is most objectionable. It narrows the street most, and cars backing out are dangerous. It is sometimes said that a car parked all day causes less trouble than the one which stays a short time—the maneuvering of the latter blocks the street.

Cleveland found that 80 percent of collisions are caused by parking. This city adopted an ingenious scheme. Parking is prohibited during the morning rush hour on the inbound side of the street, and during the afternoon rush on the outbound side. The change resulted in cars hitherto 15 minutes late being restored to schedule.

Following the lead of New York many towns have installed light signals, sometimes attempting to handle several blocks of traffic synchronously. Opinions differ as to the result—opponents of the scheme charge that the conditions of Fifth Avenue, New York, are not duplicated in other cities, and that delay rather than expedition results. Especially is this complaint made against signals timed automatically, since the interval often does not fit the traffic.

Things the Railway Company Can Do

There are some things that the street railway can itself do immediately to relieve the situation and from which to derive profit and prestige. These items have all been covered in previous chapters. Briefly, they are:

Trim out superfluous stopping points—six hundred feet is not too long an interval. Stopping points may be chosen without regard to street intersections, but they should be plainly marked and possibly provided with raised loading platforms.

Street collectors will help even with two-man cars.

Cars should be double berthed.

Through routes are valuable if traffic can be fairly balanced on either end.

If possible, turns in heavy traffic should be avoided. Left turns are worse than right. In laying out necessary loops, strive to put at least two or three of the four turns outside of the crowded district. If practicable place the outbound side of the loop on the main outgoing street, so that, once loaded in the afternoon, no further curves are necessary.

If practicable, transfer points should be removed from the congested district.

Once the street car company has done all that it can with its own service, it can proceed to influence public action. The pedestrian is the final unit all means of transit exist in order to place pedestrians near their destination. The street car carries most of the pedestrians and the street car does not park at all. If the public is convinced of the company's good faith, much can be done. The company can make itself a defender and a spokesman for the great mass of people. With its professional experience in handling traffic, it can speak with weight. It can inquire—are the streets intended primarily for a few stores, a few thousand motor cars, or for all the population? Are the streets storage space, or are they channels for movement? If the city or the stores feel that free parking space is needed, the logical course is to erect municipal or private garage room off the street.

Among the most essential matters is to have good relations with the police, not only for their influence in the adoption of regulations, but because of the necessity of its enforcement. It is surprising too, what a difference a friendly traffic officer, or a hostile one, may make in clearing up a jam.

It is possible that the present automobile traffic problem may pass as completely as did the question of bicycle "scorchers" and riding on sidewalks thirty years ago. But it is very real now, and it offers the street railway company an opportunity and a reward for all its skill, tact and wisdom.

Chapter XV

Elimination of Non-Productive Mileage

A STREET railway company is in the manufacturing business — the manufacture and sale of urban transportation.

Among the requisites for a successful manufacturing business are many factors, certain of which bearing on this discussion are:—

1. An accurate knowledge of the demands of the market—is the product what the customer needs and wants to buy?
2. A periodical analysis of market conditions—is the product brought out at such times and in such amounts as to best meet the demands?

On any given line the traffic varies widely during different hours of the day, and at different points along the line. The ultimate ideal would be a seat for every passenger at all times, and no vacant seats at any time. It is obviously uneconomical to meet the first condition and impossible to meet the second, but these should be the aim, and actual accomplishments should be brought as close to this ideal as conditions permit.

Routes should be carefully checked from time to time to be sure that a direct journey without change is furnished to the maximum number of people. Changes in the character of development of property along the route will often so affect the riding characteristics as to require material modification in the kind and amount of service supplied.

The development of new residence districts in a growing community will require rearrangements of routes or schedules or both, in order properly to serve the district.

The officials responsible for traffic are constantly on the alert for opportunities to improve service on the one hand and on the other to furnish the necessary service at minimum cost. The traffic count is an almost indispensable tool in their work. Traffic surveys are made in many ways, differing in detail and in degree of elaborateness. The final result desired is a tabulation or graph showing the following for each line:—

1. Number of passengers boarding at each stop.
2. Number of passengers alighting at each stop.
3. Passengers on car at various points along line.
4. Variations in headway.
5. Total passengers, cars, and seats passing given points by 15- or 30-minute intervals.

In addition data is sometimes collected to show for each passenger the origin and the destination of trip, and also the density of other vehicular traffic at congested points.

On the basis of this data, headways, seating capacities, etc. can be so regulated as to meet the desired standard of service in passenger per seat ratio.

Investigation may show the desirability of "turn back" runs, to maintain a better relationship between seats per hour and passengers per hour on the outer ends of lines and thus to avoid running more cars than the service requires. Extreme care must be used, however, in inaugurating such practice. While apparent economy might indicate such a course, care must be taken to avoid making the headway too long on the outer end of the lines. Experience with one-man operation and the safety car indicated the improvement in riding brought about by frequent service, and if the headway is made too long the loss of income and good will may offset the gain in reduced car miles.

Public Confidence Essential

The inauguration of such service should be preceded by a campaign of education to acquaint the patrons of the line with the purpose of the step, and with the details of the plan. The turn back point might well be made at some natural traffic exchange point as this will usually be found to suit best the personal convenience of a large number of riders. Crossover or turning facilities should be so arranged as to interfere as little as possible with other traffic. The general objection to long headways may sometimes be at least partly overcome by insisting on regularity of service and close adherence to schedule. A passenger might not mind so much waiting ten, fifteen or twenty minutes for a car if he knew that the car *always* passed his corner a certain number of minutes past the hour.

Destination signs of short route cars should be particularly clear to avoid annoyance to passengers from boarding the wrong car, and crews should be instructed to be particularly careful in handling the occasional irate passenger who finds that he has to change cars.

Under certain conditions it may be possible to effect economies by the operation of light-weight one-man cars or safety cars on the outer sections of lines, running the larger capacity two-man cars only as far as the passenger density determined by the traffic survey requires. In such cases public good will can be cultivated by the provision of suitable waiting shelters at the point of change and by so arranging the schedules as to require the minimum waiting time for passengers.

The whole problem resolves itself into a balance between short headways with excess seat miles and long headways with resultant economy but loss of traffic. Care and experienced judgment are necessary in making a choice, and beyond that, everlasting vigilance to provide rapid, regular service to encourage patronage and public good will.

Philadelphia-Paoli Electrification

Typical All-Steel Electric Passenger Train

Reprint No. 27

Heavy Traction Department

Westinghouse Electric & Manufacturing Company

East Pittsburgh, Pa.

Reprinted from Electric Railway Journal *November 13, 1915*

PHILADELPHIA-PAOLI ELECTRIFICATION—VIEW OF CROSS-CATENARY OVER NINE TRACKS, SHOWING READY VISIBILITY OF SEMAPHORE SIGNALS NOTWITHSTANDING COMPLICATED OVERHEAD CROSS-OVERS

Philadelphia-Paoli Electrification

Operation of This 20-Mile Suburban Electric Zone Has Been Begun by the Pennsylvania Railroad with Multiple-Unit Cars Having Repulsion-Starting, Series Motors, Single-Phase Power Being Supplied from an 11,000-Volt Catenary Contact System Carried on Cross-Wire Bridges

The recent establishment of regular suburban service over the single-phase electric zone of the Pennsylvania Railroad at Philadelphia marks an important step in the electrification plans of that company and makes a description of the details of the installation particularly timely. At present the electrified tracks extend only from the Broad Street Station in Philadelphia to Paoli, 20 miles to the west on the main line, but work is under way also on the electrification of the Chestnut Hill line, 12 miles northward from Philadelphia. Both of these installations, as outlined in an account that was published exclusively in the ELECTRIC RAILWAY JOURNAL for April 18, 1914, were projected primarily to relieve congestion at Broad Street Station, and it is expected that they will take care of the normal growth of traffic for the next seven or eight years.

For the Philadelphia-Paoli service the rolling stock consists of ninety-three standard all-steel cars, eighty-two of which are for passenger service, nine for combined passenger and baggage service and two for combined baggage and mail service. All are motor cars, as no trailers are operated on the electrified suburban runs. Trains of from two to seven cars are operated regularly, the average acceleration on a straight level track being approximately 1 m.p.h.p.s. up to 30 m.p.h., with a balanced speed of 60 m.p.h.

ELECTRICAL EQUIPMENT ON CARS

The equipment of each car consists of two 225-hp. Westinghouse single-phase, air-blast-cooled motors mounted on one truck. Automatic acceleration is provided with the control, and automatic multiple-unit electro-pneumatic brake equipment has been installed. All of the main pieces of the electrical apparatus are mounted on one end of the car and the brake equipment is mounted at the other end. This gives an uneven weight distribution on the two trucks, approximately 60 per cent of the total car weight being on the driving wheels. The cars are designed for double-end operation.

The motors, which are connected in series, are started and operated up to approximately 15 m.p.h. as repulsion motors, with the auxiliary or compensating field, the armature and the main field in series. With these series connections the armature is short-circuited through resistance. Resistance is also inserted in series with the motors on the first step and is cut out on the second step. The third step changes the connections to energize the auxiliary field from one portion of the transformer and the armature and main field, connected in series, from another portion of the transformer, thus affording doubly-fed connections. The armature short-circuit is removed when the motors are operating with the double feed. Subsequent steps in the control are obtained by increasing the motor voltages.

Power for the control system is supplied in the usual manner from a motor generator set, which is in parallel with a storage battery, through a control plug, and the movement of the master controller handle to the right or to the left energizes the proper control circuit for forward or for reverse movement of the train. The closing of the unit switches is governed by a current limit switch which has two settings, one for repulsion

PHILADELPHIA-PAOLI ELECTRIFICATION—STANDARD CATENARY CONSTRUCTION ON TANGENTS

connections of the main motor and the other for the doubly-fed connection, the change in the limit setting being obtained by energizing a battery coil on the limit switch. All of the switches are interlocked through the No. 9 switch, so that no switches can close until the No. 9 switch is closed. A small knife-switch is placed in the control circuit of the No. 9 switch, and opening this switch cuts off the supply of current to the main motors. Ten control wires between cars are necessary to operate cars in trains, one of these wires performing in the dual function of the third operating wire and the "trolley unlock" wire.

Each motor has a continuous rating of 200 hp. when ventilated with 1200 cu. ft. of air per minute. The armature is of standard construction, the commutator and the laminations being mounted on the spider, the former being undercut 1/16 in. The armature is wave-wound and cross-connected, and no resistance leads are used between the windings and the commutator. The field windings consist of two entirely independent sets of coils, one being the main field circuit for producing the effective magnetic field, and the other the auxiliary or compensating winding which balances the armature reaction on the field. In addition, the latter has a neutralizing effect on the sparking voltage. The field consists of six poles, the coils being of copper bars suitably insulated, connected at the ends by straps.

Flexible gears are used, the gear ratio being 24 : 55. Each gear is made up of a rim on which the teeth are cut, a center, a cover plate and spring details. The rim is spring-mounted on the center, the periphery of the center and the cover plate acting as the bearing surfaces for the rim.

One pantograph of especially light construction is installed on each car. The springs which raise it are designed to give flexibility to the framework, so that in operation a slight dragging of the contact shoe takes place, resulting in its following the wire much closer than with a rigid framework. In addition, the shoe is spring-mounted on the framework. The pantograph is provided with four insulators suitable for 11,000-volt service, and the whole mechanism is mounted on a base provided with insulators similar to those of the pantograph, thus providing double insulation. The pantograph is lowered and unlocked by air at 70-lb. pressure, a small hand pump being provided for unlocking it when no air pressure is available.

The safety-first principle has been carried out in the provision of a grounding device of novel design. Steps for mounting to the roof are provided at one corner of the car only, and a lever is placed on the roof at this corner. When one climbs to the top of the car this lever is thrown up, thus locking the trolley in the down position and grounding the entire framework.

Repulsion Motor Principles

In American practice it has not been customary heretofore to short-circuit the armature of an a.c. commutator motor, and the adoption of this plan in the Philadelphia-Paoli electrification for even a part of the accelerating period is of more than passing interest.

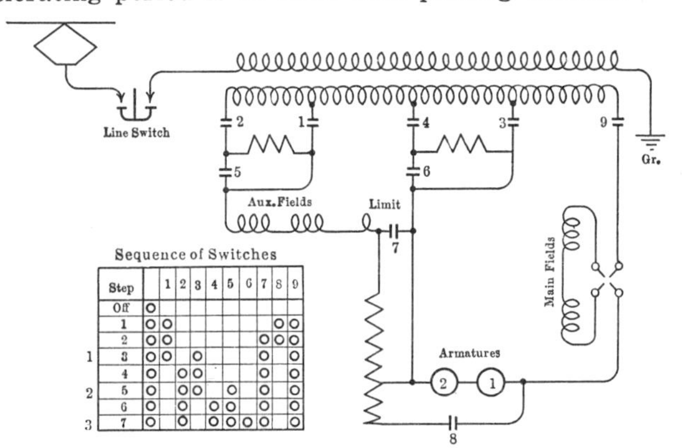

	Step		1	2	3	4	5	6	7	8	9
	Off	O									
	1	O	O							O	O
	2	O	O						O	O	O
1	3	O	O		O				O		O
	4	O		O	O				O		O
2	5	O		O	O		O		O		O
	6	O		O		O	O		O		O
3	7	O		O		O	O	O	O		O

PHILADELPHIA-PAOLI ELECTRIFICATION—SCHEMATIC DIAGRAM OF REPULSION MOTOR CONNECTIONS

The invention of the repulsion motor is due to Prof. Elihu Thomson, who more than twenty-five years ago developed the scheme shown diagrammatically at *A* in the diagram on page 5. This diagram represents a two-pole motor having field coils wound on laminated field cores and connected to an a.c. supply line, an armature with brushes and coils connected to the commutator bars as in a d.c. motor, and a connection short-circuiting the brushes. The brushes are shifted from the normal position as shown.

If the brushes were placed under the middle of the field poles the magnetic field flux would pass trans-

PHILADELPHIA-PAOLI ELECTRIFICATION—VENTILATED REPULSION MOTOR—FLEXIBLE GEAR WITH COVER PLATE REMOVED TO SHOW SPRINGS

versely through the spaces inclosed by the armature coils and a current would flow between brushes, because the passage of alternating magnetic flux through a loop of wire sets up current flow in the wire, in accordance with the well-known principle on which a transformer operates. The short-circuiting of the brushes would permit a current that would be limited only by the impedance of the electric circuit. This flow of current in the short-circuited coils would set up torque, because each conductor would be within a magnetic field. No net torque would be produced, however, because the current would flow in the armature conductors in opposite directions on opposite sides of the brushes and the torques produced on the two sides of the armature would balance each other. But if the

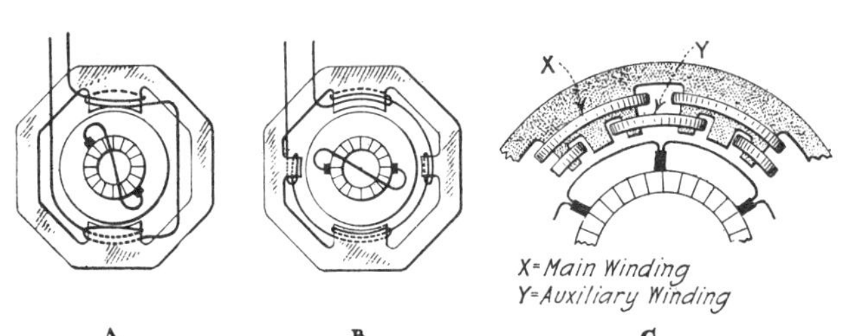

PHILADELPHIA-PAOLI ELECTRIFICATION—DIAGRAMS ILLUSTRATING REPULSION MOTOR PRINCIPLES

brushes are shifted as shown in diagram *A*, the balance is disturbed and net torque is produced, because the torque of the windings on one side of the brushes overcomes the lesser torque of the windings on the other side.

The principle of the repulsion motor can be explained also by means of the second diagram, shown at *B*. This motor is equivalent in every way to the preceding. Here the winding is represented as divided into two parts, main and auxiliary coils, one furnishing the field for the production of torque and the other inducing the armature current. It is evident that the compensating winding used in the a.c. series motor, shown crudely in principle in *C*, could be used as an auxiliary winding, to produce a shifting of the magnetic field equivalent to a shifting of the brushes. A series motor can therefore be readily adopted to repulsion starting.

In the Philadelphia-Paoli motor the original scheme of Professor Thomson is still further modified by the addition of the series connection of armature and field. The armature and field windings are therefore both "conductively" and "inductively" connected. To the extent, however, that current is induced in the short-circuited armature because of the transformer effect of the auxiliary winding it is a repulsion motor. On account of the low resistance of the armature short-circuit the induced current during the starting period has a high value.

Electro-Pneumatic Brake

The air-brake equipment on the cars is designed to be used either in steam or electric service, and differs from the ordinary pneumatic brake in that the brake pipe reduction is made on each car by means of electric control instead of being made entirely with the engineer's brake valve. The addition of electric control to the pneumatic brake does not change its function in any way, but shortens the time required to get the brakes applied on all cars.

The motorman's brake valve contains both electric contacts and pneumatic parts, the electric portions being mounted above the pneumatic portions. There are six positions: (1) The release and running; (2) the electric holding; (3) the handle off; (4) lap; (5) service, and (6) emergency. The first-named position is to the left and in this position all train brakes are released and the system charged. The "electric holding" position, as the name implies, holds the train brakes through the electric control system, but recharges the system. Pneumatically this position is identical with the release and running position. All ports are closed in the "handle off" position, and the handle may be removed, and in the "lap" position the ports are also closed. The "serv-

PHILADELPHIA-PAOLI ELECTRIFICATION—PANTOGRAPH WITH DOUBLE INSULATION ON CAR ROOF—END VIEW OF STANDARD CAR

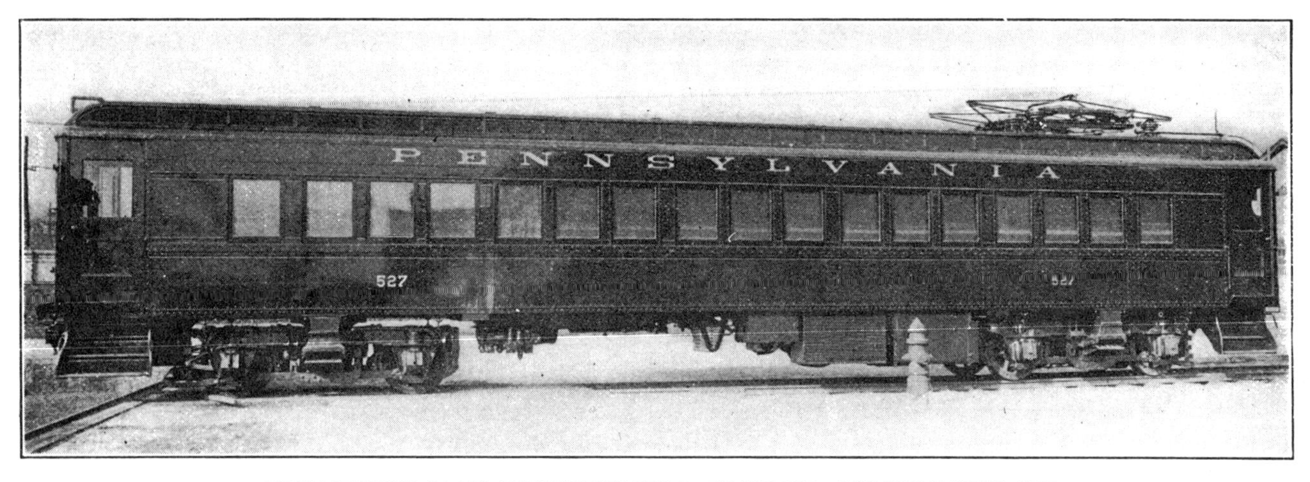

PHILADELPHIA-PAOLI ELECTRIFICATION—STANDARD MULTIPLE-UNIT CAR

ice" and "emergency" positions are the regular ones for setting in service or emergency applications. For the service application a limiting valve is provided in conjunction with the brake valve, and this allows a maximum reduction of 20 lb. in the brake pipe. A small cut-out plug is provided for cutting out the electric operation when desirable.

The universal valve is built up of five different portions: (1) The pipe bracket, which contains a quick-action chamber and a quick-action closing chamber for use in emergency application; (2) the equalizing portion, which contains the moving parts employed in service and emergency operation; (3) the quick-action portion, which contains the moving parts employed in producing quick action; (4) the high-pressure cap, which contains the parts employed in securing a high cylinder pressure in an emergency application, and (5) the magnet brake portion, which contains the magnet valve for electric control of brake operation, an emergency switch, and a cut-out cock. This universal valve controls the charging of reservoirs, the application of brakes and the release of brakes. It is mounted at the side of the car near the trailer truck.

The main reservoir pressure is 100 lb. and the brake-pipe pressure is 70 lb. To permit the operation of these equipments in steam service, where the brake-pipe pressure is 110 lb., without making adjustment, a main reservoir by-pass and limiting valve is employed. By its use the same cylinder pressure is secured in making an emergency application in either steam or electric service, although the operation of the universal valve is the same for either. In steam operation the pipe line, which is used as a main reservoir line in electric service, is used as a signal line.

Eight wires, including the battery plugs, and ground wires are required for the electric control of the brakes, for governor synchronizing and for train signaling. Since the two battery wires are common to the brake control and the unit switch control, a seven-point receptacle and jumper is used to carry the brake control wires. Two receptacles are mounted on each end of the car and on each side of the coupler, and these are con-

PHILADELPHIA-PAOLI ELECTRIFICATION—SHOP AND INSPECTION BUILDING AT PAOLI

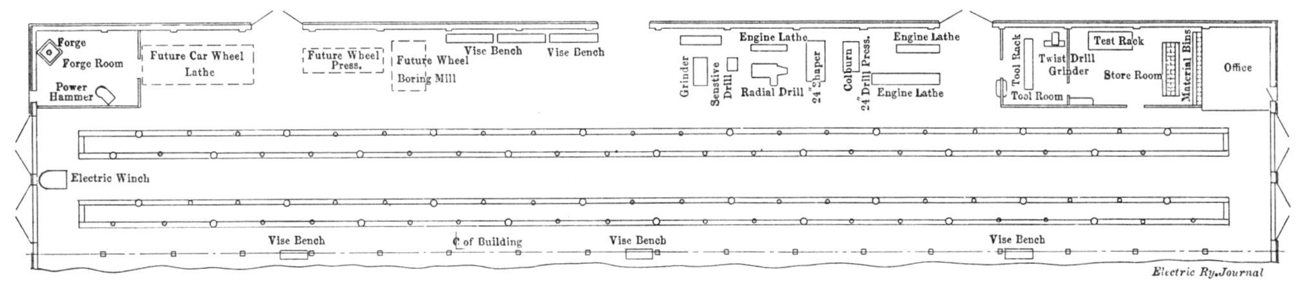

PHILADELPHIA-PAOLI ELECTRIFICATION—HALF PLAN OF INSPECTION BUILDING SHOWING ARRANGEMENT OF SHOP SECTION

nected in multiple in the same way as the nine-point receptacles for the unit switch control.

Car-Inspection Building

A very substantial and completely-equipped car-inspection building has been constructed at the Paoli yard. This has been planned to serve not only the cars required for the present electrification, but also for the cars required by other divisions when electrified. Adjacent to the inspection building proper is a small service building, which contains boilers for heating, locker and wash rooms, air compressors and motor generators for supplying power for the tools and signals. Current for the operation of the motor generator sets is obtained from the Paoli substation, in which are located two 11,000/2200-volt transformers.

Power Supply and Transmission

Power for traction purposes is purchased from the Philadelphia Electric Company and is generated in its main power station at Christian Street on the easterly bank of the Schuylkill River about 1 mile south of the West Philadelphia passenger station. It is delivered to the railroad company at a substation at Arsenal Bridge, on the westerly bank of the Schuylkill River, opposite the main generating station, the connection be-

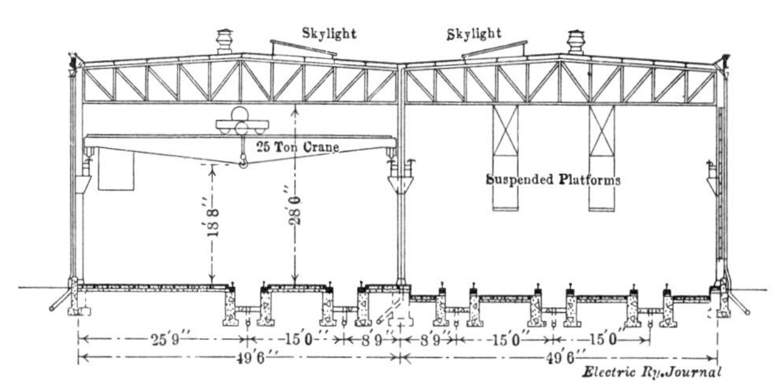

PHILADELPHIA-PAOLI ELECTRIFICATION — CROSS-SECTION OF INSPECTION BUILDING

tween the power house and the substation consisting of armored submarine cables under the river.

While the present service is on one phase only of the power company's three-phase generating system, the plan is to supply succeeding electrifications from the remaining phases. The power is delivered to the railroad company's substation at 25 cycles and 13,200 volts. Here it is stepped up to 44,000 volts, and by means of duplicate single-phase overhead circuits is transmitted to the step-down substations. Special provisions have been made by the Philadelphia Electric Company to balance this single-phase load as well as to

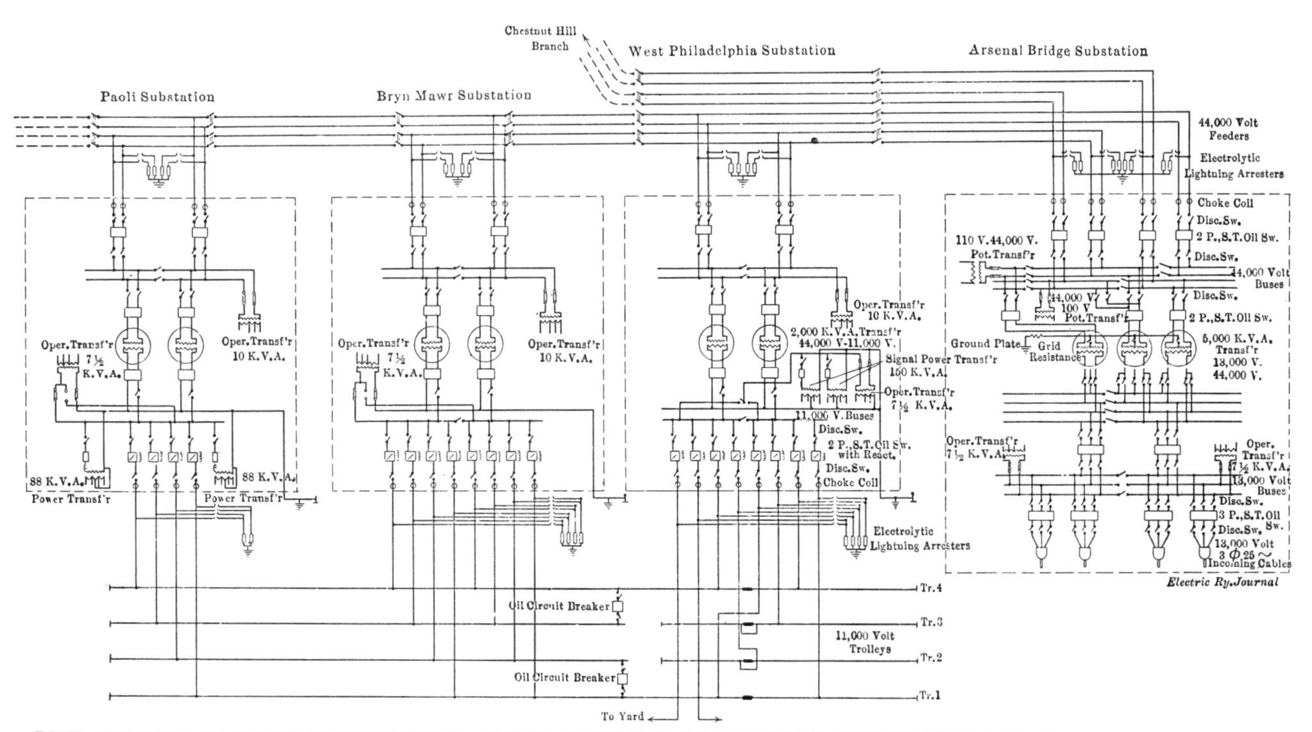

PHILADELPHIA-PAOLI ELECTRIFICATION—GENERAL WIRING DIAGRAM FOR SUBSTATIONS AND TRANSMISSION SYSTEM

correct for the relatively low power factor in order that the full three-phase capacity of the generating units may be available.

From the Arsenal Bridge substation four 44,000-volt single-phase transmission lines extend to the West Philadelphia substation. These four lines will tee into the West Philadelphia substation and two of them continue on to the Bryn Mawr and Paoli substations. The other two lines will later go to the Chestnut Hill substation. The four transmission lines are carried on brackets on the side of the elevated structure between the Arsenal Bridge substation and the West Philadelphia substation, but beyond the West Philadelphia substation they are carried on the catenary supporting structures. Along the right-of-way the lines are carried on both sides of the tracks.

The transmission lines are No. 00, seven-strand, hard-drawn copper wires. They are spaced 5 ft. apart where the two wires of a single-phase feeder are on the same cross-arm, and where there is more than one circuit on a pole the vertical spacing is 3 ft. 6 in. The lines are protected by a 3/8-in. steel ground wire on the top of the poles. Where the transmission lines pass under highway bridges the ground wire is dead-ended on the bridge structure and the wires carried on post-type insulators.

Where the lines pass under the Belmont and Girard Avenue bridge, which approaches tunnel conditions on account of its width, they are covered with rubber and varnished cambric insulation. The joints between the bare and insulated wire either side of the bridge are made outside of porcelain sleeves filled with compound. The wire is continuous throughout the sleeve, the insulation being tapered off inside the sleeve.

At the Arsenal Bridge substation the lines are protected by relays which operate on overload and on an unbalanced load on either side of the circuit caused by a ground. In the other substations the relays operate only differentially, and in case of a ground between substations, the circuit on which the trouble occurred would be cut out first in three of the substations, and finally at the Arsenal Bridge substation. Overload relays are provided in the 13,200-volt lines at the Philadelphia Electric Company's power station, and there are reverse current relays in these feeders in the Arsenal Bridge substation.

PHILADELPHIA-PAOLI ELECTRIFICATION—LIGHTNING ARRESTERS AND HIGH-TENSION WIRING ON SUBSTATION ROOF

The pin-type porcelain insulators used on the transmission lines are 8 7/16 in. high and 12 in. in diameter, made up of four parts. These insulators withstand tests for dry flashover at 165,000 volts, for wet flashover at 120,000 volts, and for puncture at 250,000 volts. After erection the complete transmission lines were tested out to ground at a potential of 66,000 volts, or far in excess of the working pressure.

SUBSTATIONS

Transformer substations are provided at suitable points along the 93 track-miles of electrified railroad for stepping up the voltage for transmission and for reducing it to that required in the contact conductors. These are substantial fireproof brick buildings adjacent

PHILADELPHIA-PAOLI ELECTRIFICATION—VIEW OF SUBSTATION SHOWING ARRANGEMENT OF 44,000-VOLT TRANSMISSION CABLES AND 11,000-VOLT FEEDERS

PHILADELPHIA-PAOLI ELECTRIFICATION—INTERIOR VIEW OF SECOND FLOOR OF TYPICAL SUBSTATION SHOWING BUSBAR AND SWITCHING EQUIPMENT

to the tracks. Electrolytic lightning-arrester equipment and high-tension-feeder sectionalizing switches are located on the roof, the busbars and switching equipment on the second floor, and the transformers on the ground floor. Space is provided in all substations for 100 per cent increase in capacity. The installed capacity of the substations are as follows:

Arsenal Bridge	Three 5000 kva. step-up transformers
West Philadelphia	Two 2000 kva. step-down transformers
Bryn Mawr	Two 2000 kva. step-down transformers
Paoli	Two 2000 kva. step-down transformers

The transformers in all substations are of the 25-cycle, single-phase, oil-insulated, water-cooled type, with the usual voltage taps on the primary and secondary coils. They were furnished, together with the switching equipment, by the Westinghouse Electric & Manufacturing Company. The cases are mounted on wheeled trucks, and in each substation a transformer truck and chain hoist are provided for handling the transformers and cores. Thermostats have been provided to operate an alarm bell in case of high temperature in the transformers. Oil filtering and drying apparatus is located in each substation.

The neutral point of the high-tension winding of the step-up transformers is grounded through a grid resistance, thus limiting the potential to ground from either side of the 44,000-volt transmission lines to 22,000 volts. All 44,000-volt circuits are connected to the buses in the various substations through oil circuit breakers with boiler-plate tanks arranged along the floor without barriers.

The circuit breakers of the 11,000-volt and 13,200-volt circuits are of the oil type, those on the 11,000-volt trolley circuits having two poles with a reactance connected across one pole. The circuit breakers are automatic and remote controlled.

Open buses mounted on insulators and carried by pipe framework are used throughout. In general, all power wiring is bare, and copper tubing or solid wire is used, sufficient clearance having been provided so that no barriers are required between buses or wires except in the case of the incoming cables from the Philadelphia Electric Company in the Arsenal Bridge substation. Control, instrument and lighting wires are rubber insulated and run in conduits. The 44,000-volt and 11,000-volt buses are sectionalized in each station, normal operation being carried on with these bus-disconnecting switches closed, and in the step-down substations one transformer is connected to each side of the bus. One leg of the 11,000-volt side of all step-down transformers is connected to a bus through a disconnecting switch, this bus being connected to the track rails.

Low-voltage power for the opening and closing of oil circuit breakers is obtained from the 25-cycle buses, but 60-cycle power is also provided in all substations to trip circuit breakers in case of the loss of the 25-cycle traction power.

Except in the West Philadelphia substation, where the power director, or system operator, is located, there are no attendants. A switchboard with the necessary instruments, controllers and indicating lamps is provided in signal towers near the Arsenal Bridge, Bryn Mawr and Paoli substations. This board is connected with the board in the substation through a control cable, and the opening and closing of circuit breakers is done by the interlocking plant operator in the tower. Telephones are provided in all substations and in the signal towers controlling them, so that the power director is in constant touch with all substations and tower men. An alarm bell connected to the thermostat on the transformers is located in each signal tower.

PHILADELPHIA-PAOLI ELECTRIFICATION—TYPICAL CATENARY CONSTRUCTION ON SHARP CURVE

Catenary System

One of the notable features of the installation is the use of what are called the "tubular cross-cantenary bridges" for carrying the contact wires. This construction was described in detail in the Electric Railway Journal for April 18, 1914. In brief, it consists of National tubular steel poles on either side of the tracks grouted into concrete foundations and anchored by double guys that are made of steel rods with turnbuckles. Spanning the tracks between the poles are two cross wires that form a cross-catenary support for the longitudinal wires.

The cross wires are of Roebling's extra-high-tension galvanized-steel strand, the upper strand usually being ¾ in. and the lower one ½ in. in diameter. Both are socketed at each end, and at one side a turnbuckle is installed to permit of adjustment. The top and bottom cross-wires are joined together by means of a vertical ¾-in. rod and suitable malleable iron clamps at the points where insulators carrying the longitudinal wires are located. Each insulator consists of three-suspension-type units made by the Locke Insulator Manufacturing Company, the porcelain being 8 in. in diameter and the flashover value of the three being many times that of the line voltage.

The cross-wire bridges are located about 300 ft. apart on tangents, but are closer on curves, the exact spacing depending upon the degree of track curvature. Insulators are suspended over the center of each track, being offset toward the outside of the curves. The main messenger wire, which is strung out and suspended from the insulators, is a ½-in., extra-high-tension, seven-strand, double-galvanized steel cable, having a sag of 5 ft. in a span of 300 ft. At intervals of approximately 1 mile this messenger is socketed and dead-ended on one of the heavy structural signal bridges which are spaced about ½ mile apart. The messenger is insulated from the signal bridge by using two or more sets of three-unit suspension type insulators, these insulators being similar to those used for suspending the messenger from the cross-wire and other bridges.

Every 15 ft. on curved track and 30 ft. on tangent track a hanger supports the lower two wires from the messenger wire. The top one of these two wires, called the auxiliary messenger, is of No. 0 round copper, and its purpose is to give suitable current capacity to the system. The contact or trolley wire is a No. 000 grooved phono-electric conductor furnished by the Bridgeport Brass Company. Both wires are carried in a vertical plane, generally about 22 ft. above top of rail, except where they drop down to pass under an overhead highway bridge having insufficient clearance to permit this height.

In the terminal division, which includes the first 5 miles from Broad Street Station, and where the dense steam locomotive traffic causes a great deal of smoke and corrosive gas, a non-corrodible tube hanger is used. The hanger tube, which is 9/16 in. outside diameter and made of No. 18-gage metal, is fastened to a casting at each end by rolling or crimping the tube into grooves turned in the shank of the casting. Some of the tubes are of Monel metal, while others are composed of a bronze mixture containing 90 per cent copper.

On the Philadelphia division, where there is relatively less steam traffic, wrought-iron strap hangers 1 in. wide by 3/16 in. thick are used. The main messenger cable at the hanger clip is protected from corrosion by a collar of zinc inside of the annealed brass or Monel metal clip, which is bolted to the hanger strap. The flat-strap hangers, which have a quarter turn in them to minimize the area exposed to the wind in the direction crosswise with the tracks and to resist bending when

PHILADELPHIA-PAOLI ELECTRIFICATION—TYPICAL ANCHOR AND SIGNAL BRIDGE LOCATED AT END OF CURVE; TRACK-BOOSTER TRANSFORMERS MOUNTED AT BOTH SIDES

placed on curves, are bolted to the castings that clamp the auxiliary messenger and trolley wires.

On tangents the castings at the bottom of the hangers hold only the auxiliary messenger, and the trolley wire is, in turn, supported from this auxiliary messenger every 15 ft. at points equidistant from the hanger. This insures a very flexible or smooth riding trolley wire.

On curves the two lower wires do not hang directly beneath the messenger, but the whole system swings into a curved plane until a balance is reached between its weight and the tension in the wires. The tensions in both the auxiliary messenger and trolley wires are selected so that in extreme hot weather there will be enough tension to prevent sagging, and yet in extreme cold weather the contraction will not cause stresses beyond the elastic limit.

The catenary system over each of the four main tracks is separated electrically from those over the other tracks, and trolley sectionalizing points with switches are provided at all cross-overs so that sections of the line may be temporarily cut out of service for repairs. On the main running tracks sectionalizing is of the "air-break" type, wherein the two ends of the trolley wire are spread apart, each end being lifted up at a different point, and an insulator is placed in each wire at a point above that where contact is last made with the pantograph. Thus, while the pantograph is making contact with one wire the other is lifted up and sectionalized.

At cross-overs and in yards the trolley wires are sectionalized by means of wood stick insulators that have runners or gliders on each side, these being so arranged that, while the pantograph always makes contact with at least one of the runners, they are separated electrically. The switches are of the disconnecting knife type, mounted on top of the wood section insulators, and are operated from the ground by means of a long impregnated wooden switch stick.

An interesting detail in the erection of this catenary work was the use of cars, the top platforms of which could be readily raised or lowered by means of chain hoists. The cars were also equipped with removable outriggers so that in the four-track section the work could be completely erected over one of a pair of tracks without in any way interfering with the regular steam traffic on this track.

The electrified route is crossed in many places by overhead highway bridges of restricted height, and in such cases, where it was impracticable to raise the highway bridge, the trolley wires had to be carried underneath, each catenary system being steadied by being held with post-type insulators supported by brackets on the bridge structure. The transmission wires on either side of the main line tracks are also carried down underneath the bridge and supported from the bridge structure by insulators, the metallic brackets carrying the insulators being carefully bonded together and earthed by means of ground plates. To prevent pedestrians on the bridges from contact or interference with the wires, there have been erected solid wooden fences, either vertical or inclined, of sufficient height to shut out all view of the wires. To protect the trainmen, general orders have been issued that no employees are allowed on top of any car in the electrified zone.

Track-Bonding, Signals and Telephones

Throughout the electric zone the rails are bonded with pin-type expanded-terminal bonds furnished by the American Steel & Wire Company, the Electric Service Supplies Company and the Ohio Brass Company. One end of the bond has a terminal solidly welded to the bond while the other end has a soldered terminal. This enables the bond to be installed by being slipped back of the splice plate without the necessity of removing the plate. Each rail joint has two No. 0 bonds, but through the interlockings only one rail of each track is bonded, although all of the traction rails are connected together.

To minimize the inductive effect of the traction currents on adjacent telephone and telegraph wires, series or track-booster transformers have been mounted on the signal bridges at approximately 1-mile intervals. Details covering the method of operation of this scheme were published in the Electric Railway Journal for May 2, 1914. The need for this equipment, however, applies only to commercial circuits in towns along the route, because the railway company has had all wires along the right-of-way installed underground for some time, this action having been taken subsequent to the heavy sleet storm of March, 1914, which did a great deal of damage to overhead wires throughout the Eastern States.

In the electrified portion of the Philadelphia division, about 15 route-miles of four-track line, the old form of semaphore signals have been replaced by position-light signals made by the Union Switch & Signal Company. With these different rows of five lights each indicate the various positions of clear, caution and stop, the mechanical operating device being replaced by relays and Kerite wire. The signals are operated by 60-cycle current track circuits in the usual manner, but to nullify the induction effect of the traction current in the signal circuits of adjacent tracks, resonant shunts have been installed which permit the local induced currents to be shunted around the track relays and thus avoid disturbing the signal circuits.

In addition to the usual telephone facilities between substations and between the electric power director and the train dispatchers, there are permanent telephone boxes located at every signal bridge, approximately ½ mile apart, throughout the electrified zone. In consequence prompt and reliable intercommunication by telephone is possible between any parts of the whole system.

Engineering and Construction

The design and construction of the electric installation was carried out by Gibbs & Hill, consulting electrical engineers for the railroad company, in co-operation with the engineering department and the officials of the railway. All construction except that of the substation buildings and inspection building, which were covered by outside contracts, was carried out by a specially organized force. The mounting of the multiple-unit car equipment on the cars was carried out by the railroad forces at the Altoona shops under the direction of the motive power department, and the signal equipment and the changes in telegraph and telephone lines were designed and installed under the direction of the signal and telegraph departments respectively.

PENNSYLVANIA RAILROAD

ELECTRIFICATION

WESTINGHOUSE ELECTRIC & MANUFACTURING CO.

EAST PITTSBURGH **PENNSYLVANIA**

1905

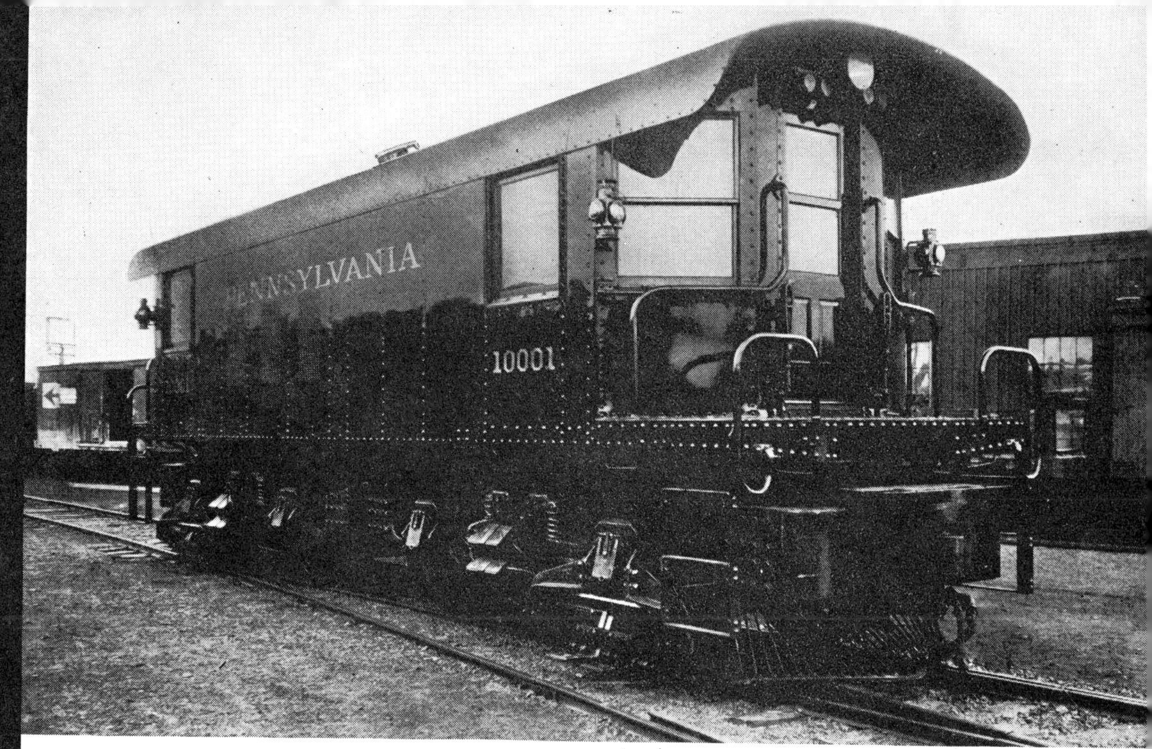

1935

F O R E W O R D

In an address at Union Station, Washington, January 28, 1935, on the occasion of the inauguration of through electric service over The Pennsylvania Railroad between Washington and New York, M. W. Clement, then vice-president of The Pennsylvania, and now president, said:

"The electrification of The Pennsylvania Railroad between Washington, Baltimore, Philadelphia, and New York, marking an evolution in railroad transportation, is one of the outstanding achievements of a long career of one of the Country's really great railroad executives, General W. W. Atterbury. ¶"It was he who conceived it, but more than the conception was his resourcefulness in bringing it to completion in these difficult times. I am glad of the opportunity to pay this tribute to him, coming from those who have been associated with him in this project."

Fig. 1—Early Long Island Commuter Train

LONG ISLAND SUBURBAN ELECTRIFICATION

The Long Island was the first steam railroad in the United States to operate extensively by electricity.

In 1905, this service was inaugurated out of the Flatbush Avenue Terminal to the suburban territory and resorts on Long Island. One hundred thirty-four motor cars, each equipped with two 200-hp. Westinghouse motors and multiple unit control, and 55 trail cars were placed in operation. By 1925 the number of motor cars had increased to more than 700.

To date, this electrification has been extended to cover a total of 141 route miles with 448 miles of electrified track, operating also from the 33rd Street Station in New York through tunnels to Long Island. To take care of this service, 1044 motor cars and 262 trailer cars are in the Railroad's present service with more than 1000 trains in daily operation—carrying the heaviest suburban traffic in the U. S.

At the time of the inception of this electrification with heavy concentration of passenger traffic, and the state of electrical development, then existing, the 650-volt, third rail, D-C. system was adopted to handle this service.

Power for this electrification was furnished by The Pennsylvania Railroad's Power Station constructed specially for this purpose at Long Island City. This power house was laid out with room provided for six 5500-kw. turbo-generator units for traction service. The initial installation consisted of three Westinghouse-Parson units delivering three-phase, 25-cycle, 11,000-volt power to substations located along the electrified lines where it was converted to 650-volt D-C. for the third rail.

Five permanent substations were initially constructed, all equipped with Westinghouse converters and transformers as well as two portable substations mounted on flat cars which could be moved at will to points of excessive load concentration.

Fig. 2—Long Island City Power House

The completion of the Pennsylvania Station in New York City in 1910 witnessed the extension of the Long Island service through the East River tunnels to the heart of Manhattan. Some idea of the growth of Manhattan traffic can be visualized by the fact that, in 1911, 6,334,429 Long Island passengers were handled in and out of the Pennsylvania Station in New York. By 1930 this figure had increased to 54,203,239.

Obviously, this vast increase in traffic necessitated the construction of additional power substation facilities with the result that at that time the Long Island System represented the heaviest power load for electrification in the U. S.

Thirty years ago, the Long Island Railroad looked to the Westinghouse Company to solve its problems. Since that date its electrification developments have been executed largely in cooperation with Westinghouse engineers.

Fig. 3—Long Island Multiple-unit Train

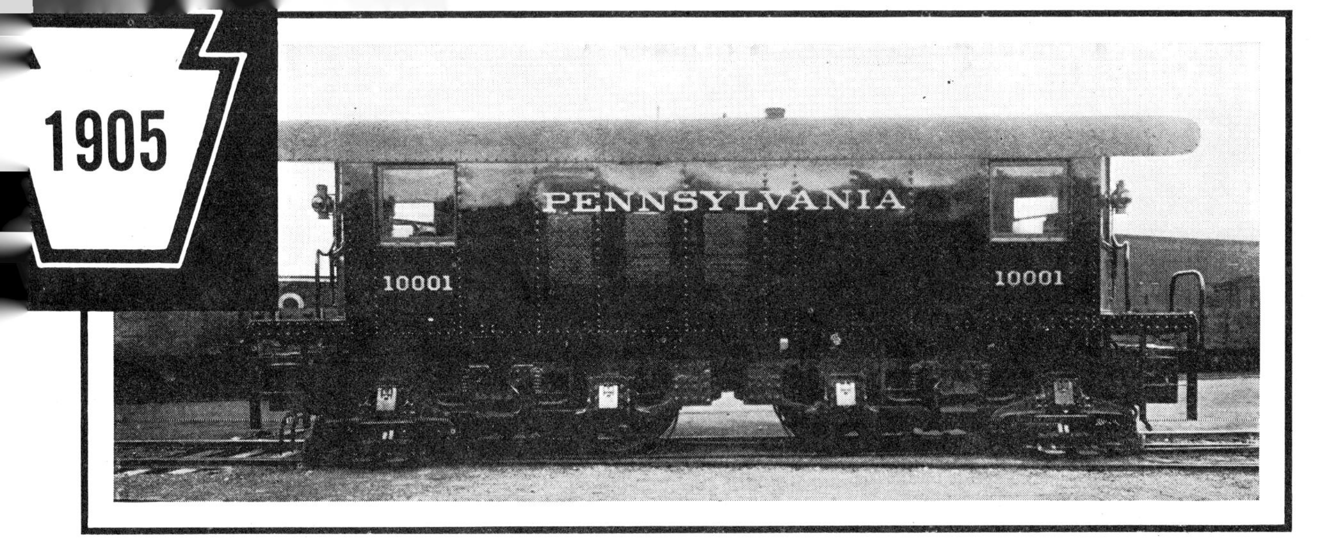

Fig. 4—First Direct-current Locomotive

FIRST LOCOMOTIVE BUILT . . . Forerunner of New York Terminal Electrification

For years, prior to the beginning of the present century, officials of The Pennsylvania Railroad had dreamed of a passenger terminal in the heart of New York City. To provide this, required a means of going underneath the Hudson River, which was not possible with the steam locomotive, the only form of power then in use.

But the rapid strides in the development of electrical apparatus during the last few years of the nineteenth and in the early years of the twentieth century soon made the use of tunnels for electrified train service into New York entirely feasible.

Anticipating electrification of the North River tunnels, The Pennsylvania Railroad, as early as 1905, in cooperation with the Westinghouse Electric and Manufacturing Company, built and tested two experimental locomotives, each having two 4-wheel, center-articulated trucks with motors on each axle. These locomotives were given Road numbers 10001 and 10002.

Locomotive 10001, still in operation, had its motors geared to the axles in a manner more or less conventional. On the other hand, locomotive 10002, still available for service, had four motors concentric with the axles which were driven by quills and flexible drive. The design for two of these axles was the forerunner of the present cup drives used on all modern Pennsylvania electric locomotives.

These locomotives were operated from a third rail at 650 volts, direct-current.

A third locomotive, Road number 10003, designed for single-phase, 11,000-volt operation, was also constructed, having a four-wheel guiding truck and two driving axles, receiving power from gearless motors through quills. This locomotive underwent exhaustive tests on a special track on Long Island. Two of these locomotives placed back to back would have had the same wheel arrangement as the locomotives ultimately adopted for the New York Terminal service.

The final result of this early development work was the design and building of the Class DD-1 locomotives, shown in Fig. 7, for use in the service into New York Terminal.

Fig. 5—First Single-phase Locomotive

Fig. 6—Original Cup Drive

Fig. 7—Class DD-1 Locomotive awaiting Train at Manhattan Transfer

NEW YORK TERMINAL ELECTRIFICATION

In 1910, following the completion of two single-track tunnels under the Hudson River and four single-track tunnels under the East River, as well as the Station in New York City, electrified operation was inaugurated by The Pennsylvania Railroad between Manhattan Transfer and the Coach Yards at Sunnyside, Long Island.

Power for this electrification was obtained from the Long Island City Power House.

After extensive tests with two Class DD locomotives, twenty-four Westinghouse-equipped Class DD-1 locomotives were used for initial operation of passenger trains on this division. Increased traffic shortly necessitated the addition of 7 more locomotives of this type to take care of the service with the result that a total of 33 Westinghouse-equipped locomotives were soon in operation.

Serious consideration was given to the adoption of the single-phase, alternating-current system for this service. All the tunnels were laid out so that the overhead trolley could be used. However, in view of the fact that the Long Island Railroad, utilizing the same tunnels under the East River, had been laid out for the 650-volt, D-C., third rail system, and furthermore, in view of the fact that extensions to the electrification of the Railroad west of Manhattan Transfer were not contemplated for some years, locomotives used in this shuttle service were equipped for the same power supply as that utilized on the Long Island.

Fig. 8—Passenger Train hauled by DD-1 Locomotive leaving New York Terminal

Further traffic expansions of course followed, which were taken care of by locomotives of a different type, which will be mentioned later. Suffice it to say that, as a result of the foresight of The Pennsylvania Railroad and the close cooperation with Westinghouse, the Class DD-1 locomotives were among the most successful electric passenger locomotives placed in service up to that time. More than half of these locomotives are still in daily service, largely on the Long Island Railroad, and the remainder are all available for service.

Fig. 9—Manhattan Transfer where interchange was made between steam and electric power

Fig. 10—Multiple-unit train approaching Broad Street Station, Philadelphia

PHILADELPHIA SUBURBAN ELECTRIFICATION

By 1915, traffic in and out of the Broad Street Station in Philadelphia had reached a point where it was practically impossible to handle the necessary train movements through the approaches to the station and on the station tracks.

The solution consisted in the electrification of the suburban service on the main line of the Railroad, west as far as Paoli. This reduced the movements per train round trip through the "bottle neck" from eight train movements to two, and greatly relieved the congestion.

However, before inaugurating the suburban electrification, the Railroad engineers, in co-

Fig. 11—Class MP-54 Suburban Multiple-unit Coach

operation with Westinghouse engineers, made a very careful and complete analysis of the power supply problem, having in mind the ultimate extension of electrification. The result was the adoption by the Railroad of the 11,000-volt, single-phase trolley system.

Power for the Paoli electrification was purchased by the Railroad from the Philadelphia Electric Co. and distributed by the Railroad to 3 Westinghouse-equipped substations at West Philadelphia, Bryn Mawr, and Paoli.

Initial operation called for the use of 93 motor cars, each equipped with 2 Westinghouse 200 continuous hp., single-phase commutator type motors and electro-pneumatic control. As traffic increased into the Philadelphia Terminal, additional congestion necessitated the extension of this electrification to other branches—Chestnut Hill line in 1918, White Marsh in 1924, West Chester and Wilmington in 1928, Trenton and Norristown in 1930—thus completing all-electric suburban operation in this area.

Characteristic of the progress in the design of electrical equipment by Westinghouse is the fact that in the later car equipments a motor of 370 continuous hp., but weighing somewhat less than the original 200-hp. motor, was designed to fit into the same space in the truck as the original motor. This motor was further designed in such a way that its speed characteristics enabled cars equipped with two of these motors to haul a trailer and operate in the same train with the cars equipped with the original 200-hp. motors.

Today 431 electrically-equipped multiple-unit cars are in service on The Pennsylvania Railroad 11,000-volt system.

Fig. 12—{Upper} Early totally-enclosed type substation {Paoli}
Fig. 13—Modern outdoor substation

Fig. 14—Multiple-unit train on Bridge over Schuylkill River

Fig. 15—Two-car suburban train passing Allen Lane Outdoor Substation onto Chestnut Hill Branch

Fig. 16—Class FF-1 Freight Locomotive, single-phase, three-phase; 10-20 mph., 4000 hp.

EXPERIMENTAL LOCOMOTIVES FOR MAIN LINE SERVICE

The highly successful operation of the A-C. electrification in the Philadelphia suburban area confirmed the advantages of this system for main line electrification, and showed beyond doubt that a system was available which could meet all of the requirements of Pennsylvania traffic for future expansion.

With the idea of electrifying its main line definitely in mind, The Pennsylvania Railroad undertook, over a period of years, a program of development work during which a number of experimental locomotives were constructed. These experimental locomotives followed an orderly development consistent with the progress taking place in electrical design which resulted in the locomotives now used for the New York to Washington service. All locomotives from this time on were designed for A-C. service, or for conversion to A-C. by minor changes.

FF-1 FREIGHT LOCOMOTIVE—1917

With the prospect of extending main line electrification, a 4000-hp., A-C., experimental freight locomotive especially adapted to heavy grades was built from a joint design of the Railroad and Westinghouse.

The wheel arrangement was 1C+C1 and the drive was by geared jack shaft and side rods. Each jack shaft carried two flexible gears which meshed with pinions on the shafts of two motors.

The motors were of the three-phase induction type and ran at one or the other of two constant speeds regardless of grade conditions drawing power from the system on the upgrade and returning power automatically on the downgrade. Single-phase, 25-cycle power was collected by the pantograph from an overhead wire.

Fig. 17—Class L-5 Universal Freight-Passenger Locomotive. 204 tons, 3070 hp. continuous

Fig. 18—Class O-1 Single-phase Passenger Locomotive. 150-tons, 2500 hp. continuous

This design at that time represented the most powerful locomotive ever built and after completion of tests it was placed in freight service between Philadelphia and Paoli.

L-5, L-5A LOCOMOTIVES—1924

Looking further toward the ultimate main line electrification, The Pennsylvania Railroad felt the need of a more flexible variable speed motive power unit and conceived the idea of building a universal locomotive, adapted to the entire railroad, which would be good for either freight or passenger service with merely a change in gear ratio.

The Railroad by this time had definitely confirmed their confidence in the 11,000-volt single-phase system. Westinghouse cooperated with the Railroad in the design of this universal locomotive and in 1924 there was placed in experimental work, one locomotive (L-5) for A-C. operation and two locomotives (L-5A) for D-C. operation, all equipped with single-phase traction motors and other rotating apparatus. The two L-5A locomotives arranged for 650-volt D-C. operation were capable of conversion for A-C. service by comparatively simple and minor changes. Since motors of sufficient hp. which would go between a pair of driving wheels were not then available, the motors were placed ahead of the driving wheels, and jack shafts and side rods were employed to transmit the motor horsepower to the driving wheels.

In 1926, ten additional L-5A Westinghouse-equipped locomotives were placed on the New York Terminal Division.

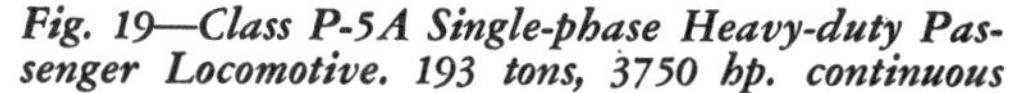

Fig. 19—Class P-5A Single-phase Heavy-duty Passenger Locomotive. 193 tons, 3750 hp. continuous

Fig. 20—Class L-6 Single-phase Freight Locomotive. 152.5 tons, 2500 hp. continuous

PIONEER LOCOMOTIVES—NEW YORK TO WASHINGTON

The next step in locomotive development was a single-phase twin-motor capable of delivering 1000 hp. per axle or 500 hp. per armature, which was sufficient capacity to develop 25% adhesion at start and attain speeds up to 100 mph., and yet small enough to go between the wheels, a feat never before performed in the design of a single-phase commutator type motor. This was a radical improvement and made possible the elimination of side rods.

Early in 1927, Westinghouse pioneered and built such a motor, and at the end of the same year it was exhibited to officials of The Pennsylvania Railroad as conclusive proof that a motor of this capacity was attainable. This one accomplishment alone gave tremendous impetus to the Railroad's plan for electrification from New York to Washington and demonstrated the feasibility of building motors of greater horsepower per axle.

Almost immediately three new type locomotives were ordered built. These were:

The Class O-1 for light-duty passenger service, a two-driving axle locomotive, having a four-wheel guiding truck on either end.

The Class P-5 for heavy-duty passenger service, a three-driving axle locomotive, having a four-wheel guiding truck on either end.

The Class L-6 for freight service, a four-driving axle locomotive, having a two-wheel guiding truck on either end.

The motor and control parts were made interchangeable for the several classes of locomotives, the only difference being in motor frames and methods of mounting.

Twin motors, with such further improved design, as to rate at 1250 hp., transmitted power to each axle of the passenger locomotives through a quill-type cup drive. The axles of the freight locomotive were each driven by one 625-hp. motor through flexible gearing.

All three of the locomotives were so designed that the maximum number of mechanical parts were interchangeable and the general design similar.

Several of these pioneer locomotives were built in the period between 1930 and 1933, Westinghouse supplying equipment for three locomotives of the O-1 Class, and one equipment each for the P-5 and L-6 Classes. At this time a large number of road locomotives, referred to later, were ordered and built.

In 1934, studies were made of larger size motive power units. In order to obtain more power and somewhat lighter axle loadings than incorporated in the P-5A locomotive, the Class R-1 locomotive was designed and built, using Westinghouse equipment. Four driving axles were used in a rigid wheel base, each having an axle loading of 60,000 pounds carrying 1250-hp. single-end, quill-drive motors, resulting in a total locomotive rating of 5000 hp. at 100 mph.

The outstanding feature of this motor design is that, although it has the same hp. rating as the motor used in the P-5A locomotive built 3 years earlier, it weighs only about one half.

Fig. 21—Class R-1, Single-phase, High-powered Passenger Locomotive. 202 tons, 5000 hp. continuous

Fig. 25—Train between New York and Philadelphia hauled by P-5A Locomotive

ELECTRIC SERVICE
New York to Philadelphia

As a result of the experimentation with the trial locomotives and the survey of main line electrification which promised attractive economies, The Pennsylvania Railroad announced in the fall of 1928 plans for the most extensive electrification project in transportation history. A further announcement projected complete electric train operation between New York City and Washington.

Inauguration of electric service between New York and Philadelphia early in 1933 marked the completion of the first step of this project. A further extension of this main line electrified service to Wilmington, Delaware, and Paoli, Pa., was made a short time later and permitted the elimination of D-C. operation on the New York Terminal Division between Manhattan Transfer through the New York Terminal to Sunnyside Yards for through trains. This was equivalent, in effect, to moving Manhattan Transfer to Wilmington and Paoli where the change from electric engines to steam was then made.

LOCOMOTIVES

The Pioneer P-5 type of heavy-duty passenger locomotive mentioned previously was chosen for this service.

A few modifications in design were made and sixty-two of the modified units, classed as P-5A locomotives, equipped with electrical apparatus of design fully coordinated between Westinghouse and General Electric were built and placed in service. Thirty-seven of these are Westinghouse-equipped.

The box-type cab is used with a 2-C-2 wheel arrangement. Each driving axle is driven by twin motors through single-end quills. Each motor is rated at 1250-hp. at the driver rim, thus providing 3750-hp. (at 63 mph.) for the three driving axles. This engine is capable of developing a peak of 6500 hp. for short periods of time. Also, it is suitable for continuous sustained operation at 90 mph. at full capacity.

Fig. 22—Class B-1, Single-phase Switcher Locomotive—Philadelphia Coach Yard

NEW TYPE LOCOMOTIVES FOR SWITCHING SERVICE

To provide an electric switching locomotive, the Class B type was built in 1924, having an O-C-O wheel arrangement.

A thorough study of conditions by Pennsylvania Railroad engineers indicated that the resistance lead type of single-phase series motor was the most suitable for this type of service due to characteristics inherent in such a design. Westinghouse had a large number of this type motor in service on the New Haven, Canadian National (Port Huron, Mich.), and other railroads. The engines at Port Huron comprised three driving axles and since this size of locomotive was suitable for the Pennsylvania switching service, this type of locomotive was selected.

Using the same type motor with either A-C. or D-C. control, 42 engine units, the majority of which were Westinghouse-equipped, were built for A-C. operation in the Philadelphia and Long Island Yards, and for D-C. operation in the Sunnyside Yards.

Fig. 23—Class BB-2, D-C. Switcher Locomotive for New York Terminal Service

Fig. 24—Class BB-3, Single-phase, Long Island Switcher Locomotive

Fig. 26—GG-1 Streamlined Single-phase Locomotive. 230 tons, 4620 hp. continuous

ELECTRIC SERVICE New York to Washington

February 10, 1935, was a memorable day for all those connected with this great electrification project. At that time, through scheduled electrified passenger service was inaugurated between New York and Washington, D. C.

Headed by the world's most powerful streamlined locomotives, honors for the first run were given to the two "Congressionals", outstanding trains on this run, one leaving each of the terminal cities about the same time.

What a climax this was to a program visioned years ago—the completion of a truly great project, consummated by fine cooperation between the Federal Government, the Railroad, and Industry. An Electric Highway, serving America's densest traffic, affording high-speed mass

Fig. 27—P-5A Streamlined Single-phase Locomotive. 197 tons, 3750 hp. continuous

transportation of passengers and freight between these two famous cities, on schedules unsurpassed by any other type of motive power in comparable service.

LOCOMOTIVES

A steeple-type cab had now been adopted as standard. This places the operator in the center of the locomotive and further affords streamlining. The remaining twenty-eight P-5A locomotives on order, 17 Westinghouse-equipped, were redesigned and completed to include this feature.

These later P-5A streamlined locomotives are equipped with the same motors and electrical equipment as previously supplied on the sixty-two P-5A engines used in the New York to Philadelphia service, and hence rate 3750 hp. at 63 mph. or at 6500 hp. for short periods of time.

Coincident with the building of the Class R-1 locomotive previously mentioned, the Railroad, Baldwin, General Electric, and Westinghouse also coordinated the design of another type of locomotive—Class GG-1. Fifty-eight of these locomotives, 34 Westinghouse-equipped, consisting of a 2C + C2 articulated wheel arrangement with approximately 50,000 pounds loading per driving axle, were placed in service.

The driving axles of the GG-1 locomotive are powered by twin motors through double-end quill drive. Each motor is rated at 385-hp. per armature, thus providing 770-hp. per axle, or 4620-hp. (at 90 mph.) for six driving axles. This locomotive, designed for speeds in excess of 100 mph., can develop 8000 hp. for short periods of time.

POWER SUPPLY SYSTEM FOR NEW YORK TO WASHINGTON ELECTRIFICATION

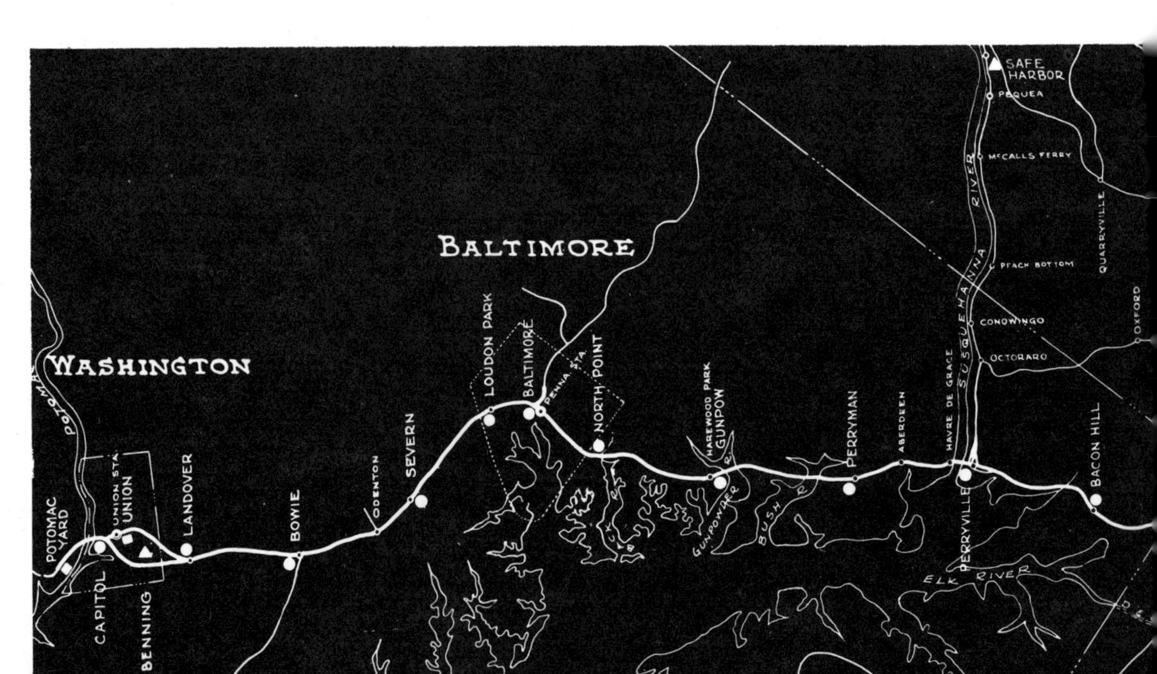

Fig. 28—Map showing New York-Washington Electrification with Power Supply and Substation Locations

TWENTY-FIVE cycle electrical energy for the motive-power equipment is supplied through the 11,000-volt, single-phase trolley system from 48 step-down transformers and switching stations. The transformer stations are, in general, supplied by 132,000-volt transmission lines usually located on the railroad right of way. The comprehensive extent of the power supply system may readily be visualized by reference to Fig. 28 which shows the electrified tracks together with the location of the power sources and step-up stations, the step-down substations, and principal switching stations.

Power for the electrification is obtained from seven stations; namely, Long Island City, L.I.; Metuchen, N.J.; Richmond and Arsenal Bridge in Philadelphia and Lamokin, in Chester, Pa.; Safe Harbor with transmission line connections at Perryville, Md.; and Benning near Washington. The ratings of the 25-cycle generators and of the step-up transformers are given in Table I. The Long Island City generating station has three 20,000 kv-a. turbo-generators arranged to supply three-phase power to the direct-current electrified lines of the Long Island Railroad or single-phase power to the New York end of the A-C. electrification. Arsenal Bridge station, which originally provided power for the Philadelphia-Paoli electrification, is now used as an auxiliary source. Lamokin station was installed for the Wilmington suburban electrification, Richmond station for the extension to New York, and the remaining stations for the through-electrified service to Washington.

One of the more recently added sources is from the Safe Harbor hydro-electric station illustrated in Fig. 29, which in the foreground shows the frequency-changer installation. This Westinghouse frequency changer is

TABLE I—POWER SUPPLY SYSTEMS

Station	Source	Single-Phase, 25-Cycle Rating	Step-Up Transformers	Connection
Long Island City	Turbo-Generator	3—13,000 Kv-a.	3— 7500 Kv-a.	P.R.R. Generating Station
Metuchen	Frequency Changer	1—35,700 Kv-a.	2—15,000 Kv-a.	Public Service Electric & Gas Co.
Richmond	Frequency Changer	3—42,900 Kv-a.	4—20,000 Kv-a. (East) 2—20,000 Kv-a. (West)	Philadelphia Electric Co.
Arsenal Bridge	Auxiliary	21,000 Kv-a.	1—20,000 Kv-a.	Philadelphia Electric Co.
Lamokin	Frequency Changer	3—21,500 Kv-a.	4—15,000 Kv-a. (North) 2—20,000 Kv-a. (South)	Philadelphia Electric Co.
Safe Harbor	Frequency Changer WW Generator	1—31,250 Kv-a. 1—37,500 Kv-a.	4—20,000 Kv-a.	Pennsylvania Water & Power Co.
Benning	Frequency Changer	1—31,250 Kv-a.	2—15,000 Kv-a.	Potomac Electric Co.

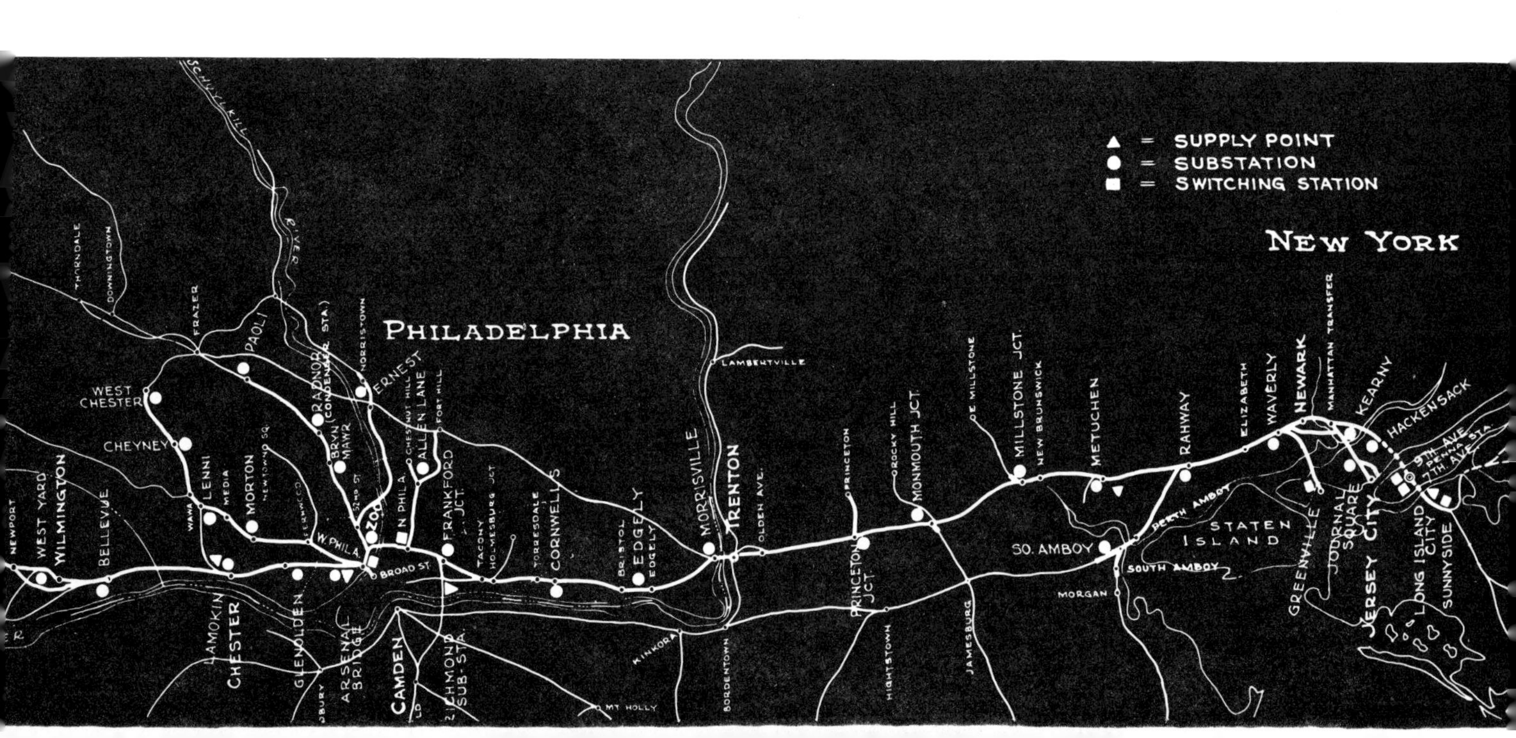

of the steel-enclosed type for outdoor service. Because of the large size and single-phase rating of the generator, see Fig. 30, the rotor is provided with special damper windings to minimize rotor heating and the stator is provided with enclosed spring mounting, which absorbs the pulsating torque due to single-phase loading. The 60-cycle motor is provided with stator frame-shifting mechanism to facilitate synchronizing and load division.

Fig. 29—Safe Harbor Generating Station with Steel-enclosed Outdoor-type Frequency Changer Set in Foreground

Fig. 30—Interior View showing 31,250 kv-a. Single-phase, 25-cycle Generator of the Safe Harbor Frequency Changer Set

TRANSMISSION SYSTEM AND CONNECTION OF POWER SOURCES

The general scheme of connection of power sources, step-up transformers, transmission lines and step-down stations is indicated in Fig. 31. The individual generator is connected to the bus through low voltage circuit breakers, such as shown in Fig. 32 for the Richmond station.

At each supply station, the step-up transformers are directly connected to the transmission line and the combination is energized or de-energized through low-voltage breakers. The principal transmission

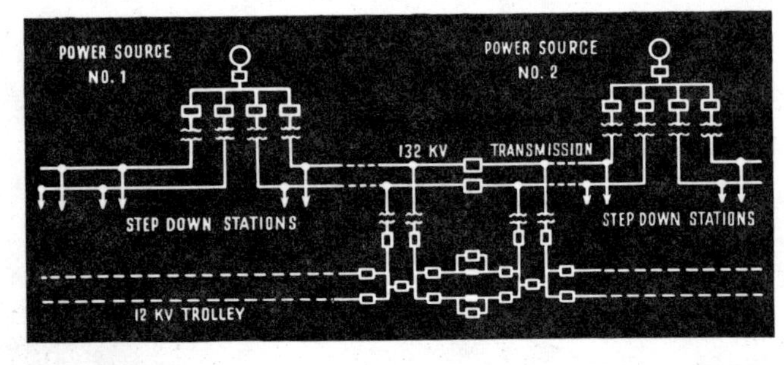

Fig. 31—Schematic diagram of connection of power sources showing the "looped-in" arrangement of transmission circuits and sectionalizing breakers

Fig. 32—One of the 5,000 amp., 13,200-volt, 2,500,000 kv-a. Oil Circuit Breakers between the Generators and Transformers at Richmond

lines are located on the railroad right of way and are supported on the same structures that also support the catenary system. Four 132,000-volt transmission lines are provided between Hackensack and Richmond, two lines between Richmond and Washington (except 4 lines between Zoo and Arsenal Bridge) and two lines each for the West Chester and Norristown suburban branches at Philadelphia. 44,000-volt transmission is retained for the Paoli and Chestnut Hill electrifications at Philadelphia. It will be noted from an examination of Fig. 31 that the transmission arrangements employed insure continuity of power supply to trolley sections without the use of 132,000-volt breakers for step-up and step-down transformers. Another feature of the scheme of connections is the "looped-in" arrangement of transmission circuits which is employed at Richmond, Lamokin, and Safe Harbor stations. Normally all the power sources are operated in parallel. In the event of trouble the system may be separated into sections at many places, but special provision has been made for isolating the system at Hackensack, Zoo, and Perryville stations. The circuit breakers for the 132,000-volt system are of Westinghouse manufacture, the one shown in Fig. 33 being for the Zoo station.

STEP-UP STATIONS

A general view of the Lamokin step-up station is shown in Fig. 34. The principal equipment in the step-up station is, of course, the step-up transformer for which two ratings have been standardized, namely, 15,000 kv-a. and 20,000 kv-a., 12,000 to 132,000 volts. Fig. 35 shows the 15,000 and 20,000 kv-a. transformers as built by West-

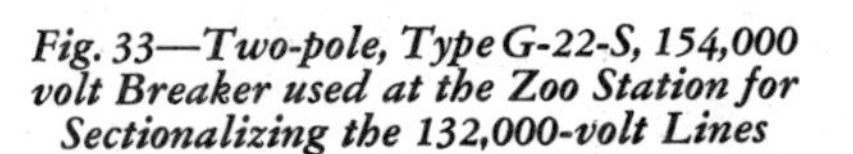

Fig. 33—Two-pole, Type G-22-S, 154,000 volt Breaker used at the Zoo Station for Sectionalizing the 132,000-volt Lines

Fig. 34—General View of the Lamokin Step-up Station

inghouse. These transformers were among the first to incorporate the new Westinghouse design of high voltage insulation for protection against lightning. They are arranged with mid-point tap brought out for grounding the 132,000-volt transmission line through resistance.

STEP-DOWN SUBSTATIONS

Power to the trolley-rail system is supplied through 48 step-down substations and principal switching stations. Of these, 37 are supplied from the 132,000-volt transmission system. The location and transformer capacities of the step-down stations are listed in Table II.

Fig. 35—15,000 kv-a., 12,000–32,000-volt Step-up Transformer

The step-down substations (except the initial ones) are of the outdoor type and are located adjacent to the railroad right of way. The typical substation, illustrated in Fig. 36, is laid out for an ultimate capacity of 4—4500 kv-a. transformers and the main power circuit connections are indicated in Fig. 37. The step-down transformers have been standardized with a rating of 4500 kv-a., 132,000–

TABLE II—STEP-DOWN SUBSTATIONS

Substation	Transformers	Transmission Voltage
Hackensack	4—4500 Kv-a.	132 Kv.
Kearny	3—4500 Kv-a.	132 Kv.
Journal Square	3—4500 Kv-a.	132 Kv.
Waverly	5—4500 Kv-a.	132 Kv.
Rahway	4—4500 Kv-a.	132 Kv.
South Amboy	2—4500 Kv-a.	132 Kv.
Metuchen	3—4500 Kv-a.	132 Kv.
Millstone Junction	3—4500 Kv-a.	132 Kv.
Monmouth Junction	3—4500 Kv-a.	132 Kv.
Princeton Junction	4—4500 Kv-a.	132 Kv.
Morrisville	4—4500 Kv-a.	132 Kv.
Edgely	3—4500 Kv-a.	132 Kv.
Cornwells	3—4500 Kv-a.	132 Kv.
Frankford Junction	6—4500 Kv-a.	132 Kv.
Allen Lane	2—3000 Kv-a.	44 Kv.
Zoo	8—4500 Kv-a.	132 Kv.
Ernest	2—4500 Kv-a.	132 Kv.
Bryn Mawr	4—2000 Kv-a. 2—3000 Kv-a.	44 Kv.
Paoli	2—2000 Kv-a. 2—3000 Kv-a.	44 Kv.
Arsenal Bridge	5—4500 Kv-a.	132 Kv.
Morton	2—4500 Kv-a.	132 Kv.
Lenni	2—4500 Kv-a.	132 Kv.
Cheyney	1—4500 Kv-a.	132 Kv.
West Chester	2—4500 Kv-a.	132 Kv.
Glenolden	4—4500 Kv-a.	132 Kv.
Lamokin	4—4500 Kv-a.	132 Kv.
Bellevue	3—4500 Kv-a.	132 Kv.
West Yard	4—4500 Kv-a.	132 Kv.
Davis	4—4500 Kv-a.	132 Kv.
Bacon Hill	4—4500 Kv-a.	132 Kv.
Perryville	5—4500 Kv-a.	132 Kv.
Perryman	4—4500 Kv-a.	132 Kv.
Gunpow	4—4500 Kv-a.	132 Kv.
North Point	4—4500 Kv-a.	132 Kv.
Baltimore	2—4500 Kv-a.	132 Kv.
Loudon Park	4—4500 Kv-a.	132 Kv.
Severn	4—4500 Kv-a.	132 Kv.
Bowie	4—4500 Kv-a.	132 Kv.
Landover	4—4500 Kv-a.	132 Kv.
Capitol	4—4500 Kv-a.	132 Kv.

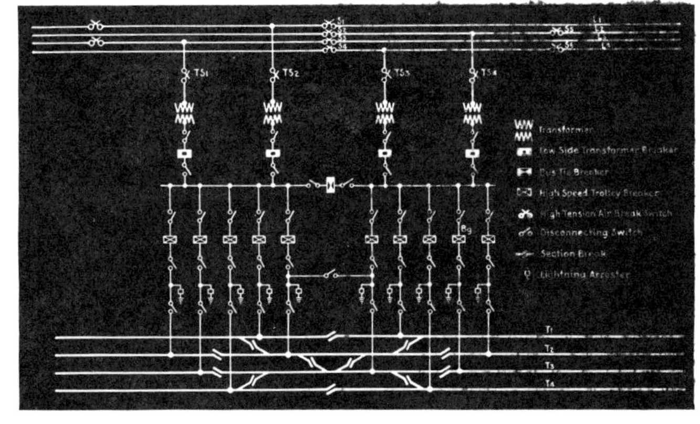

Fig. 37—(Above) Diagram of Power Circuit for one of the Step-down Substations

Fig. 36—(Left) General View of Davis Step-down Substation

Fig. 38—(Right) Typical Installation of 4500-kv-a., 132,000–12,000-volt Step-down Transformer and Associated Low Voltage Circuit Breaker

12,000 volts. These transformers have design features similar to those of the step-up transformers previously described. Additional features include the unusually low reactance, low losses and high overload ratings. Special bushing potential devices are provided which are used in the operation of relays to disconnect the transformer on the low voltage side in the event of a ground fault on the transmission line supplying the transformer. A total of 61—4500 kv-a. step-down transformers have been supplied by the Westinghouse Company for the New York to Washington electrification. See Fig. 38.

The step-down transformers are normally arranged to supply two bus sections for feeding the trolleys in either direction from the substation, being connected through a bus tie circuit breaker, such as shown in Fig. 39. The individual trolleys are fed through high-speed circuit breakers rated at 1500 amperes with an interrupting capacity of 50,000 amps. Westinghouse has supplied 64 high-speed oil circuit breakers of the type shown in Fig. 40, and 46 high-speed "De-ion" circuit breakers of the type shown in Fig. 41. In certain terminal areas individual low capacity circuit breakers have been used as feeder breakers with a high-speed circuit breaker as the master breaker. Fig. 42 illustrates the "De-ion" type of breaker supplied for such service in the Washington area.

Fig. 39 — 34,500 - volt, 1500 Amp., "De-ion" Grid Oil Circuit Breaker

The switching control equipment for step-down substations have a number of unusual features. In general, the substation equipment may be controlled either in the substation, Fig. 43, or remotely from the signal tower or by supervisory control. See Figs. 44 and 45.

Fig. 40—1500 Amp., High-speed Oil Circuit Breaker with Interrupting Capacity of 50,000 Amps.

Fig. 41—1500 Amp., High-speed "De-ion" Circuit Breaker, 50,000 Amp. Interrupting Capacity

The Pennsylvania Railroad operates a very extensive system of supervisory control, the equipment in the New York and Philadelphia areas being supplied by Westinghouse. See Figs. 45 and 46. The dispatcher's equipment for the Philadelphia area installed in the Power Director's Office is shown in Fig. 45 and the receiving equipment for one of the controlled substations, Glenolden, is shown in Fig. 46. The substation equipment also includes frequency-changer sets and associated switching equipment for supplying power from the 11,000-volt system to the 100-cycle signal supply lines. The principal part of this equipment for the New York to Washington electrification has been supplied by Westinghouse. The switching equipment is arranged for control either in the substation or from the signal tower. A number of three-panel control boards similar to the ones shown at the left of Fig. 43 have been supplied.

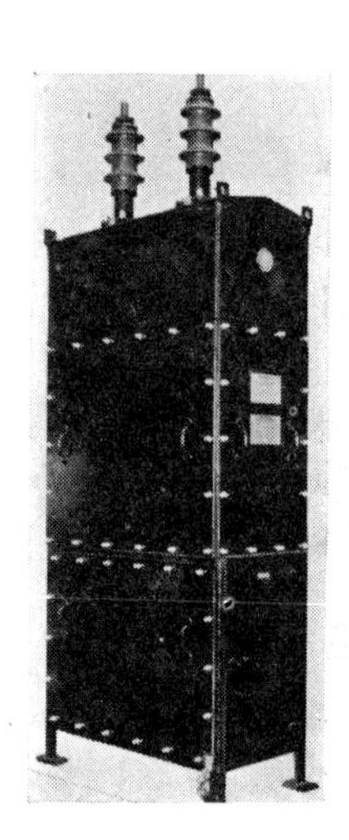

Fig. 42—Low Capacity "De-ion" Type Trolley Feeder Breaker used in Terminal Areas

SPECIAL ENGINEERING PROBLEMS

The power supply apparatus for the New York to Washington electrification has required a considerable amount of pioneering work. Because of differences in rating or in application requirements, there has been in the design of many pieces of apparatus no counterpart in either the central station or railway

ig. 43—Substation ontrol Board at rsenal Bridge Station. The Three Panls at the Left are for ignal Power Supply

fields. The rating and quantity of apparatus has been specified by The Pennsylvania Railroad, but many features of design have been worked out cooperatively with the electrical manufacturers.

One of the special apparatus problems was the development of the A-C. trolley circuit breaker whose requirements included the capacity to interrupt the very heavy short circuit currents in a sufficiently short period of time as to minimize the damage to the contact system and the disturbances to the operation of adjacent motive power equipment and inductive effects in adjacent communication circuits. Westinghouse pioneered in this development and in cooperation with the Railroad tested the first high-speed trolley circuit breaker at the Radnor station in 1925. Not only was it necessary to build circuit breakers involving radical departures from prior art, but it was also necessary to introduce a similar development in the relay field. The combination of high-speed circuit breakers and high-speed trolley relays is capable of interrupting heavy short circuit currents in .04 second which includes both breaker and relay time. The trolley relays are not only capable of high-speed action, but are highly selective and can distinguish between a fault and a load of higher current value.

The relaying of the 132,000-volt transmission system constituted another engineering problem of considerable complexity because of the number of power sources and of the step-down substations and particularly because of the fact that the layout adopted restricted the use of the 132,000-volt circuit breakers. Practically all of the transmission line relaying equipment has been supplied by Westinghouse.

Fig. 44—Typical Signal Tower Station Control Board

Fig. 45—The Dispatcher's End of Supervisory Control Equipment located in the Power Director's Office at the Thirtieth Street Station, Philadelphia

Protection of the transmission circuits and electrical apparatus against interruption due to lightning has constituted another important type of problem. In this field Westinghouse engineers have made important contributions because of novel concepts as to the nature of lightning and of the practicability of various measures for providing protection against direct strokes. All the insulation elements of the system in lines and apparatus were studied by the Railroad in cooperation with principal electrical manufacturers with a view of coordinating the insulation to best advantage.

Fig. 46—Receiving Station Supervisory Control Equipment installed at Glenolden

In addition to the development of the many kinds of apparatus and coordination of their characteristics, considerable work was done in connection with special engineering problems incidental to electric operation. These included parallel operation of power sources, inductive effects in communication and supervisory control circuits and problems incident to the operation of A-C. and D-C. services over the same tracks in the New York area. Also methods were developed for predicting the electrical performance of a complicated A-C. railway system. Originally it was necessary to make all such calculations analytically, but it is now possible to carry on much of this work by means of calculating boards, developed by Westinghouse. See Fig. 47.

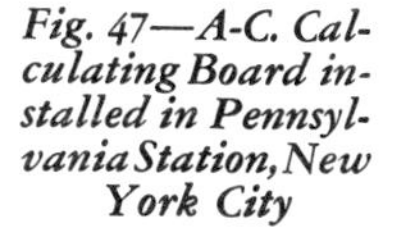

Fig. 47—A-C. Calculating Board installed in Pennsylvania Station, New York City

TERMINAL FACILITIES KEE

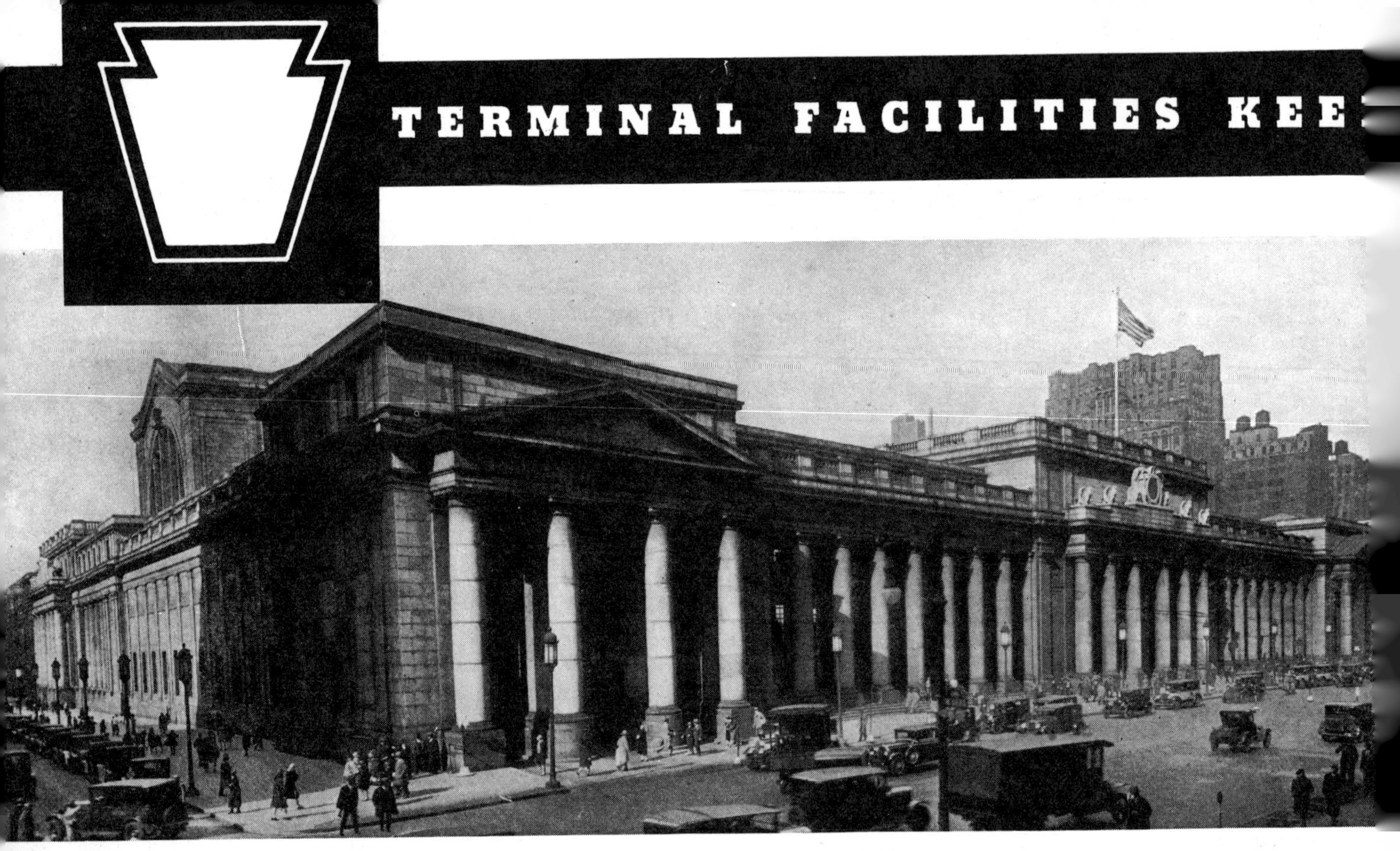

Fig. 48—Pennsylvania Station, New York City

NEW YORK STATION

September 7, 1910, marked a new era for Pennsylvania and Long Island Railroad commuters working in New York City. This day recorded the opening of the Pennsylvania Station in the heart of New York City. Previously, passengers had to cross either the North or East River ferries and even then found themselves far from the center of the city.

Electric service through the under-river tubes made such a station possible and was the fulfillment of a long-cherished desire of the Railroad. Naturally, a terminal of this magnitude required many kinds of power service. Adjacent to the station, therefore, the Railroad built a power plant—known as the Service Plant. This was completely Westinghouse-equipped.

PHILADELPHIA STATION

Another important link in the electrification was the improvement of suburban and through passenger facilities in the Philadelphia zone. In addition to the downtown suburban station, there stands a magnificent edifice at 30th Street, Philadelphia's new through and suburban station . . an enduring facility to a great transportation system.

Throughout these stations, Westinghouse electrical equipment provides convenience and personal comfort for patrons, and helps make possible the most efficient and economical operation of terminal facilities. Illustrated on these pages are but a few of the numerous applications of Westinghouse apparatus.

Fig. 49—Pennsylvania 30th Street Station, Philadelphia

Fig. 50—(Left) Westinghouse 500-hp. Synchronous Motor driving I-R compressor

Fig. 51—(Right) Three Westinghouse-Hertner m-g. sets charge car lighting and air conditioning batteries

Fig. 55—This 150-hp. Westinghouse Wound-rotor Induction Motor drives forced-draft fans for the boilers

Fig. 54—(Below) Three Westinghouse 667-kv-a. Power Transformers are used to step down the voltage from 13,200 to 133 volts

Fig. 52—(Left) Westinghouse Nofuze Lighting Panel eliminates fuses. A mere flip of the handle restores service

Fig. 56—(Below) Truck-type Westinghouse Oil Circuit Breakers. The movable truck feature simplifies maintenance and inspection

Fig. 53—(Left) In the steam heating plant, the main Switchboard is made up of Westinghouse "De-ion" and CL Circuit Breakers

NEWARK STATION

In view of the eventual elimination of Manhattan Transfer, due to all-electric operation, a new station was constructed for Newark, New Jersey, to facilitate the interchange of passengers to and from connecting trains at this point. Westinghouse equipment was selected to perform many of the duties that make for passenger comfort and economy of operation in this new station.

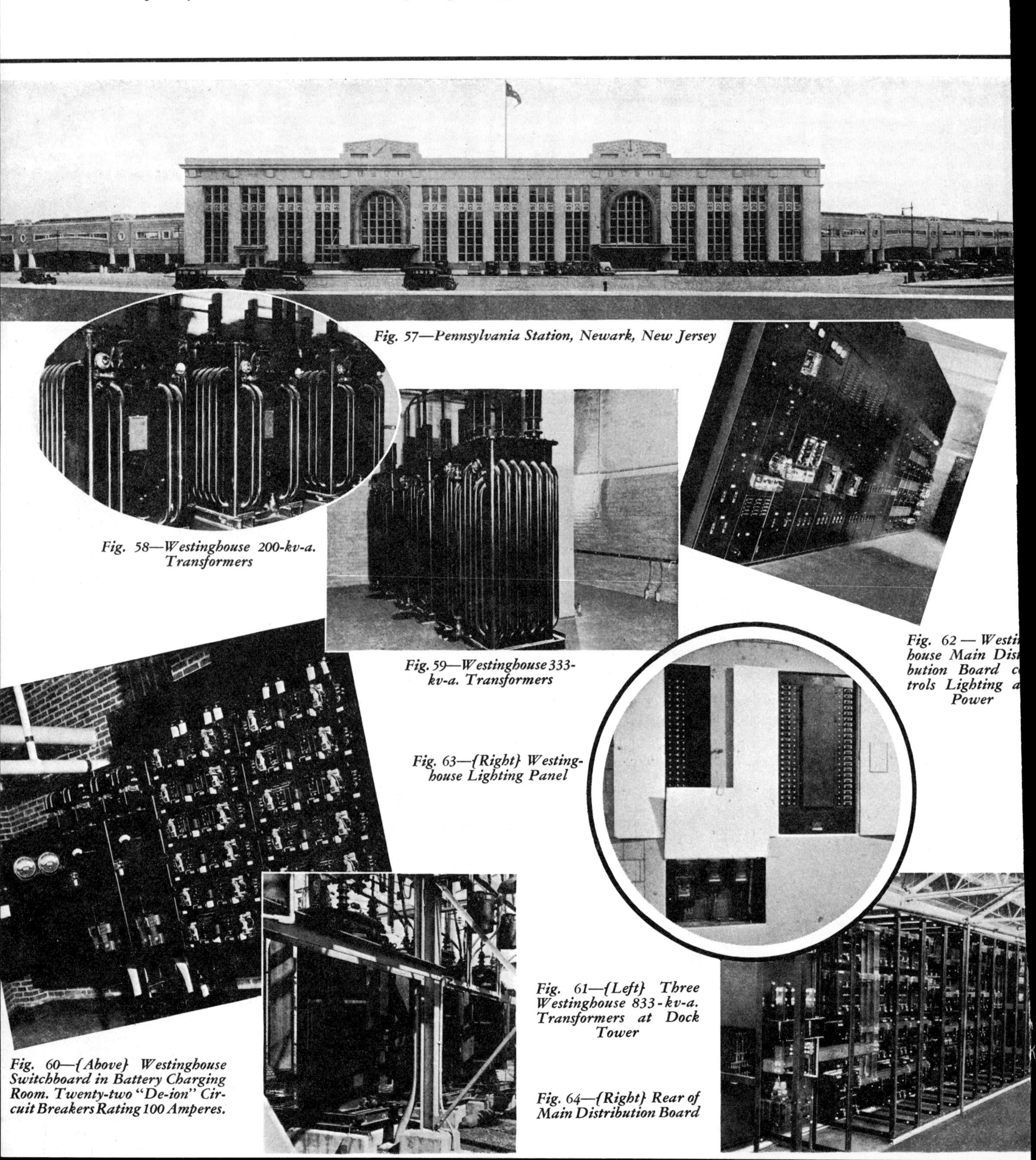

Fig. 57—Pennsylvania Station, Newark, New Jersey

Fig. 58—Westinghouse 200-kv-a. Transformers

Fig. 59—Westinghouse 333-kv-a. Transformers

Fig. 62 — Westi[nghouse] Main Dis[tri]bution Board c[on]trols Lighting a[nd] Power

Fig. 63—{Right} Westinghouse Lighting Panel

Fig. 60—{Above} Westinghouse Switchboard in Battery Charging Room. Twenty-two "De-ion" Circuit Breakers Rating 100 Amperes.

Fig. 61—{Left} Three Westinghouse 833-kv-a. Transformers at Dock Tower

Fig. 64—{Right} Rear of Main Distribution Board

THE PENNSYLVANIA ELECTRIFICATION

The electrified main line of The Pennsylvania Railroad between New York and Washington is one of the world's great arteries of commerce and travel. Over this portion of The Pennsylvania System, the greater percentage of the north and south traffic of the Eastern Seaboard is moved, augmented by the heavy traffic converging on this territory from the central and far western regions. In the zone served by the electrified lines lie many of the Country's largest centers of population . . . New York, Philadelphia, Baltimore, Washington, Newark, Trenton, Wilmington. Thus, an enormous suburban traffic is superimposed on the through traffic flowing to and from these centers.

The Pennsylvania Railroad wisely appreciated that a railroad is, and should always be, the primary artery of commerce for handling traffic efficiently and expeditiously. Moreover, in so doing, it must always provide the highest possible standards of service and performance. Hence, The Pennsylvania Railroad has shown its faith in rail transportation by embracing electrification for its New York to Washington line so that this territory now and in the future may be served as it should be served.

While electric service affords many operating advantages, its greatest accomplishment has been to provide for this section a standard of operation not approached by any other transportation system. Large fleets of heavy passenger trains, sometimes running 15 cars or more, are operated on fast schedules, the minimum time between Washington and New York now being 3¾ hours, which gives promise of further reduction. Comparable schedules with such dense traffic and heavy trains are operated on no other railroad. Similarly, in freight service, maximum tonnage trains are being moved at average road speeds between terminals in excess of 40 mph., corresponding to normal passenger schedules on many railroads and a speed which is greatly in excess of the average steam operating speed for freight throughout the U. S.

The operation in this electrified territory comprises 686 daily passenger and freight trains, operating 29,000 train miles over 1405 miles of track. To move this traffic requires 191 electric passenger, freight, and switching locomotives and 431 multiple-unit cars. The locomotives employed to handle this traffic are capable of maximum outputs of 6500–8000 rail horsepower, and when needed, two or more locomotives can be operated in multiple. Thus, passenger and freight traffic is moved on schedules calling for peak outputs at the rail up to 13,000 horsepower.

In this booklet there has been presented a short history of a most remarkable development, which has culminated in the greatest installation of its kind in history. By electrification, one of the Nation's greatest transportation arteries has been provided with a standard of operation which would have been economically unattainable with any other form of motive power. The Pennsylvania has set the standard for operation that the future growth of the United States will require.

With the fruition of this New York-Washington electrification involving heavy freight superimposed on dense passenger traffic, operating figures indicate confirmation of the savings expected when this electrification project was approved by Pennsylvania Railroad Executives.

Westinghouse is proud to have played a leading part over these years in the continued development and progress of this project. Westinghouse is confident that with this Pennsylvania Railroad record of accomplishment, further and extensive Steam Railroad electrification is inevitable.

ELECTRIC RAILWAY TRANSPORTATION

The Light-Weight Double-Truck Car

The Safety Car

The Trolley Bus

Their Field and Application

SPECIAL PUBLICATION 1655
OCTOBER, 1922

Westinghouse Electric & Manufacturing Co.

EAST PITTSBURGH, PA.

PREFACE

GREATER changes in the design and construction of electric transportation devices and equipment have been made during the past twelve years than in any other similar period. Many experiments were started early in this period, which led to the first substantial departures from the old standard forms and customs. Among these innovations were reduced step heights and the introduction of enclosed platforms, center entrances, scientific methods of loading and unloading passengers, safety devices, etc., and finally, equally important—lighter weight.

Due to this foresight on the part of individual operators and to the cooperative efforts of their allied associations and the manufacturers, these developments have become standardized so that car designs of today lend themselves to adjustment of details to include all features of value that may be required for local conditions and at the same time adhere to the all-important principle of light weight.

The industry as a whole is exceptionally fortunate in having so many developments completed and available for application at the present time when there has been created such widespread appreciation of the importance of improved transportation.

It is for the purpose of emphasizing the essential merits and possibilities of each of the three outstanding electric transportation vehicles, and to assist in the selection of suitable equipment to meet local conditions whatever they may be, that this publication has been prepared. All three vehicles, the light-weight double-truck car, the safety car, and the trolley bus, have passed the experimental stage and are now performing with full satisfaction in their respective classes of service. The results in actual operation have definitely verified the economies that they render and the classes of service to which they are best suited.

Light-weight double-truck cars of from 12 to 20 tons with seating capacity for 40 to 60 passengers include all the practical requisites for handling modern mass transportation problems. They are being successfully used on many roads to handle traffic conditions previously impossible with former equipment.

Although the single truck safety car has made a nation-wide reputation for itself during the past six years, it was regarded in the early days as suitable only for light service and for smaller cities. It has since proved of great importance and value in congested districts of many of the country's largest cities, such as Brooklyn, Boston, Baltimore, Los Angeles, Seattle, and many others. The safety car in reality is one of the most successful of any of the distinctive types of cars ever produced.

The trolley bus, the most recent development, in this country, for transportation purposes, has proved itself an economical method of transit. The trolley bus, like the gas bus, has certain economical fields of operation which as yet are not very well defined. Their application depends largely upon operating conditions encountered.

It is of vital importance to the electric railway transportation industry that the correct application and use of all modes of transportation be made. Also, it is imperative that the industry adopt the most economical types of vehicle and methods of propulsion for the varied services involved if it is to keep abreast with its allied industries in rendering service to the public.

One of the First Light-Weight Double-Truck Low-Floor Cars Placed in Service. The Pittsburgh Railways

Light-Weight Double-Truck Cars

For Interurban and City Service

LIGHT-WEIGHT double-truck cars have been successfully performing both city and interurban service during the past few years. The necessity for lower operating costs has been an important factor in promoting equipment replacements by lighter and more efficient cars whenever possible. The greatest saving in operating expense will naturally occur with the lightest car that can be operated satisfactorily, but this does not necessarily mean the highest return on the investment. Comfort and safety must be given equal consideration with weight as the proper balance between these three factors is essential to obtain the most beneficial results.

There are a number of different types of light-weight double-truck cars in city and interurban service in various parts of the country. No fixed rule can be made as to the exact type of car which will be suitable for a given service as each individual application must be studied and its particular needs considered. Some of the most successful of the city cars range in weight from 24,000 to 28,000 pounds complete without load. Among these cars will be found types for double and single-end one-man or two-man operation, Peter-Witt cars, two-man cars with center entrance and front and center exit, and one-man, two-man cars with multiple-unit control for service in trains. The advantage of light cars in maintaining better schedules due to higher acceleration and braking rates is being realized in handling heavy rush hour traffic. Also, train operation with light cars is replacing the motor car and trailer as a better means of meeting peak load demands, and of obtaining economical operation at other periods.

Light-weight double-truck interurban cars have been developed for both moderate and high-speed service. To meet the requirements of short suburban or interurban lines where moderate speeds are satisfactory, a 13-ton car (complete excluding load) with or without safety features and operated by one or two men has been showing some remarkable operating economies. A number of properties are obtaining very pleasing results from cars of this type, and undoubtedly there are many other places where such cars can be successfully applied. For high-speed service, light-weight cars ranging from 25 to 30 tons complete, excluding load, are proving their economic value. Many interurban roads that are now using heavy equipment can profitably make replacements with lighter cars of this type. A careful study of service requirements will often show that operating expense can be materially reduced without impairment to traffic; in fact, an increase in revenue is attracted by the use of new, more comfortable, better-riding cars.

The two principal savings that will be realized by operating light-weight cars are lower power consumption and reduced maintenance. The reduction in maintenance will apply to both way and structures and cars and equipment. Of course in the case of a new road, light cars would require less investment for feeder copper, machines in substations, and power house units. Also, it may be possible to remove feeder copper on properties where replacements are made with light-weight cars. However, the former case does not occur often; and when considering the latter, it will usually be found more desirable to retain the feeder and permit schedules to benefit from the increase in line voltage. Power savings will be realized almost in direct proportion to the reduction in car weight since the power required per ton mile is only slightly affected by train resistance. Maintenance of cars and equipment will be less with the lighter, better constructed cars and modern ventilated motors. The lighter cars will also be less severe on the tracks and roadbed thus reducing tie renewals, rail renewals and the cost of ballasting. In some cases tie spacings have been increased, which gives a more direct saving.

In many localities bus competition has made inroads into the earnings of electric railways, both interurban and city. New light-weight cars will prove attractive to the public since faster schedules can be maintained due to improvement in voltage and accelerating and braking rates. Better service on interurban roads will be a means of securing business taken by cross-country bus routes. Lower operating expense will often permit more frequent headways,

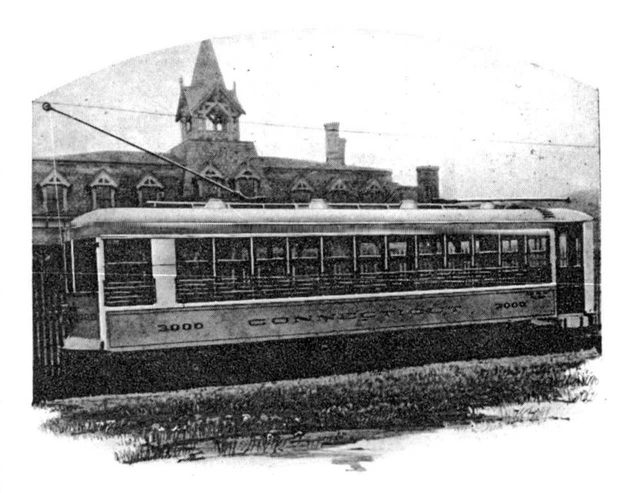

Light-Weight Double-Truck One-Man Interurban Car
The Connecticut Company

particularly in city service, thus regaining business lost to buses as well as promoting additional car riding.

Below is a typical example based on certain assumptions as indicated, showing the savings that may be expected in city service.

	Old Cars	New Cars
Weight of car without load, tons	24	14
Seating capacity	50	50
Number of motors per car	4	4
Horsepower of motors	50	25
Miles one way	4.02	4.02
Round trip time including layover, minutes	60	60
Headway, minutes	6	6
Number of cars	10	10
Schedule speed, miles per hour	8.04	8.04
Stops per mile	8.5	8.5
Average length of stop, seconds	8	8
Annual cost of power	$30,900	$18,000
Annual way and structure maintenance	9,500	6,010
Annual equipment maintenance	16,450	9,400
Annual cost of labor	66,900	44,300
Totals	123,750	77,710
Annual saving in above items		46,040
Cost of 11 new cars at $10,000 each		110,000
Return on investment		41.9%

From the foregoing typical example it will be seen that by changing over to light-weight cars the net saving will pay off the investment in a comparatively short time.

Pittsburgh Railways Company

One of the first light-weight double-truck low-floor cars for city operation was placed in service by the Pittsburgh Railways. This car, which is remarkable for its large capacity and saving in weight, seats 59 passengers, and weighs 33,200 pounds complete with all equipment exclusive of load. It was designed for and has successfully performed heavy city service where schedule speeds are high and grade conditions severe.

Light-Weight Double-Truck Low-Floor Car of The Pittsburgh Railways

The electrical equipment includes four 40 hp. motors suitable for 24 or 26 in. wheels. The car is center entrance, with a low step obtained by the use of a ramp and small wheels, and is operated either center or front exit. The length over bumpers is 45 feet and the distance between truck centers is 21 feet 8 inches.

United Railways Company, St. Louis

Another light-weight double-truck city car that is giving very satisfactory service was built by the United Railways of Saint Louis. This car is similar in design to the Peter-Witt type that has been used in Cleveland so successfully for many years. The wheel diameter is 26 in. and four 35 hp. motors are used. While the grades are not so severe as encountered in Pittsburgh, the traffic is extremely heavy and high schedule speeds are maintained. In actual service the car has shown a decided improvement in operation over the older type equipments which were replaced. The loading and unloading facilities are exceptionally good on account of wide doors and aisles. This car weighs 38,000 pounds complete with all equipment but without load, and seats 59 persons.

Chicago Surface Lines

The Chicago Surface Lines has recently built light-weight double-truck city cars adapted for double-end one-man operation and equipped with four 35 hp. motors. With this equipment, but without load, the weight of the car is approximately 30,600 pounds. The over-all length of the car is 39 feet 7 inches and the seating capacity is 50 persons. The schedules in Chicago are unusually severe when the number of stops is considered. The schedule speed is 12 miles per hour with seven stops per mile, each of six seconds duration. An interesting feature of the Chicago car is separate entrance and exit doors with selective control that permits either door to be open or closed as desired. Wide aisles, comfortable seats and attractive interior construction add to the riding qualities and the appearance.

Cincinnati Traction Company

One of the late applications of light-weight double-truck cars to city service is that made in Cincinnati. The Cincinnati Traction Company is placing in service 75 cars which will weigh 28,000 pounds each, complete with all equipment, but without load. These cars will seat 50 persons each, and will have 26 in. wheels. The cars are specially designed to meet the heavy service and severe grade requirements of Cincinnati. The service requires operation on grades as high as 12 per cent on some of the lines. The cars are equipped with four 35 hp. motors and multiple-unit HL control for cabinet mounting. Provision is made for one-man, two-man and train operation. Trains of two and three cars are to be operated during rush hours to assist in speeding up the schedules.

Light-Weight Cars in Interurban Service

There is a wide field for the application of light-weight double-truck cars in moderate and high-speed interurban service. Many heavy equipments that are extremely wasteful of power and expensive to maintain are in operation in a service that can be performed by lighter and more efficient cars. For high-speed interurban service, cars weighing 25 to 30 tons can be successfully applied in many cases. In this service

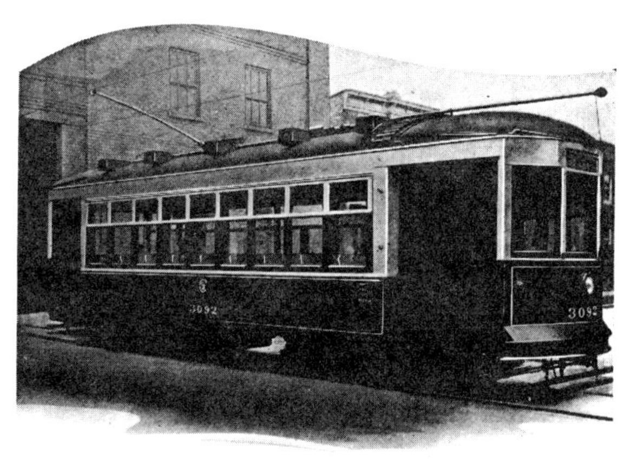

Light-Weight Double-Truck Car for One-Man Operation Chicago Surface Lines

two-man operation is a necessity on account of safety, schedules and public relations. A conductor is necessary to flag railroad crossings, to assist in obtaining train orders, to handle baggage at non-agency stations and to give information to passengers. The demands of such service are too many for the time of only one man.

In moderate speed interurban service the 13-ton one-man double-truck car has proved very successful when properly applied. Where the time between terminals is relatively short, say not more than two hours, where the stops are frequent, that is, one or two miles apart, and where maximum running speeds of 35 miles an hour are sufficient for maintaining schedules, this type of car has demonstrated its ability to handle the service. One-man operation does not materially offset the schedule in service where these cars are used. Heavy loading is usually at terminals when there is a layover and therefore causes no delay. Improvement in acceleration and braking is ordinarily more than sufficient to offset any additional time spent in stops. It should not be overlooked that the saving incident to one-man operation is less for interurban service than for city service as the crew expense forms a smaller part of the operating cost. Platform labor cost per car mile decreases as schedule speeds increase. However, the maximum saving of crew expense on interurban lines will be realized where the traffic is suitable for the application of the 13-ton car as schedule speeds are not high in this class of service. It is evident that one-man operation has its place in the interurban field as well as in that of city service, but care should be taken not to

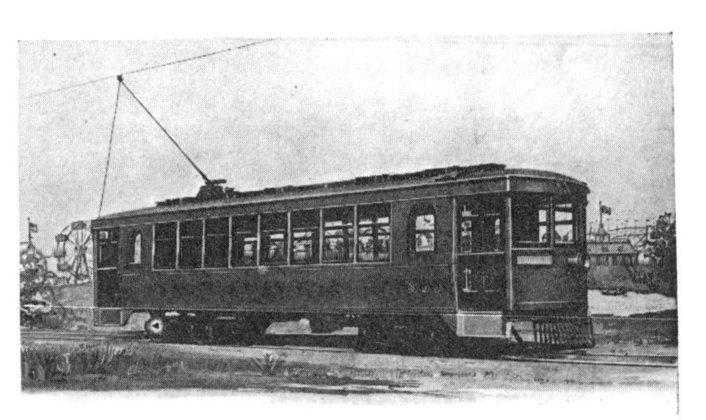

Typical Light-Weight Double-Truck Interurban Car

use it when the class of traffic demands two men to maintain the standard of service.

Below is a typical example based on certain assumptions as indicated, showing the savings that may be realized in moderate speed interurban service by replacing old equipments with new light-weight cars.

	Old Cars	New Cars
Weight of car without load, tons	30	13
Seating capacity	46	46
Number of motors per car	4	4
Horsepower of motors	75	25
Miles one way	15	15
Single trip time exclusive of layover, minutes	50	50
Headway, minutes	60	60
Number of cars	2	2
Schedule speed, miles per hour	18	18
Stops per mile	1	1
Average length of stop, seconds	15	15
Annual cost of power	$ 7560	$ 3270
Annual way and structure maintenance	5515	4260
Annual equipment maintenance	5360	3840
Annual labor cost	10980	5850
Totals	29415	17220
Annual saving in above items		12195
Cost of 3 new cars at $10,000 each		$30000
Return on investment		40.6%

This service shows a very attractive return. It is sufficient to clear the entire investment in a very short time.

Light-Weight Double-Truck One-Man Car of the Houston Electric Company

The Connecticut Company

The Connecticut Company recently replaced heavy double-truck cars, operating on a 12 mile suburban line between Torrington and Winsted, Connecticut, with light-weight double-truck one-man cars. These cars operate in service where heavy loads are handled, thus requiring more capacity than can be secured with safety cars. The new equipment is also improving schedules, and business lost on account of "jitney" competition is being regained. This car weighs 26,000 pounds complete with all equipment, excluding load, and seats 52 persons. Considerable aluminum has been used in piping and fittings in order to reduce the weight to a minimum. The length over-all is 40 feet 3 inches and the distance between truck centers is 22 feet 11 inches. The equipment consists of four 25 hp. motors mounted on 26 in. wheels. These cars have performed excellently and have proved a very satisfactory investment.

Princeton Power Company

A light-weight double-truck interurban car that has been in service for some time is operated by the Princeton Power Company. This car represents exceptional weight economy and has rendered very satisfactory service. It operates on a line where stops occur every two miles and a schedule speed of 20 miles an hour is necessary. The car is equipped with a baggage compartment and weighs 26,450 pounds complete with all equipment, excluding load. It has four 25 hp. motors and is geared for operation at moderate speeds. It is of the low floor type with 26 in. wheels and seats 44 persons. By using a ramp extending from the bolster line to the platform, the step heights are reduced to 16 inches from track to step and 13 inches from step to platform. The length of the car over bumpers is 41 feet 6 inches and the distance between truck centers is 25 feet.

Wheeling Traction Company

The Wheeling Traction Company is operating an interurban car which is especially interesting when comparing weight and capacity. This car is of the low-floor center-entrance type and has seats to accommodate 72 persons. The car weighs 49,500 pounds without load but including all equipment. It is 57 feet 10 inches over bumpers and the distance between truck centers is 25 feet 8 inches. The wheel diameter is 33 inches. The car operates in service that requires a schedule speed of 18 miles an hour with one 20 second stop every three fourths of a mile. Aside from lightness and large capacity, it is noteworthy as being extremely comfortable and smooth riding and is well adapted for operation at higher speeds than the service requires.

Milwaukee Electric Railway & Light Co.

The light-weight double-truck one-man, two-man cars operated by the Milwaukee Electric Railway & Light Company offer some superior advantages in the way of speed, safety and comfort. This car was built by the operating company and the fundamental ideas and details were worked out by its engineers. The distinctive features are adaptability to light or heavy traffic, unusually light-weight, separate exit and entrance, special design of trucks, and advantages of safety, comfort and improved service to patrons. The car weighs 31,900 pounds complete with all equipment but excluding load.

It has specially constructed arch-bar trucks with 26 in. wheels. The length over buffers is 45 feet and the truck center distance is 20 feet 9 inches. A seating arrangement to provide the maximum number of cross seats is used with accommodations for a total of 58 persons. The electrical equipment is composed of four 25 hp. motors and double-end type K control. All standard safety features are included in the equipment. Arrangement for either one-man or two-man operation has made the car readily adaptable for heavy traffic service, as it can be used as a one-man car during the non-rush hours and as a two-man car during the rush hours. The use of separate entrance and exit doors assists materially in speeding up loading and unloading. The operation of this car has been very satisfactory in service where exceptionally heavy traffic is handled.

EQUIPMENT FOR LIGHT-WEIGHT DOUBLE-TRUCK CAR

Single Car Operation, Four 25 Hp. Motors

Apparatus	No. Req'd	Type	Weight Pounds
Motor Items:			
Motors	4	508-A	4200
Gear cases	4	Pressed steel	
Axle bearings	8	Bronze 4-in.	
Axle collars	4	4-in.	
Gears—Helical 74-tooth	4	Forged steel BP	
Pinion—Helical 13-tooth	4	Forged steel BP	
Main Circuit Control Items:			
Trolley base with 13 ft. pole	2	No. 15	210
Trolley harps	2	No. 25	4
Trolley wheels	2	No. 40-A	7
Lightning arrester	1	M.P.	8
Circuit breaker	2	611-A	48
Controllers	2	K-35-JJ	540
Grid resistor	1 set	Single 3 pt.	150
Equipment Details:			
Main cable	1285 ft.		187
Knuckle joint connectors	16		4

Apparatus	No. Req'd	Type	Weight Pounds
Lighting Details:			
Keyless wall receptacles	27		
Mazda lamps	29		
Snap switches	2		38
Transfer switches	1		
Cable	450 ft.		
			5396

EQUIPMENT FOR LIGHT-WEIGHT DOUBLE-TRUCK CAR

Single Car Operation
Four 35 Hp. Motors

Apparatus	No. Req'd	Type	Weight Pounds
Motor Items:			
Motors	4	510-A	
Gear cases	4	Pressed steel	
Axle bearings	8	Bronze 4-in.	
Axle collars	4	4-in.	5970
Gears—			
Helical 69-tooth	4	Forged steel BP	
Pinions—			
Helical 13-tooth	4	Forged steel BP	
Main Circuit Control Items:			
Trolley base with 12 ft. 4 in. pole	2	No. 13	312
Trolley harps	2	No. 25	4
Trolley wheels	2	No. 40-A	7
Lightning arrester	1	M.P.	8
Circuit breakers	2	611-A	48
Controllers	2	K-35-JJ	540
Grid resistor	1 set	8 in. 3 pt.	276
Equipment Details:			
Main cable	1520 ft.		
Knuckle joint connectors	16		7
Lighting Details:			
Keyless wall receptacles	27		
Mazda lamps	29		
Snap switches	2		38
Transfer switch	1		
Cable	450 ft.		
			7210

EQUIPMENT FOR LIGHT-WEIGHT DOUBLE-TRUCK CAR

Train Operation
Four 25 Hp. Motors

Apparatus	No. Req'd	Type	Weight Pounds
Motor Items:			
Motors	4	508-A	
Gear cases	4	Pressed steel	
Axle bearings	8	Bronze 4-in.	
Axle collars	4	4-in.	4200
Gears—			
Helical 74-tooth	4	Forged steel BP	
Pinions—			
Helical 13-tooth	4	Forged steel BP	
Main Circuit Control Items:			
Trolley base with 13 ft. pole	2	No. 15	210
Trolley harps	2	No. 25	4
Trolley wheels	2	No. 40-A	7
Lightning arrester	1	M.P.	8
Main switch	1	496-I	24
Main fuse box and fuse	1	220-A	11
Control and reset switch	2	494-B	14
Master controllers	2	15-B	120
Line switch	1	8063-C	100
Switch units	6	806	174
Overload trip	1	371-G10	12
Reverser	1	184-A	37
Motor cutout switch	1	X-D-1	10
Grid resistor	1 set	5-in. 3 pt.	113
Control resistor	1	197-J	35
Junction boxes	2	426-D	54
Train line receptacles	2	448-D	34
Train line jumpers	1	449-F	17
Equipment Details:			
Cabinet control details	1 set		2
Insulating details	1 set		8
Pneumatic details	1 set		40
Main cable	410 ft.		73
Control cable	460 ft.		45
Knuckle joint connectors	16		4
Lighting Details:			
Keyless wall receptacles	27		
Mazda lamps	29		38
Snap switches	2		
Transfer switch	1		
Cable	450 ft.		
			5394

EQUIPMENT FOR LIGHT-WEIGHT DOUBLE-TRUCK CAR

Train Operation
Four 35 Hp. Motors

Apparatus	No. Req'd	Type	Weight Pounds
Motor Items:			
Motors	4	510-A	
Gear cases	4	Pressed steel	
Axle bearings	8	Bronze 4-in.	
Axle collars	4	4-in.	5970
Gears—			
Helical 69-tooth	4	Forged steel BP	
Pinions—			
Helical 13-tooth	4	Forged steel BP	
Main Circuit Control Items:			
Trolley base with 12 ft. 4 in. pole	2	No. 13	312
Trolley harps	2	No. 25	4
Trolley wheels	2	No. 40-A	7
Lightning arrester	1	M.P.	8
Main switch	1	496-I	24
Main fuse box and fuse	1	220-A	11
Master controllers	2	15-B	120
Control and reset switch	2	494-B	14
Line switch	1	806-C-3	100
Switch units	6	806	174
Overload trip	1	371-C-10	12
Reverser	1	184-A	37
Motor cutout switch	1	X-D-1	10
Grid resistor	1 set	8 in. 3 pt.	211
Control resistor	1	197-J	35
Junction boxes	2	426-D	54
Train line receptacles	2	448-D	34
Train line jumper	1	449-F	17
Equipment Details:			
Cabinet control details	1 set		2
Insulating details	1 set		8
Pneumatic details	1 set		40
Main cable	445 ft.		110
Control cable	460 ft.		45
Knuckle joint connectors	16		7
Lighting Details:			
Keyless wall receptacles	27		
Mazda lamps	29		
Snap switches	2		38
Transfer switch	1		
Cable	450 ft.		
			7404

THE SAFETY CAR

THE safety car was first operated in 1916 and met with immediate success. On account of its small size and its one-man operation, it was regarded for a time, by the majority of operators, as suitable for only lightly traveled lines in small cities. It had a certain physical resemblance to the single-truck "dinkeys" of unpleasant memory. One-man operation began in horse-car days and had been gradually abandoned partly because of objections of the platform men, and partly because the use of only one man on existing cars resulted in slow schedules and generally unsatisfactory service. The modern safety car, however, eliminates all of these objections.

The standard safety car is approximately 28 feet in length and 7 feet 10 inches in width. The double-end car seats 32 persons and the single-end car seats 35 persons. The weight of the car completely equipped but without passenger load is 16,000 pounds. Twenty-four or 26 in. wheels are used on practically all cars.

The safety features which are included on the standard car are of primary importance. They actually do prevent accidents and they also are of material assistance in winning over the public when safety cars are first used in any particular town, and they contribute by labor saving to the speed of operation. With the usual safety devices it is necessary for the brakes to be applied, bringing the car to a stop before the door can be opened. Likewise, the door must be closed before the brakes can be released and the car started. In case the operator releases the handle of the controller through illness or inattention or for any other cause, the power is immediately cut off, the brakes applied, the track sanded and the doors released so that they can readily be opened by hand. Boarding and alighting accidents have been almost eliminated by the safety features and there has been a number of instances where the emergency features have been brought into action and accidents averted which otherwise would have occurred.

In view of the variety of designs of older cars and the diversity of opinions regarding almost every detail of their construction, it is not surprising that there have been various modifications of safety cars constructed for various conditions of service. Rather, it is surprising that so few changes have been made, and that essentially the same car has been found adequate to meet the variety of service in which standard safety cars are used. As a result of standardization, there is undoubtedly considerable advantage in first cost and in quickness of obtaining renewal parts.

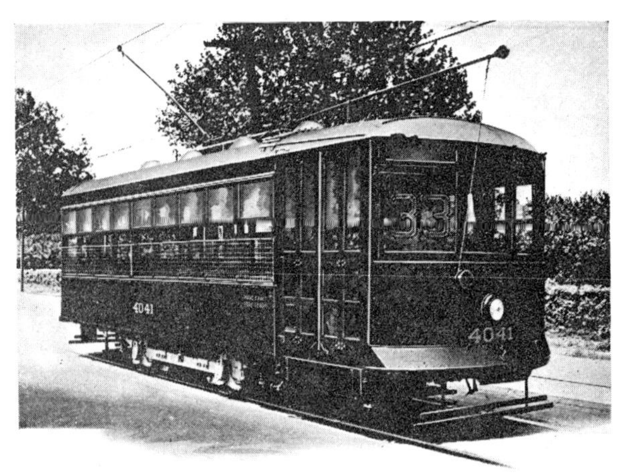

The United Railways & Electric Company, Baltimore, is One of the Largest Operators of Safety Cars

In certain cities, where physical limitations permit, wider aisles and a greater distance between seats are desirable. In some cities, loading conditions are such that double doors are considered essential. In others, the seating arrangement is changed to provide longitudinal seats on one side of the car for a short length at the front end of a single-end car or at both ends where double-end cars are used. There has been a general tendency in the northern part of this country and in Canada to strengthen various parts of the car and to protect the interior of the car better from cold. Where cars are heavier than normal and where severe snow or grade conditions are encountered, two 35 hp. motors may be required instead of the customary 25 hp. motors. When double doors are used, provision should be made for operating each door independently of the other.

The use of safety cars has resulted in the expected decrease in operating expenses. The principal item in street railway operating expense is wages of conductors and motormen. In order to share the benefits of safety car operation with the car operator, it is customary to increase his rate of pay 10 per cent above that of a conductor or motorman on two-man cars. There is thus a saving in platform labor per car mile of 45 per cent.

The safety car effects savings in maintenance of cars and equipment. On account of its light weight, repairs are easily made and renewal parts cost less. As compared with double-truck cars, there are fewer and lighter parts to maintain; and as compared with older single-truck cars, the same weight economies obtain while the improved design of truck and body eliminates many of the strains set up by the rocking prevalent in the older cars. The actual maintenance of safety cars depends largely on local conditions, but usually will be between 1.5 and 2.0 cents per car mile. On a number of properties, safety car maintenance has been found to be 40 to 50 per cent of that of the average double-truck car, on a mileage basis.

Maintenance of track and roadway will undoubtedly be decreased by safety car operation. The exact amount saved is doubtful on account of the difficulty of segregating maintenance for any particular sections of track or conditions of operation. When new track is laid or old track is rebuilt, there should be a very material saving, for less expensive construction may be used with light-weight safety cars than with heavy double-truck cars.

Energy consumption varies approximately as the ton-mileage and hence the saving per car mile will be almost directly proportional to the saving in car weight. As compared with heavier equipment, the less power drawn by the safety car results in less line drop with resulting higher voltage at the car and higher efficiency of distribution. When generating and converting equipment are loaded to capacity, safety car operation may make it possible to postpone the purchase of additional power house and substation equipment. When safety cars are used in new installations, a lower investment is necessary for complete generating, conversion and distribution systems than when heavy cars are operated.

While the safety car was first proposed as a means of reducing operating expense in com-

Safety Car of the San Francisco-Oakland Terminal Railways

bating jitney bus competition in the West and Northwest, shorter headways were operated and it was found that the increase in receipts due to the improved service outweighed the reduction in operating expense. In the last five years it has been demonstrated frequently that the most successful safety car applications from the standpoint of financial return and popular favor, are those where the service is considerably increased. The result has been that although the original applications were on the basis of not more than 25 per cent increase in service, the later applications which have been most successful have been made on the basis of 50 to 100 per cent increase in service.

The 1920 report of the American Electric Railway Association, Committee on Safety Car Operation, says of earnings: "The averages of the companies which increased their schedules with safety cars show an average of 40.5 per cent increase in safety car mileage compared to former two-man car mileage on the lines affected, with a corresponding decrease in the car mile gross earnings of but 0.93 per cent. That is, the gross earnings with safety cars show substantially the same percentage of increase as the car miles operated."

The following table shows the present and proposed service and the results to be obtained by replacing double-truck cars on a base headway of 10 minutes with safety cars on a base headway of 6 minutes.

SAFETY CAR OPERATION

	Present Operation	Proposed Operation
Car Type	Double Truck	Safety
Weight of car with av. load, tons	18.65	9.25
Seating capacity	40	32
Number of motors per car	2	2
Type of motor	No. 532-B	No. 508-A
Round trip distance, miles	10.32	10.32
Round trip time, including layover:		
Non-rush service, minutes	60	60
Rush service, minutes	64	60
Normal headway, non-rush, min.	10	6
Maximum number of cars, non-rush	6	10
Normal headway, rush, minutes	5	3
Maximum number of cars, rush	12	20
Total car hours per day	150	260
Total car miles per day	1505	2680
Kw. hrs. per car mile at sub-station	2.73	1.255
Power cost per day, 1.25c per kw. hr	$ 51.40	$ 42.00
Labor cost per day, 47c per man-hour	141.10	....
Labor cost per day, 52c per man-hour	....	135.10
Car maintenance per day, 2.98c per car mile	44.90	....
Car maintenance per day, 1.5c per car mile	....	40.20
Total cost of power, labor and maintenance per day	$ 237.40	$ 217.30
Total cost of power, labor and maintenance per year	$86650.00	$ 79315.00
Saving in operating expenses per year	....	$ 7335.00
Increase in car mileage with safety cars	....	78%
Estimated increase in earnings		39%
Estimated increase in annual earnings, 39% of $169215		$ 65995.00
Total increase in net annual revenue		73330.00
Cost of 22 new cars, $6000 each		132000.00
Return on investment		55.5%

United Railways & Electric Company Baltimore

The United Railways & Electric Company of Baltimore is one of the larger operators of safety cars in congested districts. The original installation on the Fremont Line was made in July, 1920 with 33 cars. The Fremont Line skirts the congested section of the city, and crosses a comparatively large number of the traction lines. Heavy transfer service results. Actual operation during a period of several years has fully demonstrated the ability of this type of transportation to meet most exacting conditions.

The car employed is 30 feet long and is provided with a wide aisle. A longitudinal seat is placed at each corner of the car to provide more standing room, and also to afford the least possible resistance to ingress and egress of passengers who are mostly short distance riders.

Los Angeles

Safety car operation was initiated in Los Angeles, in July, 1920. The first line was a shuttle which served an outlying section of the city and connected two lines running to the principal business district. The second line connects a number of main lines and provides a cross-town service outside the business district. Operation of safety cars was started in the business district in August, 1921. Greatly improved service was furnished, twelve safety cars superseding nine older cars of the same seating capacity.

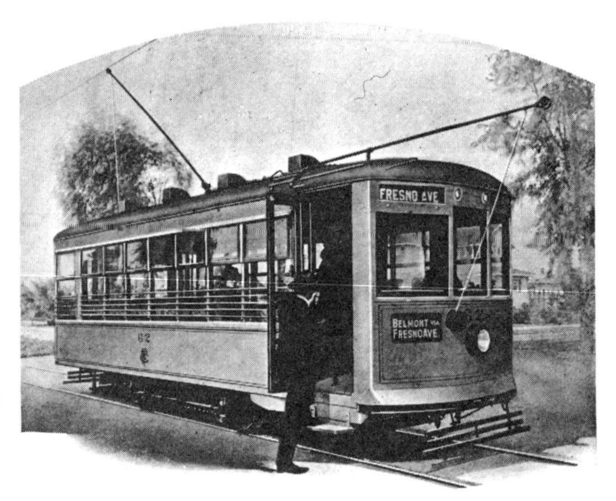

Safety Cars are Operated by the Fresno Traction Company

Brooklyn Rapid Transit Company

One large company, the Brooklyn Rapid Transit Company, uses safety cars extensively for heavy traffic feeders and for congested city service in Brooklyn. Their successful operation is an example of the usefulness of this type of car in a large city. Other examples of the extensive use of safety cars are the Eastern Massachusetts Street Railway Company and the Stone & Webster Corporation. These companies have hundreds of safety cars in service in various towns served by their systems, where in many cases safety cars are operating in congested traffic. The original safety cars met these requirements so satisfactorily and with such marked improvement in service that large numbers of safety cars are being added by these two companies from time to time.

Pennsylvania-Ohio Electric Company

The Pennsylvania-Ohio Electric Company operates safety cars in Sharon, New Castle and Youngstown, all of which cities are busy industrial centers. Safety car operation was started in September, 1919. On the initial installation the number of seats furnished was increased 84 per cent by providing a six-minute service instead of ten-minute service. Safety cars were used on other lines in rapid succession with the most gratifying results. On lines where the service remained the same, riding decreased during the recent industrial depression; but on the safety car lines receipts increased very materially.

Mr. Clinton D. Smith, general superintendent of the company, summarizes the features of the safety car which contribute to the merchandising of transportation as follows:

"(1) The economical operation of the car permits more frequent headway.

"(2) The car, being the last word in modern mechanical electrical equipment, appeals to the people in the same manner as the modern department store, and encourages its frequent use.

"(3) Appreciation of the fact that the car is fully equipped with safety devices tends to increase the number of rides per capita.

"(4) Finally, our analysis indicates conclusively that there are fewer accidents with the safety cars, as compared with two-man cars. There has been a practical elimination of boarding and alighting accidents."

Safety Car of the Union Traction Company of Santa Cruz

The Safety Car in Congested Service
Kansas City Railways

EQUIPMENT FOR STANDARD DOUBLE-END SAFETY CAR

Two 25 hp. Motors

Motor Items:

Apparatus	No. Req'd	Type	Weight Pounds
Motors	2	508-A	2100
Gear cases	2	Pressed steel	
Axle bearings	2	Pinion end	
Axle bearings	2	Comm. end	
Axle collars	2	M.I.	
Gears, solid tr. st.	2	74-tooth	
Pinions, forged st.	2	13-tooth	
Main Circuit Control Items:			
Trolley base with 14-ft. pole	2	No. 15	......
Trolley harps	2	No. 25	220
Trolley wheels	2	No. 40-A	......
Lightning arrester	1	M.P.	8
Circuit breakers	2	611	48
Controllers	2	K-63-BR	268
Reverse handle	1	K-63-BR	......
Grid resistors	1 set	5 in. 3 pt.	75
Equipment Details:			
Main cable	695 ft.	7x".064	90
Knuckle joint connectors	8	Pivot	1
Lighting Details:			
Keyless wall receptacles	20	....	29
Mazda lamps	20	....	
Snap switch	1	....	
Transfer switch	1	....	
Snap switch (d.p.d.t.)	1	....	
Cable	400 ft.	19x".0142	
			2839

For Single-End Operation—Omit:

1 Trolley	165 Ft. Main cable 7x".0545
1 Controller	1—D.P.D.T. snap switch
1 Circuit breaker	1—Transfer switch

EQUIPMENT FOR DOUBLE-END SAFETY CAR

Two 35 hp. Motors

Motor Items:

Apparatus	No. Req'd	Type	Weight Pounds
Motors	2	510	2985
Gear cases	2	Pressed steel	
Axle bearings	2	Pinion end	
Axle bearings	2	Comm. end	
Axle collars	2	M.I.	
Gears, solid tr. st	2	69-tooth	
Pinions, forged st	2	13-tooth	
Main Circuit Control Items:			
Trolley base with 14-ft. pole	2	No. 15	....
Trolley harps	2	No. 25	220
Trolley wheels	2	No. 40-A	....
Lightning arrester	1	M.P.	8
Circuit breakers	2	611	48
Controllers	2	K-63-BR	268
Reverse handle	1	K-63-BR	....
Grid resistors	1 Set	5 in. 3 pt.	113
Equipment Details:			
Main cable	695 ft.	7x".064	111
Knuckle joint connectors	8	Pivot	1
Lighting Details:			
Keyless wall receptacles	20	....	29
Mazda lamps	20	....	
Snap switch	1	....	
Transfer switch	1	....	
Snap switch (d.p.d.t.)	1	....	
Cable	400 ft.	19x".0142	
			3693

For Single-End Operation—Omit:

1 Trolley	165 ft. main cable 7x".064
1 Controller	1—D.P.D.T. snap switch
1 Circuit breaker	1—Transfer switch

For magnetic line switch, omit circuit breakers and add the following:

FOR DOUBLE-END:

Control switch with fuses	2	494-C	14
Line switch	1	801-E	95
Insulating bolts	4	....	6
Ratchet switches for:			
Controller	2	....	15
Control cable	150 ft.	19x".0142	5

FOR SINGLE-END:

Control switch with fuses	1	494-C	7
Line switch	1	801-E	96
Insulating bolts	4	....	6
Ratchet switch for:			
Controller	1	....	7½
Control cable	100 ft.	....	3½

A Safety Car of the Pacific
Electric Railway

The Trolley Bus

A FEW trolley buses were operated a number of years ago in the United States. These experimental lines were abandoned and until the last two or three years, the activity in trolley bus development was confined to foreign countries. Foreign experience, however, could not be applied directly to American conditions because of the many double-deck buses operated by two men, and because of speeds being lower than desirable in this country. The trolley bus, in order to be a success here, must be operated by one man and must make schedule speeds which compare favorably with safety car schedules.

The operator of a safety car is able to perform the duties of both motorman and conductor without being overworked, but he admittedly is kept busy. With the trolley bus (or gasoline bus) he has the added duty of steering and the operating of doors which, up to the present time, have been manually instead of pneumatically operated as on the safety car.

The higher speeds necessary in this country increase the difficulty of current collection. One of the major problems encountered has been in connection with current collectors. The bus body and chassis, as well as the electrical equipments, are essentially adaptations of existing designs. After extensive trials of various devices, experience indicates that the most satisfactory method of current collection in this country to date has been obtained from two independent trolleys having swivel harps. Probably the most important advantage of this form of current collection is the fact that standard overhead construction can be employed. The use of two

Control Apparatus for Trolley Mounted Under Hood

trolley wires requires well maintained overhead construction and a good roadway since delays caused by a trolley leaving the wire are more liable to occur than when a single wire ground return system is employed. It must also be remembered that the bus must have a reasonable cruising range, and that to provide economic operation it must be operated by but one man.

Considerable work has been done in connection with the development of a single-pole current collecting device for the trolley bus; while great strides have been made and there is every promise as to the feasibility of the scheme, further work must be carried on to make it more satisfactory than the two-pole current collection.

The higher speed of operation requires greater motor capacity in this country than is used abroad. In general, two 25 hp. motors, practically the same as those used on safety cars, have been used. The motors are mounted in tandem on the bus chassis and drive the automobile type rear axle through universal joints and a propeller shaft. Single-motor equipment has been tried on certain experimental buses, but owing to lack of capacity in the units tested, two motors have been used on commercial installations. One motor of sufficient size could be used. When one motor is used, it may be desirable to have it arranged for field control in order to reduce resistor losses and to provide two economical running speeds. The single-motor equipment has the advantage of low maintenance and simpler control circuits. On the other hand, a motor failure of almost any kind means a dead bus and a "tow-in." This feature is not so important in a bus as in a railway car because the bus can be moved to one side of the street or road and does not tie up the other traffic following. A disabled railway car always delays the cars following unless it can be moved quickly.

The subject of control also has been given serious thought and various types have been applied. If the trolley bus is ever to be used extensively in such service as that for which it seems to be economically suited, it should not be handicapped by adding the manual operation of a controller to the already numerous duties of the bus operator. Most automobiles are now equipped with a foot accelerator because of the inconvenience of the hand throttle. It seems reasonable to assume that the trolley bus operator who must collect fares, make change, issue transfers, operate a door, and keep the vehicle within range of the trolley wires, should be provided with a simple control equipment. The development of the automobile has carried with it many simplifications in the control of a vehicle, and their economical use and general adaptation must not be overlooked whenever trolley bus or gas bus operation is involved.

Profiting by the extended experience of the automotive industry in the control of vehicles, it appears to be the concensus of opinion that the trolley bus should be equipped with a foot-operated control mechanism or throttle.

Non-automatic control would ordinarily be favored for operation under the varied conditions met on city streets, for the same reasons that

Unit Switch Panel and Sequence Switch for Trolley Bus

have made its use almost universal on street railways. In addition to the inability of the usual forms of automatic control to meet traffic emergencies, it has the additional handicap on buses of having to overcome a wide variety of road conditions. It is necessary, therefore, with the usual automatic control, to provide means of overcoming the automatic feature in order to meet certain traffic conditions; that is, the automatic control is made non-automatic.

One form of automatic control which includes a time element feature and permits foot operation, is peculiarly fitted to trolley bus service. This equipment consists of a small number of switches which are actuated by a motor-driven "sequence switch" or "controller." When a heavy current is drawn by the motor driving the bus, the "sequence switch" motor operates at a lower speed than when smaller currents are drawn. Thus, within certain predetermined time limits, the more severe the accelerating condition, the greater the length of time required to close the switches and the lower the rate of acceleration. The movement of the sequence drum continues, however, at a slow speed under the most severe conditions of operation and thus insures against stalling provided the overload capacity of the motor is not exceeded. This scheme of automatic acceleration has been employed most successfully on trolley buses operated in regular service in Minneapolis, Toronto (Ont.), Windsor (Ont.), and Baltimore.

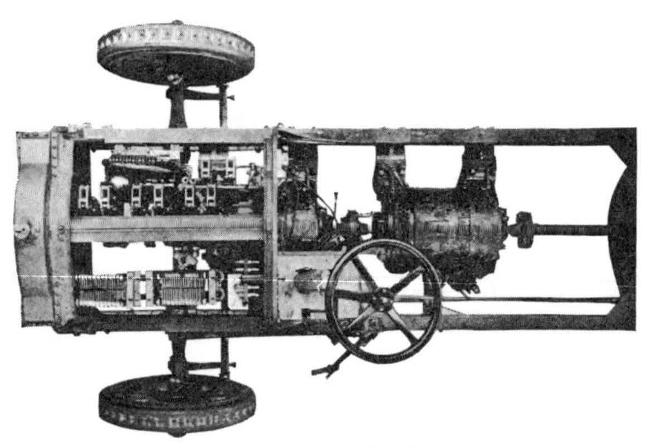

Electrical Equipment for Trolley Bus Mounted on an Automotive Chassis

Type 508 Motor Mounted on Chassis for Car-Type Trolley Bus

Two general classes of buses are built. One type resembles a safety car and the other resembles the automotive type of bus. The advocates of each type feel that the resemblance is of psychological advantage. It is argued that the companies operating trolley buses are primarily in the street railway business and the car-like appearance advertises the street railway business. Those that favor the "bus" appearance feel that the public like to ride in an automobile and that they therefore will ride more often if the trolley bus resembles the automotive or "gas bus." The car-type of bus has the advantage of a shorter wheel base, while the "automobile" type has the advantage of permitting a somewhat better location of the electrical equipment and possibly better riding qualities.

Much has been written about the economic field of the trolley bus and about trolley bus operating expenses. Operating and financial conditions are so different in Europe that European operating records cannot be applied directly to American conditions. No installations in this country have been in operation a sufficient length of time to furnish reliable data on cost of operation. A study of European operating costs for street railway cars, gasoline buses and trolley buses, and the operating costs of street cars and gasoline buses in this country, permits the making of close estimates for trolley buses. The following costs for safety cars and gasoline buses are actual costs for average conditions based upon records of a large number of

Trolley Buses are Operated as Feeders at Windsor, Ontario, Canada

properties, and the trolley bus costs are estimated from the best information available.

	Cents per Car Mile		
	Safety Car	Trolley Bus	Gas Bus
Way and structure (includes road tax for buses)	2.5	1.7	1.0
Equipment	2.0	5.0	10.0
Power	2.5	2.5	6.0
Conducting transportation	9.0	9.0	9.0
General and miscellaneous	3.5	3.5	3.5
Total operating expense	19.5	21.7	29.5

The density of traffic determines which form of transportation is most suitable for any particular case. While the safety car has the lowest operating costs and the gas bus the highest, the investment required is generally highest for safety cars and lowest for gas buses. The following estimated first costs apply to a three-mile extension where ten-minute service is to be provided. Four cars are necessary to perform the service. One spare safety car and two spare trolley buses or gas buses are allowed. The life of the safety car is taken as 15 years, the trolley bus as 10 years, and the gas bus as 5 years. It is assumed that tracks can be laid in unpaved roadway, but that a well paved parallel route is available for bus operation. The necessary investments are as follows:

	Safety Car	Trolley Bus	Gas Bus
Cars or buses	$ 30000	$45000	$36000
Tracks	105000	......	......
Overhead construction	11100	21000	......
Total investment	$146100	$66000	$36000

Assuming annual deposits in a sinking fund bearing compound interest at 5 per cent, expecting to retire the equipment at the end of its useful life, and taking interest and taxes on the investment at 7 per cent, the annual fixed charges and depreciation are as follows:

Safety car	$11,960
Trolley bus	8,850
Gas bus	9,000

On the basis of twenty hours' operation per day, the operating and fixed charges per car mile are:

Operating expenses	19.5	21.7	29.5
Fixed charges and depreciation	4.5	3.4	3.4
	24.0	25.1	32.9

It will be seen from these estimates that for the special example chosen, the safety car is most economical. If conditions were such that there would be difficulty in financing the safety car installation while the trolley buses could be financed relatively easily, the trolley bus would be better. If there were a chance that the need for transportation might not be permanent, the trolley or gas bus would be preferable. The trolley bus would have lower total operating costs and fixed charges where track construction costs were materially more than the values used and where a headway of 20 minutes would be sufficient. The advantage of the trolley bus as compared with the safety car becomes greater as the frequency of the service decreases. For very infrequent and temporary service the gas bus is more economical than the trolley bus and provides a means of rendering some service at a low first cost to districts where the present traffic does not warrant a great outlay but where increased traffic may ultimately develop.

Toronto Transportation Commission

The Toronto Transportation Commission is operating four trolley buses on a short suburban line acting as a feeder for the street railway system. The buses are of the "automobile" type, with a wheel base of 192 inches. Two type 508 25 hp. motors and foot control are used, with provision for arresting the control on the "switching", "full series", "full parallel" running positions. Very satisfactory performance is obtained, the operator manipulating only two foot pedals for starting and stopping the vehicle.

Two standard trolley bases and poles, with swivel harps and wheels are used for current collection. Standard overhead line material and construction are used. It has been demonstrated that turn-outs and cross-overs can be used successfully with trolley buses. As there is no space available for looping at the ends of the line, all turns are made by "wye-ing", spring type frogs and non-insulating cross-overs, sectionalized from one side of the line, being used in the overhead construction.

Twin City Rapid Transit Company

The Twin City Rapid Transit Company is operating a bus of the safety car type in Minneapolis. The total length of the bus is a little more than 23 feet, with a wheel base of 137 inches. Automatic foot-operated control, giving series-parallel acceleration of the main motors, is employed. The control equipment is mounted beneath the bus, fully enclosed for protection against dust and water. The master controller has three positions, permitting running for short periods with series connections of motors and all resistance in the circuit for switching, an economical low running speed with motors in series and all resistance cut out, and full parallel connections of motors for normal service. Upon starting, the foot controller may be at once operated to the desired position, and proper acceleration will be accomplished regardless of route or load conditions.

Hydro-Electric Power Commission

The transportation system of Windsor, Ontario, Canada, is augmented by three trolley bus lines in non-congested districts. Two lines are now in operation and a third is contemplated. The routes connect with the street railway lines, serving as feeders to these lines, and transfers are issued between the trolley bus and railway lines, a small charge being made for the transfer. The bus lines are independent of one another.

The buses are of the automobile type, each equipped with two type 508, 25 hp. motors mounted in tandem. The automatic foot control apparatus is mounted beneath the hood of the bus where all parts are readily accessible. The bus has an over-all length of 25 feet 2 inches, with a wheel base of 194 inches.

One bus is used on each line and since there is, therefore, normally no passing of buses en route, only one pair of overhead wires is required. A wye is used to turn the bus at one place and at other places loops are provided on adjacent streets. Spring frogs are used in the overhead construction so that after looping, the collectors return to the original line without the necessity of the operator stopping to manipulate them.

United Railways & Electric Company Baltimore

To supply the demand for service in the lightly populated district along Liberty Heights Road, the United Railways & Electric Company of Baltimore has established a trolley bus line between Gwynn Oak Junction and Randallstown. Three trolley buses are giving service on this line, which has a route length of approximately five and one-half miles. Buses of the safety car type are used, each equipped with two type 508, 25 hp. motors connected in tandem, and automatic foot-operated control. Two trolley bases and poles with swivel harps and wheels are used for current collection.

EQUIPMENT FOR TROLLEY BUS
Two 25 Hp. Motors—Automatic Foot Control

Apparatus	No. Req'd	Type	Weight Pounds
Motor Items:			
*Motor	2	508-T-2	1570
Main Circuit Control Items:			
Trolley base with 18 ft. pole	2	Special for Trolley bus	230
Trolley harp	2		
Trolley wheel	2		
†Unit switch panel	1	Magnetic	70
Reverser	1	XR-21	20
Grid resistance	1 set	5 in. 3 pt.	75
Control Circuit Control Items:			
Fuse blocks	4	Porcelain	4
Fuses	8	10 ampere	1
Snap switch	1	Double pole, 600-V	2
Master controller	1	XM	14
Control operating mechanism	1	Foot type	15
†Auxiliary panel	1		58
‡Overload trip relay	1	729	
§Reset button	1		1
‡Sequence relay	1	728 B-2	
Motor operated sequence drum	1	108-G	48
Equipment Details:			
Control cable	125 ft.	19x".0142	5
Main cable	150 ft.	7x".0545	15
Knuckle joint connectors	8	Pivot	2
Lighting Details:			
Keyless wall receptacles	10		12
Mazda lamps	10		
Snap switch	1		
Cable	150 ft.	19x".0142	
			2142

**508-TN-2 Ball bearing motor may be substituted.*

†Arranged for mounting under hood or in box under body.

‡For underneath mounting these items are included on the auxiliary panel.

§Used only when apparatus is mounted under body.